THE MEASUREMENT OF SAFETY PERFORMANCE

GARLAND SAFETY MANAGEMENT SERIES

Dan Petersen - Series Advisor

THE MEASUREMENT OF SAFETY PERFORMANCE

William E. Tarrants, *Ph.D., P.E., CSP*

Published under the auspices of the
American Society of Safety Engineers
Park Ridge, Illinois

Garland STPM Press
New York & London

Library of Congress Cataloging in Publication Data

Tarrants, William Eugene.
 The measurement of safety performance.

 At head of title: American Society of Safety
Engineers.
 Includes papers originally presented at a symposium
held in Chicago on Sept. 14–17, 1970.
 Bibliography: p. 392
 Includes index.
 1. Industrial safety—Evaluation. I. American
Society of Safety Engineers. II. Title.
T55.A1T37 614.8′52 79-19208
ISBN 0-8240-7170-0

Published by Garland STPM Press
136 Madison Avenue, New York, 10016

Printed in the United States of America

Contents

List of Contributors

Dr. James W. Altman
President
Synectics Corp.
Allison Park, Pennsylvania

Dr. Vivek D. Bhise
Research Scientist
Auto Safety Center
Ford Motor Company
Dearborn, Michigan

Dr. Murray Blumenthal
Professor of Psychology
University of Denver
Denver, Colorado

Dr. Alphonse Chapanis
Professor
Johns Hopkins University
Baltimore, Maryland

Dr. C. West Churchman
Professor of Business Administration
University of California
Berkeley, California

Dr. John V. Grimaldi
Director of Degree Programs
Institute of Safety and Systems
  Management
University of Southern California
Los Angeles, California

Dr. Herbert H. Jacobs
President
Herbert Jacobs and Associates
Overland Park, Kansas

Dr. Thomas H. Rockwell
Professor
Department of Industrial and Systems
  Engineering
The Ohio State University
Columbus, Ohio

Dr. Harold M. Schroder
Chairman, Management Department
University of South Florida
Tampa, Florida

Dr. William E. Tarrants
Director, Manpower Development
National Highway Traffic Safety
  Administration
U.S. Department of Transportation
Washington, D.C.

Mankind has inhabited this planet for several million years. The members of this unique fraternity seem to have been endowed with an insatiable desire to improve their lot. Yesterday's "wants" become today's "needs."

The painfully slow unlocking of nature's secrets suddenly accelerated when a few members of the fraternity discovered that nature would respond to planned, systematic inquiry faster than to casual observation alone. Thus was born science. And science sired technology.

Mankind seized upon technology as its long-sought panacea, accepting its fruits often without question or regard for the consequences. One of the consequences turned out to be significant impairment to the body and loss of life. Therefore, in just the United States, approximately 15,000 lives are sacrificed each year in industry so that our needs and wants can be satisfied; approximately twice as many are sacrificed in our homes; and approximately double that number are lost in transportation. The accident death toll in the United States alone is more than 100,000 per year, in addition to injuries and permanent maimings that reach into the millions. Clearly, technological progress is not without its price!

An indignant society, eager to continue to accept the benefits of technology but abhorred by its consequences, set about to correct this state of affairs in the only way it knew how. As a result, there have been more safety-related laws passed in the last decade than in the entire history of our republic. Whether this is the response of a frustrated society, the expiation of a guilt-ridden society, or the blossoming of a truly moral and ethical society is open to argument (as is the quality of the fruits of its legislatures!). Be that as it may, these and other matters thrust safety to the forefront in technological endeavors. The safety professional was suddenly called in, handed the "ball," and told to pitch.

Alas, he was not even given time to warm up! He was somewhat aware, however, of the complexity of the game (at least more so than the manager who asked him to save it). He set about to develop winning habits. From his vantage point in the bullpen, the safety professional had observed many things wrong with the team. For example, the manager hardly knew that the safety person existed except when things got so out of hand that he (the manager) did not know where to turn. This attitude permeated the entire organization; as a result, the safety professional seldom found himself on the team when it came to developing the game plan (design, purchasing, testing, and so on).

Fortunately, about this time a set of logical procedures was developed that seemed to offer at least a partial answer to many of these problems. If properly executed, this "systems approach" would assure the safety professional's membership on the team from the conceptual phase through all stages of development and operation. It would force timely consideration of safety matters. Also, the philosophy supporting the "systems approach" readily accommodated views supporting multicausation of accidents, interactive precipitating factors, and the importance of process as well as structure. Finally, adherence to its procedures and practices would disclose the weaknesses inherent in some of the simplistic "solutions" that had been advanced regarding safety problems. Those who understood this new approach appreciated complexity and understood the difficulties that confronted those dedicated to making society safer.

The safety professional soon discovered, however, that the "systems approach" also has a price. For example, it required participants to have a valid data base, procedures that are consonant with those of the other participants, and so on. And these he really did not have.

Fortunately, individuals from many disciplines are engaged in seeking solutions to these problems. Thus, the safety professional and his colleagues now have in hand, or in the planning stage, the techniques, models, scales, and other tools that are needed in the profession. These developments hold promise for even broader involvement of other professionals whose support is necessary if improvement is to continue.

Fundamental to all of this is a need for improved measurement. This is the common language that systems engineers understand and employ. It is with this topic that this book is concerned—reliable, systematic, valid measurement—without which the safety movement can never completely succeed.

It could hardly have a more worthy or capable advocate. Dr. Tarrants is one of the very few who by education, experience, and moral persuasion is capable of turning out this book. Familiar with the past, but always with an eye to the future, he has seized this opportunity to do his share to fill the void referred to previously. The immediate beneficiaries are the members of his profession; the ultimate beneficiaries will be all members of society. Examine this offering with a sentient, searching mind. There is much here that will excite you and help you in your quest for a means to implement and express your individual contributions to safety.

Julien M. Christensen, Ph.D.

The problem of safety performance measurement has existed since the very beginning of organized attempts to control accidents and their consequences. In its most elementary form, measurement has been defined as "the process of assigning numerals to objects according to rules" (Stevens, 1951). As this definition is applied to the safety field, problems concerning what "objects" to measure and what "rules" to follow are revealed quickly. The safety professional must continually ask himself, "Am I really measuring what I think I am measuring?" and "What are the benefits of measurement?"

As more is learned about the accident phenomenon, traditional concepts of describing it change. Since measurement is primarily a descriptive process, it is often erroneously believed that the description is the real thing; thus, the nature of the phenomenon to be described is forgotten. A particular group of familiar measures is accepted, and this frequently prevents a search for better measures. A few safety professionals have been known to retain a measurement system because it produces good results for them. Within the limits of the accepted definition, they are able to manipulate the evaluative process to their own advantage.

Measurement is the backbone of any scientific approach to problem definition and solution. Without measurement the state of our operations is unknown. Sound measurement is an absolute prerequisite for control, and both are necessary for prediction. In accident control and prediction valid and reliable measures of safety performance are essential in order to (1) locate and describe problem areas, (2) identify causal relationships, (3) make decisions concerning the optimum allocation of accident prevention resources, (4) evaluate the effectiveness of applied countermeasures, and (5) detect when the system is deteriorating toward unacceptable limits of control.

Unfortunately, none of the traditional measures of safety performance permit achievement of these objectives effectively. It is dangerous to assume that present measures are really descriptive of the safety level within an organization. It is even more precarious to assume that they will permit inferences about future problem areas and allow selection of effective solutions. On more pragmatic grounds, the measures of safety performance are inadequate because members of top and middle management often simply do not believe them, as evidenced by the frequent complaint of the safety director, "If only I could convince my managers to accept and act on my recommendations."

A number of leaders in the safety field have recognized the safety performance measurement problem and have made substantial contributions toward its solution. In 1937 the Bureau of Labor Statistics of the U.S. Department of Labor began compiling work injury statistics under the criteria established by the American Standards Association, which led to the present American National Standards Institute's *American Standard Method of Recording and Measuring Work Injury Experience: Z-16.1.* With some minor changes over the years, this standard is used today by many organizations. A new method for collecting national statistics on work injuries and illnesses was initiated by the Bureau of Labor Statistics (BLS) as the result of authorizations contained in the Occupational Safety and Health Act (OSHA) of 1970 and subsequent amendments. While the scope of recordable injuries and illnesses has been widened to include a more comprehensive picture of losses incurred, this method of measurement also has numerous limitations as an internal index of safety performance. Both the Z-16.1 standard and the BLS–OSHA safety and health statistics program are discussed in detail later in this book.

Numerous authors have addressed various aspects of safety performance measurement in past years. A book entitled *Selected Reading in Safety* (Widner, 1973) contains chapters authored by 36 international safety professionals, many of whom have addressed safety measurement in various forms. Charles L. Gilmore (1970) in his book *Accident Prevention and Loss Control* emphasizes measurement techniques as a means of accomplishing "loss prevention management control." His book "offers circumstantial analysis as a replacement for loss occurrence," and his program "employs the existing accepted measurement and statistical techniques of other disciplines and adapts these for loss prevention performance measurement." Books written by Daniel C. Petersen (1971) *Techniques of Safety Management;* Leo Greenberg (1972), *Quantitative Techniques in Safety and Loss Prevention with Computer Programs;* John V. Grimaldi and Rollin H. Simonds (1975), *Safety Management: Accident Cost and Control* (3rd edition), and W. G. Johnson (1973), *The Management Oversight and Risk Tree (MORT),* all address the subject of safety performance measurement in varying levels of detail.

A series of two symposia on safety performance measurement was conducted in 1966 and 1970 by the Industrial Conference of the National Safety Council. The first was held in May 1966 in Chicago "for the purpose of studying the current methods and needs for the measurement of safety performance by employers and establishing measurement methods that lead to total accident prevention." The workshop sessions each approached the subject of safety performance measurement from the viewpoint of special-interest disciplines or users of safety measures. These included specialists in industrial hygiene, insurance, cost accounting, safety engineering, and industrial medicine. Summary reports were published in the *Proceedings* of the 1966 National Safety Congress following presentations at a congress subject session on Measurement of Industrial Safety Performance. It was agreed at the conclusion of the first symposium that further exploration of the safety performance measurement problem would be desirable.

The major outcome of the first symposium was the identification of the state of safety performance measurement at that time. At least two fairly new approaches to safety performance appraisal were introduced. These and other measurement topics were discussed at a general, introductory level. In essence, the first measurement symposium was exploratory in nature. It established some parameters of the industrial safety measurement problem and provided some guidelines for future work and study in this field.

A second symposium was held in Chicago on September 14–17, 1970. It approached the safety measurement problem from a different point of view. The first symposium provided an overview of current measurement practice. Those in attendance were primarily safety, medical, and industrial hygiene practitioners who discussed the problem from the vantage point of their particular fields of specialization. The second symposium concentrated on various aspects of measurement and safety performance evaluation as viewed by persons primarily outside the occupational safety field. The purpose of inviting professionals from other disciplines (operations research, systems analysis, human factors engineering, statistics, measurement, individual behavior psychology, group behavior, social structure, management, communications systems), as well as from safety engineering, environmental hygiene, and industrial medicine, was to bring individuals with skills and interests in basic measurement problems in other professional fields together with professionals from the safety field with fresh approaches and new measurement thinking.

Another unique feature of the second symposium was the prior arrangement with eight outstanding practitioner-scientists within these same professional disciplines to prepare individual papers on the measurement issues within their particular fields that had some relevance to the safety performance measurement problem. Each author discussed safety performance measurement from the vantage point of the measurement concepts and methodologies in his discipline.

Dr. Julien M. Christensen, who was then Chief, Human Engineering Division, Wright-Patterson Air Force Base, Ohio, served as General Chairman of the symposium. Dr. Christensen now is a safety consultant in Dayton, Ohio.

It was recognized that the advancement of knowledge and the development of new approaches to any problem can best be achieved by means of considerable private thinking by individuals who have an interest in the subject, adequate technical background in at least a major facet of the problem, and a willingness to devote significant amounts of time in search of solutions. By bringing together a group of individuals who meet these qualifications and who have prepared or reviewed working papers on the subject in advance and by synthesizing these ingredients through competent leadership, it was expected that new insights would be developed and that substantial progress would be made toward defining the essential problems of safety performance measurement and identifying new avenues for research and demonstration efforts.

The safety practitioner seeking additional measurement information will probably benefit most if he first explores the thinking and writings of individuals who are competent in the basic subject areas related to his problem. Not only will he most likely equip himself with a more substantial foundation for understanding the problem, but he also will be less inclined to adopt time-worn solutions based on the repetitive application of the traditional "tools of the trade." It is suggested that creativity most often comes from outside a particular field of specialization and that restriction of thought to internal sources will frequently lead to circular thinking, which may result in a lack of substantial progress.

By drawing upon the talents and knowledge possessed by individuals who function within areas of specialization related to the general field of accident prevention and by concentrating on the interfaces between occupational safety and other specialties, maximum use can be made of the unique contributions of these outside sources to the concepts and methodologies of measurement.

Revised and updated versions of the prepared position papers from the second symposium have been included in this book to serve as a basis for concept formation in safety performance measurement. Four introductory chapters set the stage for exploring the various issues involved in the subject by introducing the concepts of safety performance measurement and defining problem areas to be resolved. Finally, the third section integrates the spectrum of views and presents systematic approaches to safety performance measurement, including specific measurement methodologies appropriate to the safety field. These final chapters are intended to make the treatment of measurement more complete and useful to safety practitioners and students seeking new and improved techniques for measuring safety performance.

WILLIAM E. TARRANTS
WASHINGTON, D.C.

Acknowledgments

In the years following the second measurement symposium in 1970, the American Society for Safety Research (ASSR), the former research arm of the American Society of Safety Engineers (ASSE), decided to publish a version of the discussions held and papers presented at the symposium, for the benefit of safety professionals everywhere. Later, in consultation with the author, a decision was made to adapt some of the symposium material into a text-reference book and to expand the scope of the book in order to provide a comprehensive treatment of the subject.

In 1978, ASSR recommended to the ASSE, an individual membership organization of 15,000 safety professionals, that ASSE publish the book as a major contribution to the literature of safety management and engineering. Harold Barnes, then the Vice-President-Communications, and Eugene Walters, the Society's Publications Task Group Chairman, subsequently reviewed the manuscript and concurred in the ASSR recommendation.

The author is especially indebted to ASSR and its past president, William H. Griswold, for providing the support, encouragement, and patience that permitted this work to be completed. Appreciation also is expressed to the National Safety Council, which sponsored the two measurement symposia and permitted some of the material from the second symposium to be adapted for this book.

The author is also indebted to the symposium participants and paper authors who made numerous thoughtful contributions to the development of the topic.

A special "thank you" is given to Dr. Christensen, who served as chairman of the Second Measurement Symposium, wrote the foreword for this book, and provided substantial encouragement and assistance as this work progressed. The author also is indebted to Lyle R. Schauer, the Bureau of Labor Statistics, U.S. Department of Labor, for providing materials on which the section on the BLS-OSHA record-keeping in Chapter 2 is based.

Finally, grateful acknowledgment is extended to the ASSE and its Director of Communications, Dwight B. Esau, (who edited the manuscript) and to Garland STPM Press, who worked together to bring this book to reality.

Part I

INTRODUCTION TO SAFETY PERFORMANCE MEASUREMENT

Chapter 1

MEASUREMENT AND ITS APPLICATION TO THE SAFETY FIELD

The progress and maturity of a science or technology are often judged by how much they succeed in the use of measures. Measurement, perhaps more often than any other single aspect, has been the principal indicator and stimulus of progress in all fields of scientific endeavor. Measurement permits accurate, objective, and communicable descriptions that readily lend themselves to progressive thinking. Measures can serve as models for events and relationships existing in the real world because the structure of nature as we know it has properties that are parallel to, or isomorphic to, the structure of logical systems in mathematics.

The purpose of measurement is to represent the characteristics of observations by symbols that are related to each other in the same way that the observed objects, events, or properties are related. These symbols by themselves have no meaning or significance; their significance depends solely on the circumstances and events that created them. The symbols or numbers arise from a series of operations, sometimes as simple and intuitively meaningful as counting the number of first-aid accident cases and writing down the number, sometimes so complicated that a vast body of scientific and technical knowledge is necessary to understand the relation between the final number and the particular object of measurement. It is important to remember that the real world is never exactly described by any mathematical model. All such descriptions are only approximations. In some instances the

3

equivalence is excellent in detail, while in other situations it is very rough, resulting in estimates with various degrees of accuracy.

WHAT IS MEASUREMENT?

Measurement has been defined quite simply by Stevens (1951) as any process that involves "the assignment of numerals to objects or events according to rules" and by N. R. Campbell (1938) as "the assignment of numerals to represent properties." Ackoff, Guptas, and Minas (1962) define measurement in terms of its function by stating that "it is a way of obtaining symbols to represent the properties of objects, events, or states, which symbols have the same relevant relationship to each other as do the things which are represented." As Caws (1959) has stated, "The essential function of measurement is the setting in order of a class of events with respect to its exhibition of a particular property, and this entails the discovery of an ordered class, the elements of which can be put in a one-to-one correlation with the events in question." For the field of accident prevention, measurements enable comparison of the same accident-producing characteristics of the same thing at different times and describe how accident-producing characteristics of the same or different things are related to each other.

A method of generating a class of information needs to be developed that will be useful in a wide variety of problems and situations. The qualitative assignment of objects or situations to classes and the assignment of numbers to objects or situations are two systems of measurement that will generate broadly applicable information (Churchman, 1959). Measurement is essentially a decision-making activity, and the usefulness of measures must be evaluated in terms of their ability to provide information that will improve accuracy and validity of the decisions made. Measurement connects three parts of knowledge—the mathematical, the conceptual, and the practical. It imparts relevance to the first, precision to the second, and utility to the third (after Caws, 1959).

Measurement is basic to all engineering and scientific activities. Without measurement one cannot describe nature, establish control over natural phenomena, or attempt to predict future events. Measurement is often referred to as the backbone of science. As such, it influences our degree of sophistication in understanding and applying scientific concepts for the achievement of our desired objectives.

Measurement is an essential prerequisite for control and prediction. The degree to which one is able to control real-world events and predict their future occurrence is a function of the ability to measure them. Measurement is primarily a descriptive process. It allows one to qualify, order, and quantify certain events and ultimately use the results as a basis for the control and prediction of actual performance.

MEASUREMENT PROGRESS IN THE PHYSICAL AND BEHAVIORAL SCIENCES

The scientific revolution of the seventeenth and eighteenth centuries was brought about by the use of measurement instruments. Before the telescope astronomy as a science was nonexistent. Astrology became astronomy and alchemy became chemistry only when measurement became possible within these fields. Increased measurement sophistication in science has allowed a movement from subjective, qualitative speculation to more objective, quantitative accuracy with improved descriptive and predictive precision.

In general, measurement in the physical sciences has progressed much farther than measurement in the behavioral sciences. The history of science demonstrates that the development of measurement processes is dependent upon the constant interaction of both empirical procedures for measurement and theoretical concepts about what is being measured. Experiences in the physical sciences have produced numerous precise relationships among characteristics as expressed in numerical laws. An example of this is the measurement of the density of liquids on the basis of the law that expresses density as a constant function of the ratio between the weight and the volume of the particular liquid. For many of the relationships among characteristics in the social sciences, there are so far neither adequate definitions nor the precision of measurement achieved in the physical sciences.

In the measurement of human characteristics there is a wide range in the degree of accuracy possible with existing measuring instruments. Height and weight, for example, can be measured very precisely. Within certain limits, one can achieve any degree of precision desired by simply refining the scale. In contrast, the measure of emotional stability has a low degree of accuracy since the instruments available for measuring this human characteristic are not sensitive to variations in the quantity of this characteristic. In fact, there is not general agreement on what is meant by the concept of emotional stability itself. The chart in Figure 1-1 (p. 6) provides an indication of the relative degree of accuracy of various measures of human characteristics.

ERROR AND ACCURACY IN MEASUREMENT

All types of measurements are subject to error, and an estimate of the magnitude of this error is necessary in order to determine whether or not the measures obtained are usable in a practical situation. The range of uses to which a measure can be put increases with its exactness. A perfectly accurate measurement would make it possible to answer nearly any question or solve any problem involving the characteristic being measured.

There are several possible sources of error in measurement: (1) the observer, (2) the instruments used, (3) the environment, and (4) the thing or

Figure 1-1
A Scale Indicating the Relative Accuracy of Various Measures of Human Characteristics. (Adapted from Herbert H. Mayer and Joseph M. Bertotti, "Uses and Misuses of Tests in Selecting Key Personnel," Personnel, 1956.)

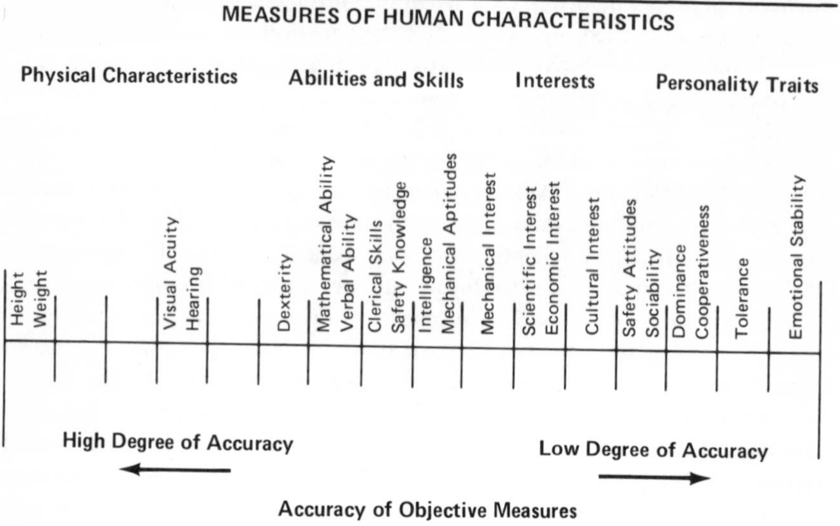

MEASURES OF HUMAN CHARACTERISTICS

Physical Characteristics   Abilities and Skills   Interests   Personality Traits

Height
Weight
Visual Acuity
Hearing
Dexterity
Mathematical Ability
Verbal Ability
Clerical Skills
Safety Knowledge
Intelligence
Mechanical Aptitudes
Mechanical Interest
Scientific Interest
Economic Interest
Cultural Interest
Safety Attitudes
Sociability
Dominance
Cooperativeness
Tolerance
Emotional Stability

High Degree of Accuracy ←   Low Degree of Accuracy →

Accuracy of Objective Measures

situation observed (Ackoff et al., 1962). The observer may not follow the required operations and thus may introduce bias into the results. He may not gather data as accurately as possible; he may read dials, meters, or rules incorrectly; he may not accurately record a response of a person he has questioned; or he may not observe all relevant factors existing in the environment. The instruments used in making observations may be biased and/or inconsistent and hence produce errors. The improvement of instruments is an important way of reducing observational error arising not only from the instruments themselves, but also from the observer, the environment, and the respondent. For example, mechanical traffic counters have reduced errors due to failure of memory or carelessness of human observers. The conditions under which the observations are made may vary so as to affect the observer, the instruments, or the thing or individual observed.

When the object observed is a human being, he may cause much difficulty, particularly when he is called on for verbal testimony as, for example, in an accident investigation. First of all, it is assumed that the respondent possesses the information that the investigator wants. If a knowledge of the information is assumed but nonexistent, then, needless to say, satisfactory results cannot be obtained. If it is assumed that the subject possesses knowledge of the information sought, then the problem becomes primarily one of communication. This means that the person understands the questions properly, that he tries to give a truthful answer, and that his answers are recorded correctly. Control of observational errors consists of periodic checking to determine whether or not the magnitude of these errors is changing over the period during which observation is made, during different observation periods, among different groups or observers, and so on. The continuous reduction of error from observation and other sources is a major objective of a science and is one of the principal gauges of its progress.

Accuracy is the measure of the degree to which a given measurement may deviate from what really exists. An estimate of accuracy should preferably accompany a measurement procedure. In statistical terms, accuracy of prediction is sometimes defined by means of a *confidence interval*. This reveals that a certain range of numbers constructed from observations has a specific probability of including the true measurement. The problem of accuracy is to develop measures that enable the user to evaluate the information contained in the measurements (Churchman, 1959).

Control is the long-term aspect of accuracy. It provides the guarantee that measurements can be used in a wide variety of contexts. A control system of measurement provides optimal information about the legitimate use of measurements under varying circumstances. Control is, in effect, the test of a good standard. If adjustments can satisfactorily be made to a standard in accordance with the criteria of control, then the standards have been sufficiently specified (Churchman, 1959).

SCALES OF MEASUREMENT

Intuitive measurements are made everyday by everyone. We constantly distinguish among objects and make differential responses to them. Many judgments require no more than the distinguishing of objects possessing qualities that are distinctive from those of others. The differences between apples and pears are noted, as are those between males and females, automobiles and motorcycles, and drivers and pedestrians. These distinctions are qualitative in nature. An object is placed in a particular class according to some predetermined characteristic that allows one to identify it as belonging to that classification and not to some other one.

Both in the sciences and everyday life it is often desirable to make distinctions of degree rather than quality. Which person possesses more knowledge about a subject? Which man is taller? How much taller is one man compared to another? In the interest of both accuracy and the discovery of constant relationships among characteristics that vary in amount as well as in kind, statements that simply affirm or deny differences are replaced by more precise statements indicating the degree of difference. These distinctions are quantitative in nature.

Suppose one is interested in studying attitudes of workers toward safety (assuming some way of accurately measuring safety attitudes exists). If one simply asserts that two workers differ in their attitudes toward safety, without specifying how great the difference or whether one is more favorable than the other, then certain attitudes must be identified as equivalent and others as not equivalent. If one wishes to state that the safety attitude of one worker is more favorable than that of another without specifying how much more favorable, he must be able to rank different attitudinal positions as being more favorable or less favorable than other positions. If the statement that the safety attitude of worker A is much more favorable than worker B's is sought or that two safety lectures have produced equal changes in safety attitudes, one must be able to determine whether the difference between two attitudinal positions is equal to the difference between two other attitudinal positions. And finally, if one wants to state that the safety attitude of worker A is twice as favorable as that of worker B, he must be able to determine the existence of an absolute zero of favorableness for the given safety attitude as well as equal units above the zero point.

These four statements correspond to four types of measurement scales described by Stevens (1951): nominal, ordinal, interval, and ratio scales. Certain formal rules and empirical operations distinguish these various scales. The scales are listed in ascending order of power. The more powerful scales presuppose the ability to perform the empirical and mathematical operations of the less powerful ones. Thus, the ratio scale implies all the operations of the nominal, ordinal, and interval scales as well as additional ones that are unique to the ratio scale. The higher-level scales require more restrictive rules. Also, the higher the level of scale, the more one can do with the numbers obtained in measurement.

Nominal Scales

Measurement at its weakest level exists when numbers or symbols are used to classify objects or individuals into two or more named categories. The only requirements for a nominal (naming) scale are that two or more categories relevant to the attribute being considered can be distinguished, that one can specify criteria for placing items in one category for an identified characteristic, and that no one can fit into two categories of the same characteristic. At least two classes must be defined. The categories need only be different from each other; there is no implication that they represent "more" or "less" of the characteristic being measured. The empirical operation is the determination of equivalence or nonequivalence with respect to the attribute in question between the given person or object and other objects or persons placed in a given category. In a nominal scale the scaling operation consists in partitioning a given class into a set of mutually exclusive subclasses.

Numbers may be used to identify the categories, but there is no empirical relationship among the numbered categories that correspond to the mathematical relations among the numbers assigned. Therefore, statistical techniques that make use of mathematical relations among numbers such as the computations of means or standard deviations, for example, are inappropriate.

Only such statistics as are appropriate to counting may be used. For example, one can count the number of cases in each class and thus obtain frequencies, or one can determine which is the most populous class and identify that class as the mode of the distribution of classes. If the same objects are classified in two ways, on the basis of two aspects or principles of classification, one can test hypotheses regarding the distribution of cases among these categories by using the chi-square statistical test. The interdependence of the two aspects can also be determined by computing a coefficient of contingency.

Examples of the use of numbers in nominal scaling include the numbers on license plates, the numbers on football jerseys, social security numbers, military service numbers, and the assignment of type or model numbers to classes (Stevens, 1959).

Ordinal Scales

If the data permit the ordering or ranking of people or objects in terms of their possession of a greater, equal, or less amount or degree of a characteristic with no implication as to the distance between positions, that is an ordinal scale. The people or objects in one category of a scale are not only different from those in other categories of that scale, but they can be placed in some kind of relation to each other. In measurements on an ordinal scale the numbers assigned utilize the property of rank order; in other words, one designates a numerical order of positions.

An ordinal scale may be compared with an elastic tape measure that is being stretched unevenly. The positions of the scales as indicated by the numbers on the tape remain in a definite order, but the numbers do not provide an indication of the distance between any two points. The distance between 16 and 17 may be equal to, less than, or greater than the distance between 56 and 57. When measurements are limited to information concerning greater, equal, or less and not how much greater or how much less, one has an ordinal scale. For example, a supervisor may rank employees in terms of their degree of cooperativeness in obeying the plant safety regulations. Although the worker who is ranked highest is given the number 1, the next highest 2, and so on, it is clear that we do not necessarily assume that the worker ranked number 1 is as much higher than number 2 as number 2 is than number 3, and so on.

The statistics that are permissible at the level of nominal scale measurement also apply to measurements on ordinal scales. The principle of order makes possible the use of additional statistics, including the median, percentiles, rank-order correlations, the sign test, the runs test, and numerous other nonparametric statistical tests (Siegel, 1956). The statistic most appropriate for describing the central tendency for frequencies in an ordinal scale is the median, since this statistic is not affected by changes of any frequencies that are positioned above or below it as long as the number of items above and below it remain the same. The median may coincide with one of the rank numbers used in the case of an odd number of classes ranked, or it may lie midway between two adjacent rank numbers in the event an even number of classes is ranked. Percentile measures may be applied to rank-order data. A value is assigned to a percentile by interpolating linearly within a class interval. Correlation coefficients, such as the Spearman Rho and the Kendall Tau, may be computed for rank-order data. These coefficients are measures of the degree of association between two variables, with the objects or individuals under study being ranked in two ordered series. Significance tests, such as the sign test and runs test, can be used in the testing of hypotheses when rank-order data are involved. The only assumptions associated with most nonparametric tests of significance used with rank-order data are that the observations are independent and that the variable under study has underlying continuity.

Examples of the use of rank-order scaling techniques include Rho's scale of hardness (the empirical relation in this case is the ability of minerals to scratch one another); street numbers; grades of leather, lumber, cloth, and the like; intelligence test raw scores; the system of grades in the military service (sergeant, corporal, private); and the judgment of the relative desirability of applicants for a job. This scale may be used by a supervisor to rank workers in terms of the degree of safety they exhibit while performing the job, from the "safest worker" down to the "most unsafe worker" in the department.

Interval Scales

When a scale has all the characteristics of an ordinal scale and when, in addition, the units or intervals of measurement are equal, one has an interval scale. In other words, the distances between any two points are known. For example, the distance between points 40 and 45 on the scale is equal to the distance between points 75 and 80. A gain of a unit in one part of the scale is equal to a gain of a unit in any other part of the scale. A number of different operations can be performed with numbers assigned to people or objects on such a scale that are precluded with the lower levels of measurement.

The placement of the zero point on an interval scale is a matter of convention. This is illustrated by the fact that a constant can be added to all scale positions without changing the forms of the scale. Because the zero point is arbitrary, relations between positions cannot be stated in terms of ratios. Thus, with data that meet the assumptions of an interval scale but not a ratio scale, one cannot state precisely that one person's safety attitude is twice as favorable as that of another person.

The interval scale is truly quantitative. All of the common statistics, such as means, standard deviations, Pearson product-moment correlations, and multiple product-moment correlations, are applicable to data appearing in an interval scale, as are the tests of significance used in hypothesis testing such as the t-test and the F-test (assuming the assumptions of these models are tenable). If measurement on an interval scale has in fact been achieved, the above statistics should be utilized since tests applicable to lesser scale levels would not take advantage of all information contained in the data.

Examples of interval scales include the ordinary temperature scales found on a thermometer, dates on a calendar, and measures of energy. The arbitrariness of the zero point is apparent when we compare the Fahrenheit and centigrade thermometer scales. In the latter the zero corresponds to the point at which water freezes; in the former the zero point is 32 degrees below the freezing point. Although relations between positions can be stated in terms of the number of scale points between them, they cannot be stated in terms of ratios. Thus, it is not true that 40° F is twice as hot as 20° F, but one can say that the temperature changes as much when it rises from 20° F to 40° F as it does when it rises from 40° F to 60° F.

Ratio Scales

When a scale has all the characteristics of an interval scale, and in addition contains an absolute zero (point of "no amount"), it is called a ratio scale. In a ratio scale one is not only able to determine equivalence—nonequivalence, rank order, and equality of intervals—but the equality of ratios as well. Since ratios are not meaningful unless there is an absolute zero point, it is only with this type of scale that one is justified in making such statements as; "A is twice the size of B," or "I have one-half of the amount that you have."

If measurements conform to the requirements for a ratio scale, any manipulation that can be carried out with numbers can be appropriately employed. All types of statistical measures are applicable to ratio scales, and only with these scales can one make transformations involved in logarithms, such as when computing the ratio of two amounts of power involved in the rise of decibels as a measure of noise intensity. In addition to the previously mentioned measures appropriate for use with interval scales, with ratio scales one can use such statistics as the geometric mean, harmonic mean, percent variation, and the coefficient of variation. All numbers in a ratio scale can be multiplied by a constant, and the ratios between any two numbers can be preserved.

Examples of ratio scales include the measure of mass or weight, length, density measures, temperature on a Rankine or Kelvin scale, loudness (sones), and brightness (brils). Ratio scales are most commonly found in the physical sciences.

Summary of Scale Characteristics

The characteristics of the nominal, ordinal, interval, and ratio scales have been summarized by Stevens (1959) and are shown in Table 1-1. Stevens has also summarized the statistical measures that may appropriately be used with measurements made on each type of scale. These are shown in Table 1-2 (p. 14).

THE IMPORTANCE OF MEASUREMENT IN ACCIDENT PREVENTION

A consideration of measurement concepts and measurement scales is important in the occupational safety context. The discovery of potential accident relationships among characteristics of the human, his equipment, and his environment is largely dependent upon the existence of measures. Numbers are assigned to observations in such a way that the numbers are amenable to analysis by manipulation or operation according to certain rules. This analysis should reveal new information about potential accident problems involving man and his work situation. The kind of measurement, and ultimately the precision of the analysis, is a function of the rules under which the numbers were assigned. In other words, when a procedure for assigning numbers to a characteristic is developed, the properties of the procedure determine the degree of scale sophistication. If sources of potential accident loss are needed, development of a system of analysis based on the assignment of numbers according to rules defined by the nature of the potential accident problem is needed. These measures must produce the type of information required to accomplish accident prevention decision-making objectives.

In accident prevention, measurement of safety performance is necessary for many reasons:

1. As a basis for causal factor detection
2. To locate and identify problem areas
3. As a basis for trend comparison
4. To describe the current safety state of an organization
5. As a basis for predicting future accident problems
6. As a basis for evaluating accident prevention program effectiveness

Table 1-1

A Classification of Scales of Measurement

Scale	Basic Emprirical Operations	Mathematical Group Structure	Typical Examples
Nominal	Determination of equality	Permutation group $x' = f(x)$, where $f(x)$ means any one-to-one substitution	"Numbering" of football players Assignment of type or model numbers to classes
Ordinal	Determination of greater or less	Isotonic group $x' = f(x)$, where $f(x)$ means any increasing monotonic function	Hardness of minerals, street numbers, grades of leather, lumber, wool, etc. Intelligence test raw scores
Interval	Determination of the equality of intervals or of differences	Linear or affine group $x' = ax + b$ $a > 0$	Temperature (Fahrenheit or Celsius) Position Time (calendar) Energy (potential) Intelligence test "Standard scores"
Ratio	Determination of the equality of ratios	Similarity group $x' = cx$ $c > 0$	Numerosity Length, density, work, time intervals, etc. Temperature (Rankine or Kelvin) Loudness (sones) Brightness (brils)

"The basic operations needed to create a given scale are all those listed in the second column, down to and including the operation listed opposite the scale. The third column gives the mathematical transformations that leave the scale invariant. Any numeral x on a scale can be replaced by another numeral x', where x' is the function of x listed in column 3."
Source: From S. S. Stevens, "Measurement, Psychophysics, and Utility," in C. W. Churchman and P. Ratoosh (Eds.), *Measurement Definitions and Theories.* New York: Wiley, 1959, p. 25.

7. As a basis for making decisions regarding the allocation of accident prevention resources
8. To assess accident costs
9. To establish long-term accident control
10. As a basis for quantifying probable risk of injury or other loss

Table 1-2

Examples of Statistical Measures Appropriate to Measurements Made on the Various Classes of Scales

Scale	Measures of Location	Dispersion	Association or Correlation	Significance Tests
Nominal	Mode	Information	Information transmitted T contingency correlation	Chi-square test
Ordinal	Median	Percentiles	Rank-order correlation	Sign test Runs test
Interval	Arithmetic mean	Standard deviation	Product-moment correlation	t-test F-test
		Average deviation	Correlation ratio	
Ratio	Geometric mean Harmonic mean	Percent variation		

Source: From S. S. Stevens, "Measurement, Psychophysics, and Utility," in C. W. Churchman and P. Ratoosh (Eds.), *Measurement Definitions and Theories.* New York: Wiley, 1959, p. 27.

Generally, measures are needed to reveal performance levels. Is the accident prevention program paying off? More specifically, measures that will enable description of the safety state currently existing throughout the entire plant or company are needed. The main function of a measure of safety performance is to reveal the level of safety effectiveness in the organization within which establishment of accident control is desired. Unfortunately, most measures presently used in the safety field require loss-type accidents to occur with a certain degree of severity before identification of accident problems is possible.

The level of safety performance within an organization involves accident situations (unsafe acts and/or unsafe conditions) that have the *potential* for producing loss but that do not necessarily produce a loss (either injury or property damage) each time they occur. In effect, then, measures of safety effectiveness that will enable us to identify accident problems that have the *potential* for producing future losses are needed as well as those that are currently producing property damage, injuries, and deaths. These measures should reveal when to expect trouble and must provide one with some insight as to what should be done about it.

A second purpose of safety performance measurement is to provide continuous information concerning changes in the safety state within an organization or operation. A valid and reliable measure of these changes permits evaluation of the effectiveness of accident prevention efforts over time. A measure of the total safety state would, of course, include accidents that have the potential for producing loss as well as those that actually result in loss during the appraisal period. It should not be assumed that the recording of only injurious or property-damaging accidents reveals a picture of the real level of safety performance. At any future point in time accidents that have a potential for loss may produce a loss. Which exposure results in loss and the degree of severity of future losses are influenced by chance factors. Therefore, measurement techniques that are more sensitive to the fundamental behavioral and conditional malfunctions that may at any time contribute to our accident loss problem are most needed.

A frustration that has long plagued those who work in the field of accident prevention has been the lack of an acceptable criterion of safety effectiveness. Safety practitioners have been forced to measure the effectiveness of their work by vague, invalid, and insensitive criteria. This complicates an already overly complex problem involving many misinterpreted, ill-defined, and unquantitative terms, such as accident proneness, carelessness, job hazard, inattention, fatigue, morale, attitudes, and motivation. Present attempts to control accidents and their consequences can best be described as trial and error chiefly because adequate measures of the effectiveness of this control do not exist in practice. The degree to which accident loss control is possible is a function of the adequacy of the measures used to identify the type and magnitude of potential injury-producing and property-damaging problem areas existing within a particular field of concern.

The safety professional, in formulating an accident prevention program, is continually confronted with this problem. Inspections, training programs, safety contests, poster programs, and the like are initiated under various levels of effort and evaluated by the only measures available, namely, lost-time accidents, recordable injuries, first-aid cases, and the extremely elusive measure of accident costs. Lacking adequate measures of what his safety program components are really accomplishing, the safety professional works in the dark. Even if he uses all recognized educational and propaganda media and effects a reduction in disabling injuries over time, he still does not know which of the many techniques produced the desired results—if, indeed, any of them were responsible.

Furthermore, he has no assurance that a reduction in disabling injuries during any one period is necessarily predictive of favorable results in the future. For example, a change in the process or a remark by the plant manager in favor of certain behavioral changes may have occurred concurrently with a new poster campaign and, consequently, may have been the actual controlling factor influencing the reduction in accidents. Safety specialists often fall into the logical fallacy trap of assuming causation from correlation when they conclude that a particular safety program "caused" a reduction in accidents. In reality, any number of variables external to the program may have intervened to produce the favorable results. In effect, the safety program analyst makes continuous use of a "machine gun with hope" when he might better use a rifle occasionally with confidence.

The real problem is to find a criterion of safety effectiveness and some way of measuring it. Because lost-time and property-damaging accidents are rare events and first-aid cases are subject to serious reporting inaccuracies, the safety specialist is faced with only an intuitive notion about the effectiveness of various accident prevention methods. Despite the fact that measurement is so critical in accident loss control, the majority of efforts in accident prevention have been concerned with the *techniques* of control, such as guard design, training, and so on. Little research has been devoted to the problem of how to evaluate the *effectiveness* of these control techniques.

CHARACTERISTICS OF EFFECTIVE MEASURES OF SAFETY PERFORMANCE

Suppose one is able to construct a perfect instrument for measuring safety performance. What would be the characteristics of this outstanding measurement technique? First, it must be recognized that the worthiness of any instrument of evaluation must be appraised in terms of the purpose for which it is constructed. Regardless of purpose, however, it is possible to postulate a number of characteristics of measuring instruments that will apply to most measurement situations. Such characteristics then become standards or criteria with which to judge any given instrument. Keep in mind that no single measure can meet all of these properties but it is useful to at least know what one should look for in a search for improved measures of safety performance.

The following are postulated characteristics of a good measurement technique without regard to their relative importance:

ADMINISTRATIVE FEASIBILITY.   The first characteristic of a good measuring instrument is its administrative feasibility. One must be able to construct and use it. The necessity for giving careful consideration to this characteristic is easily recognized. Personnel, time, and financial resources available for use in implementing a measurement system may strongly influence the type of instrument that can be used. In some cases an urgent need for immediate results may require that the measuring instrument produce practical answers in the shortest period of time. In all cases the return from the investment in a measurement system must far outweigh the various costs involved in using it.

CONSTANT UNITS OF MEASURE THROUGHOUT THE RANGE TO BE EVALUATED.
The measurement technique must provide measures that are at least on an interval scale. Statements about each interval along the scale continuum are needed. The scale should yield readings that are graduated into equal units, which means that the difference between successive points on the lower end of the scale should be the same as the differences between successive points at the upper end of the scale. For example, the difference between 12 inches and 13 inches is exactly the same as the difference between 1 mile and 1 inch and 1 mile and 2 inches on a linear scale.

Contrast this with the results obtained from an examination that measures knowledge of some subject. There is no proof that the difference between the score of 30 and 31 is the same as the difference between a score of 60 and 61 on a physics examination. In safety performance measurement more than nominal and ordinal measures are needed. The goal should be not only to know that an organization's safety performance is improving during the current month or year, but also to know how much it is improving. Having an interval scale also permits use of common statistical analysis techniques to distinguish real changes in performance levels from those resulting from only chance effects.

QUANTIFIABLE MEASUREMENT CRITERION CAPABLE OF STATISTICAL ANALYSIS.
Closely related to the interval scale requirement is the necessity for a measurement criterion to be quantifiable. A qualitative evaluation of safety performance limits statistical inference and opens the way for individual interpretation. The ideal criterion of safety performance should permit statistical inference techniques to be applied since, like most other measurable quantities dealing with human behavior, safety performance will necessarily be subject to statistical variation.

SENSITIVITY. A measurement technique should be sensitive enough to detect changes in process and performance levels in order to serve as a criterion for evaluation. Moreover, the frequency of occurrence of a measure must be large enough to permit statistical analyses to be conducted. No one would attempt to weigh a diamond on a cattle scale, since this type of scale is obviously not sensitive enough for that purpose. Similarly, one would not judge the effectiveness of a safety program by looking at a death rate alone. The ideal measure of safety performance must be sensitive to changes in environmental and behavioral conditions over time. This characteristic of a good criterion is significantly lacking in present-day safety performance measurement. Too often the measures used in accident prevention work are sluggish and fail to respond to either a successful accident prevention program or to a general deterioration of safety within an organization. For example, inspections and other prevention efforts often change worker behavior in a positive direction, but these changes are rarely reflected in the accident frequency rates used to appraise program effectiveness. With insensitive measures of safety performance, the evaluation of accident prevention methods becomes extremely difficult.

RELIABILITY.   The measurement technique must be reliable; that is, it should be capable of duplication with the same results obtained from successive application to the same situations. The changes we observe when applying the measurement technique should reflect actual changes in the criterion variable and not internal fluctuations in the measurement technique itself. In other words, it should be consistent over time. An ideal measure of safety performance should be reliable to the extent that it provides minimum variability when measuring the same condition. This quality of repeatability is related to the accuracy and precision of the measure itself. For example, in accident prevention this involves the notion of reporting. Disabling injury accidents may not be consistently reported over time, and first-aid frequency has always had dubious reliability since under identical conditions some workers will report to a dispensary for first-aid treatment while others will self-administer minor injuries.

STABILITY.   Similar to reliability is the need for a criterion to be stable. This involves the maintenance of a given range of values under repeated measures of worker behavior and environmental conditions. If a process does not change, the measure of its performance level is expected to remain unchanged. In accident prevention it is often noted that the accident frequency rate and severity rate vacillate over time even though it seems obvious that the plant's safety level has remained unchanged.

VALIDITY.   Of prime importance is the need for a measure to be valid. This means that it produces information that is representative of what is to be measured. The question of validity is a critical one in the pursuit of improved measures of safety effectiveness. Validating a measure requires the use of an outside criterion that more often than not is arbitrarily chosen. What constitutes successful safety performance? Does it involve the minimization of disabling injuries and first-aid cases or the minimization of all consequences of unsafe behavior and unsafe conditions? Near-injury or noninjury accidents constitute a dilemma in this regard since their occurrence may result in a definite loss to the company and serve as indicators of future disabling injuries, and yet it is often difficult to convince managers of their importance. Moreover, the effect of unsafe behavior on production efficiency in the absence of property damage or injury may constitute a considerable cost to the organization and yet go unmeasured. Despite the fact that there is no injury or property damage, unsafe behavior in itself requires the attention of the safety specialist and line managers, for, by definition, this type of behavior has a loss-producing potential.

ERROR-FREE RESULTS. A good measuring instrument should yield results that are free from error. The type and magnitude of these errors differ with different techniques. There is no perfect instrument in this respect since certain kinds of errors creep into the readings or results obtained from all types of instruments.

Errors may be classified into two types: *constant* errors and *variable* errors. Variable errors tend to cancel out when readings are made an infinite number of times. For example, if a coin is tossed six times the most probable number of heads and tails will be three each. There are likely to be errors from this expected one-to-one ratio in any sample of six tosses. However, as the number of sets of six tosses increases (that is, when we keep repeating the experiment), the ratio of heads to tails will approach unity. Errors of this type are called *variable errors*. If a *constant error* exists, no amount of replication will eliminate the error. If a coin is used that happens to be loaded in such a way that it will fall tails more often than heads, no amount of tossing will produce a ratio of unity between the number of heads and the number of tails.

The worth of present measures of safety performance might be evaluated in the light of these two types of errors. Variable errors in safety performance measures will tend to disappear as the number of accidents included in the rate computation is increased, as the number of individuals investigating the same accidents is increased, and as the number of instruments used to identify the same disabling injury information is increased. As all three of these factors are increased, accuracy improves. No increase in these factors, however, will eliminate a constant error.

Four types of constant errors prevail in present measures of safety performance. The first type arises from the accident investigator's failure to identify all of the causal factors associated with the accident. Perhaps this is due to weaknesses in his own problem perception capability or perhaps because he has fallen into the trap of using stereotyped, general terms such as "inattention" or "carelessness" as causes of accidents, which provide little or no usable information. The investigator's problem at this point consists of making sure that his appraisal of the accident situation yields the type of evidence that leads to the identification of all associated problem areas.

The second type of constant error is derived from the accident analyst's failure to break down the causal factors to be evaluated to the point where they are relatively homogeneous. The classification of accident causes according to the ANSI (American National Standards Institute) Z-16 or BLS (Bureau of Labor Statistics) standards, for example, results in a lack of homogeneity in the intraclass identifications since the categories included in the classification system are too broad for a precise definition of the problem.

A third type of biased error prevailing in safety performance measures arises from the accident analyst's failure to consider all behaviors and conditions that have the potential for producing future disabling injury or property damage losses. A good cross section of all unsafe acts and conditions that have this loss potential should be identified for use in appraising the true accident state of the entire system within which control is desired.

A fourth type of biased error arises from the accident investigator's failure to report all injurious accidents. In responding to a superior's pressure for improvement or in reaching for an award, there may be a temptation to stretch a point here and there to avoid reporting an injury that might serve as a deterrent to the achievement of an accident reduction objective. Manipulating data to achieve the illusion of favorable safety performance obviously defeats the purpose of measurement; the numbers are only useful if they are true representations of real-world experience.

EFFICIENT AND UNDERSTANDABLE.   Finally, a good measurement technique should be both efficient and understandable. Efficiency requires that the cost of obtaining and using the instrument is consistent with the benefit to be gained. Moreover, it should be easy to obtain with minimum disruption to the normal operations of an organization. To be understandable suggests that the criterion be understood by those charged with the responsibility for approving and using it. Without the latter property all others would be of little practical value since the measurement technique would most likely never be used.

It is obvious that these characteristics of a good performance measure can rarely if ever be achieved by any single instrument of evaluation. Usually only a combination of measures can provide even a reasonable compromise. In the field of occupational accident prevention presently used measures of safety performance are inadequate in many of the characteristics cited. New measures are needed to enhance the ability to control and predict accidents. Most probably a combination of several safety measurement techniques will be required in order ultimately to achieve worthwhile accident loss control objectives.

These then are the desirable qualities available if one waved a "magic wand" and created a perfect safety measuring instrument. Taken collectively, they represent a utopian state toward which attempts to improve safety performance measures should be directed. The next step is to examine briefly the current state of affairs and then move forward toward the ultimate state of measurement perfection that we have defined.

CURRENTLY USED MEASURES OF SAFETY PERFORMANCE

A number of safety performance indexes are now in use, such as number of disabling injuries, injury frequency rates, injury severity rates, accident costs, number of deaths, number of first-aid cases, recordable occupational illnesses, the ratio of injury severity to injury frequency, and total injury rates. A few companies have adopted a "serious injury index," which includes information about accidents resulting in certain types of injuries regardless of the degree of disability involved. Many of these indexes are recommended by the American National Standards Institute (ANSI) for use in measuring safety performance and are described in ANSI's (1967, r.1973) *Method of Recording and Measuring Work Injury Experience: ANSI Z-16.1.* Injury rates compiled in accordance with this standard are intended to show the relative need for accident prevention activities within an organization, to indicate the seriousness of the accident problem, to measure the effectiveness of safety activities in organizations with comparable hazards, and to evaluate progress in accident prevention within an organization or industry.

ANSI TERMINOLOGY FOR WORK INJURY MEASUREMENT STANDARDS

Certain definitions concerning the various aspects of disabling work injuries have been established by the American National Standards Institute as follows:

Work Injury. Any injury suffered by a person that arises out of or in the course of his or her employment. The word "injury" also includes occupational disease and work-concerned disability. Work injuries are classified as follows:

1. *Death.* Any fatality resulting from a work injury, regardless of the time intervening between injury and death.
2. *Permanent Total Disability.* Any injury other than death that permanently and totally incapacitates an employee from following any gainful occupation or results in loss of or the complete loss of use of any of the following in one accident:
    a. Both eyes
    b. One eye and one hand, or arm, or leg, or foot
    c. Any two of the following not on the same limb: hand, arm, foot, or leg
3. *Permanent Partial Disability.* Any injury other than death or permanent total disability that results in the complete loss or loss of use of any member or part of a member of the body, or any permanent impairment of functions of the body or part thereof, regardless of any preexisting disability of the injured member or impaired body function.

The following injuries are *not* classified as permanent partial disability:
a. Inguinal hernia, if it is repaired
b. Loss of fingernail or toenails
c. Loss of tip of finger without bone involvement
d. Loss of teeth
e. Disfigurement
f. Strains or sprains that do not cause permanent limitation of motion
g. Simple fractures to the fingers and toes; also such other fractures as do not result in permanent impairment or the restriction of normal function of the injured member

4. *Temporary Total Disability.* Any injury that does not result in death or permanent impairment but one that renders the injured person unable to perform a regularly established job that is open and available to him, during the entire time interval corresponding to the hours of his regular shift on any one or more days (including Sundays, days off, or plant shutdown) subsequent to the date of the injury.
5. *Medical Treatment Injury.* An injury that does not result in death, permanent impairment, or temporary total disability but requires medical treatment (including first aid).

Disabling Injury (Lost-Time Injury). A work injury that results in death, permanent total disability, permanent partial disability, or temporary total disability as defined above. These are the injuries used in calculating the standard injury frequency and severity rates.

Occupational Disease. A disease caused by environmental factors, the exposure to which is peculiar to a particular process, trade, or occupation and to which an employee is not ordinarily subjected or exposed outside of or away from such employment.

Employment
1. All work or activity performed in carrying out an assignment or request of the employer, including incidental and related activities not specifically covered by the assignment or request.
2. Any voluntary work or activity undertaken while on duty with the intent of benefiting the employer.
3. Any activities undertaken while on duty with the consent or approval of the employer.

Arising Out of and In the Course of Employment. An injury resulting from the work activity or environment of employment.

Regularly Established Job. A job that has not been established especially to accomodate an injured employee, either for therapeutic reasons or to avoid counting the case as a temporary total disability.

Total Days Charged. The combined total, for all injuries, of
1. All days of disability resulting from temporary total injuries plus
2. All scheduled charges assigned to fatal permanent total and permanent partial injuries.

The days of disability are the total of full calendar days on which the injured person was unable to work as a result of a temporary total injury. The total does not include the day the injury occurred or the day the injured person returned to work, but it does include all intervening calendar days (including Sundays, days off, or plant shutdown). It also includes any other full days of inability to work because of the specific injury, subsequent to the injured person's return to work.

Exposure. The total number of employee hours worked by all employees including those in operation, production, maintenance, transportation, clerical, administrative, sales, and other activities.

ANSI INJURY SEVERITY TIME CHARGES

In the evaluation of injury severity the American National Standards Institute has established a certain schedule of charges for death and permanent disabilities involving various parts of the body. A death resulting from work injuries is assigned a time charge of 6000 days. Permanent total disabilities resulting from work injuries are also assigned a time charge of 6000 days each. Permanent partial disabilities, either traumatic or surgical, resulting from work injuries are assigned charges according to a Table of Scheduled Charges (see Table 2-1). These charges are used whether the actual number of days lost is greater or less than the scheduled charges, or even if no days are lost at all.

Certain rules have been established for using the scheduled charges in evaluating severity:

1. *Charges for Finger and Toe Amputations.* For each finger (or toe) use only one charge—the charge shown in Table 2-1 for the highest valued bone involved. For amputations of more than one finger (or toe), total the separate charges for each finger (or toe).

2. *Charges for Loss of Use.* The charge for loss of use is the percentage of the scheduled charge, corresponding to the percentage loss of use of the member or part of member involved, as determined by the physician engaged or authorized by the employer to treat the case. An exception is made in respect to loss of hearing, which is considered a permanent partial disability by the ANSI only in the event of complete industrial loss of hearing, in which case the full-time charges apply.

3. *Impairments Affecting More Than One Part of the Body.* For permanent impairment affecting more than one part of the body, the total charge is the sum of the scheduled charges for the individual body parts impaired. If the sum exceeds 6000 days, the total charge shall be 6000 days.

4. *Permanent and Temporary Injuries in the Same Accident.* When an employee suffers a permanent partial injury to one part of the body and a temporary total injury to another part in one accident, the greater charge shall be used and shall determine the injury classification.

5. *Charges for Injuries Not Identified in the Table.* The charge for any permanent injury other than those identified in the schedule of charges (such as damage to internal organs, loss of speech, damaged lungs, back, and so on) is a percentage of permanent total disability that results from the injury, as determined by the physician engaged or authorized by the employer to treat the case.

6. *Charge for Temporary Total Disability.* The charge for a temporary total disability is the total number of full calendar days on which the injured person was unable to work as a result of the injury as defined by the temporary total disability classification.

ANSI INJURY INDEXES

The American National Standards Institute's *Method of Recording and Measuring Work Injury Experience: Z-16.1* (1967 r. 1973) suggests that

Table 2-1
Table of Scheduled Charges for Permanent Partial Disabilities

A.   For Loss of Member—Traumatic or Surgical

1. Fingers, thumb, and hand

Amputation involving all or part of bone*	Fingers				
	Thumb	Index	Middle	Ring	Little
Distal Phalange	300	100	75	60	50
Middle Phalange	-	200	150	120	100
Proximal Phalange	600	400	300	240	200
Metacarpal	900	600	500	450	400
Hand at Wrist 3000					

2. Toe, foot, and ankle

Amputation involving all or part of bone*	Great Toe	Each of Other Toes
Distal Phalange	150	35
Middle Phalange	-	75
Proximal Phalange	300	150
Metatarsal	600	350
Foot at Ankle 2400		

3. Arm

Any point above† elbow including shoulder joint	4500
Any point above wrist and at or below elbow	3600

4. Leg

Any point above† knee	4500
Any point above ankle and at or below knee	3000

B.   Impairment of Function

1. One eye (loss of sight), whether or not there is sight in the other eye	1800
2. Both eyes (loss of sight), in one accident	6000
3. One ear (complete industrial loss of hearing) whether or not there is hearing in the other ear	600
4. Both ears (complete industrial loss of hearing) in one accident	3000
5. Unrepaired hernia	50

*If the bone is not involved, use actual days lost and classify as temporary total disability.
†The term "above" when applied to the arm means toward the shoulder, and when applied to the leg means toward the hip.

injury experience be measured by means of a disabling injury frequency rate, a disabling injury severity rate, the average days charged per disabling injury and a disabling injury index (which is included as an aid in combining frequency and severity experience and is not a part of the standard). ANSI further suggests that the standard injury rates be compiled in accordance with certain rules that are included in the Z-16.1 standard.

The disabling injury frequency rate is based on the total number of deaths, permanent total, permanent partial, and temporary total disabilities which occur during the period covered by the rate. The rate relates these injuries to hours worked during the period and expresses them in terms of a million-hour unit by use of the following formula:

$$\text{Disabling injury frequency rate} = \frac{\text{Number of disabling injuries} \times 1,000,000}{\text{Employee hours of exposure}}$$

The major advantage of the injury frequency rate is that it takes into account differences in *quantity* of exposure due to varying employee hours of work, either within the plant during successive time periods or among plants within similar industry classifications. Insofar as the assumption can be accepted that the injury frequency rate is a reflection of accident prevention efforts, then this index provides a method for measuring how adequately a safety program is functioning. Also, since the injury frequency rate has been adopted by the ANSI as a standard measure of safety performance, it is available for comparison purposes among various companies within a given industry.

The disabling injury severity rate is based on the total scheduled charges for all deaths, permanent total and permanent partial disabilities, and the total days of disability from all temporary total injuries that occur during the period covered by the rate. The rate relates these days charged to the hours worked during the period and expresses the loss in terms of a million-hour unit, by use of the following formula:

$$\text{Disabling injury severity rate} = \frac{\text{Total days charged} \times 1,000,000}{\text{Employee hours of exposure}}$$

Similar to the injury frequency rate, the major value of the injury severity rate is that it takes into account differences in quantity of exposure over time. It also answers the question "How serious are our injuries?" This rate has been established as an ANSI standard and thus is available for use in making comparisons among different organizations and among various units within an organization.

The ANSI standard of average days charged per disabling injury expresses the relationship between the total days charged and the total number of disabling injuries, as defined by the Z-16.1 publication. This index may be computed by dividing the injury severity (S) rate by the injury frequency (F) rate (producing an S/F ratio):

$$S/F = \frac{\dfrac{\text{Total days charged} \times 10^6}{\text{Employee hours of exposure}}}{\dfrac{\text{Number of disabling injuries} \times 10^6}{\text{Employee hours of exposure}}}$$

Or, one may compute the average days charged per disabling injury directly by simply dividing the total days charged by the total disabling injuries:

$$S/F = \frac{\text{Total days charged}}{\text{Number of disabling injuries}}$$

In effect, this measure reveals whether or not the more severe accidents as well as those with less severity are eliminated. Thus, a sort of "third dimension" is available along with the injury frequency and severity rates. With a goal of loss control, a plotted S/F ratio hopefully will have a downward trend over time. This ratio should appear as at least a straight line during successive time periods, indicating that the prevention efforts are controlling both the major and minor losses.

An average days charged index may be calculated separately for all disabling injuries, for permanent partial injuries only, and for temporary total injuries only. To obtain the average for permanent partial injuries, divide the number of such injuries into the scheduled number of days charged for injuries appearing within this classification. To obtain the average for temporary total injuries, divide the number of these injuries into the days of disability resulting from them.

Accidents that produce injuries not resulting in death or permanent total, permanent partial, or temporary total disabilities but that require some type of treatment are classified as minor injuries. Injuries in this classification are often subdivided into doctors' cases and first-aid cases, depending on whether the injury was treated by a physician or a nonphysician. Essentially, injuries of this type result in less than 24 hours lost time and do not fall within one of the permanent injury classifications. Minor injuries provide several advantages as an index of safety performance in contrast with lost-time or disabling injury cases. They occur more frequently than disabling injuries, they are more sensitive to changes in safety performance level, and they often enable hazardous conditions and/or unsafe practices to be detected and corrected before more serious injuries occur. The main disadvantage to the use of minor injuries as a measure is that such injuries often are not reported. A decrease in reporting accuracy will have the same influence on this criterion of safety performance as would an outstanding accident prevention program. There is also a general tendency for workers to ignore or provide self-treatment for minor injuries if too much emphasis is placed on their numerical reduction.

SERIOUS INJURY INDEX

Several companies have developed a measure of safety effectiveness that includes all serious injuries involving temporary total disability, permanent total or partial disability, limited duty, fractures, lacerations requiring sutures, and eye cases requiring treatment by a physician. Injuries falling within any one of these classifications, according to the judgment of a physician, are identified as serious injuries and are included in the computation of a frequency rate regardless of the degree of lost time involved. Injuries included in the serious injury index are defined in five categories:

1. *Disabling Injuries.* Those injuries similarly labeled in the American National Standards Institute's Z-16.1 standard. In addition to deaths and permanent disablements, included as the bulk of this category are the temporary total disabilities involving more than 24 hours lost time from the job.
2. *Eye Injuries Requiring a Doctor's Attention.* The most common eye injury involves a foreign body that is removed by a nurse or dispensary attendant. Occasionally a foreign body produces a more serious result requiring removal and/or treatment by a physician. Or, a patient may have had an irritating liquid splashed into his eyes requiring diagnosis or treatment by a doctor. These latter two cases are included in the computation of the serious injury index.
3. *Injuries Resulting in Fractures.* Whenever enough force has been exerted by or applied to the body to produce a fractured bone, there is need for attention, for this force is out of control or the individual has placed himself in a hazardous position. This category is included, say the users, because it is imperative that all facts associated with such an injury be detected and problem areas corrected.
4. *Injuries Requiring Sutures.* Any laceration is serious. When the extent of the injury is great enough to require sutures, it is "out of the ordinary" and should be investigated even though the injured person immediately returns to work.
5. *Injuries Requiring Modified or Restricted Work for the Injured Person.* An injury necessitating the modification of work by work restrictions or motion limitations detracts from the efficient operation of a department. The intent here is to provide a means of discovering and correcting circumstances leading to this source of production loss.

The measure covering these classifications has been given several labels by various companies using it, including serious injury index, serious work injury index, total injury frequency rate, and serious injury frequency rate. This rate is usually calculated according to the following formula:

$$\text{Serious injury index} = \frac{\text{Number of serious injuries} \times 1,000,000}{\text{Man-hours exposure}}$$

Advantages of this index over the ANSI injury frequency claimed by its users include (1) the system provides more data, thus providing increased sensitivity, (2) it replaces the safety engineer as the interpreter of what is reportable by relying on medical opinion related to the nature of the injury, (3) it places emphasis on the nature of the injury itself regardless of whether or not sufficient lost time or permanent disablement was involved to include the accident in the standard evaluation system, and (4) it is presumably less subject to pressure for a favorable interpretation in order to win a contest or to achieve other objectives (Gilmore and Buttery, 1962; Klingel and Haier, 1956; Voland, 1962).

THE BLS–OSHA METHOD OF MEASUREMENT

A relatively new system for generating occupational safety and health statistics has been developed by the Bureau of Labor Statistics (BLS) under the provisions of the Williams–Steiger Occupational Safety and Health Act of 1970 (OSHA). Booklets available from the U.S. Department of Labor describe the OSHA recordkeeping requirements in detail (*Recordkeeping Requirements Under the Occupational Safety and Health Act of 1970,* U.S. Department of Labor, Superintendent of Documents, U.S. Government Printing Office, Washington, D.C. 20212).

The BLS–OSHA system of measuring safety and health performance is intended to serve as a nationwide survey of work injuries and illnesses. Virtually all employers throughout the United States are required to participate. The statistical universe has thus been expanded to include almost 60 million Americans at about 5 million work places. The scope of recordable injury and illness has been widened over the ANSI Z-16.1 coverage to present a more comprehensive picture of work injury and illness losses. Every work-related injury or illness that involves loss of consciousness, restriction of work or motion, medical treatment (other than first aid), or transfer to another job must be recorded. Every occupational death also must be recorded. One significant change over the ANSI record system is the elimination of the major reporting loophole of a temporary total disability requiring one or more days of disability, with "day of disability" defined as "any day on which an employee is unable, because of injury, to perform effectively throughout a full shift, the essential functions of a regularly established job which is open and available to him." This prevents masking the incidence of injury by transferring the worker to a new job or by retiring or firing him. Only the simple first-aid cases may be excluded (see definitions below). A primary use of these statistics is for standards setting and compliance activities.

The BLS–OSHA recordkeeping system provides national statistics on an industry basis for all recordable occupational injuries and illnesses occurring at the work place. In addition, the system provides OSHA and state safety inspectors with on-the-spot records of each occupational accident. The recordkeeping system involves two forms for use in recording work-related injuries and illnesses. A Log and Summary of Occupational Injuries and Illnesses (OSHA Form No. 200), a Supplementary Record of Occupational Injuries and Illnesses (OSHA Form No. 101), and a Summary of Occupational Injuries and Illnesses (OSHA Form No. 102). All forms remain at the work place and are made available to the federal or state inspectors according to the provisions of the act. Employers who fall into the statistical sample used by BLS are required to submit reports to the BLS or to the cooperating states at the end of the recordkeeping period.

Small employers with no more than 10 full- or part-time employees at any one time during the previous calendar year are exempt from the OSHA recordkeeping requirements, except that they must report any accident that results in a fatality or the hospitalization of five or more employees and must record if they are one of a few small employers who are selected to participate in the annual survey of occupational injuries and illnesses. In addition, employers of domestics and employers engaged in religious activities also are exempt.

Terminology used in the BLS system serve to highlight its scope:

Occupational Injury. Any injury such as a cut, fracture, sprain, amputation, that results from a work accident or from exposure involving a single incident in the work environment.

Occupational Illness. Any abnormal condition or disorder, other than one resulting from an occupational injury, caused by exposure to environmental factors associated with the employee's employment. It includes acute and chronic illnesses or diseases that may be caused by inhalation, absorption, ingestion, or direct contact and that can be included in the categories listed below. The following listing gives the categories of occupational illnesses and disorders that are used for the purpose of classifying recordable illnesses. The examples are considered typical and are not to be assumed to represent a complete listing of the types of illnesses and disorders included under each category.

1. *Occupational Skin Diseases or Disorders.* Examples: contact dermatitis, eczema, or rash caused by primary irritants and sensitizers or poisonous plants; oil acne; chrome ulcers; chemical burns or inflammations.
2. *Dust Diseases of the Lungs (Pneumoconiosis).* Examples: silicosis, asbestosis, coal workers' pneumoconiosis, byssinosis, silerosis, and other pneumoconioses.
3. *Respiratory Conditions Due to Toxic Agents.* Examples: pneumonitis, pharyngitis, rhinitis or acute congestion due to chemicals, dusts, gases, or fumes; farmers' lung.
4. *Poisoning (Systemic Effects of Toxic Materials).* Examples: poisoning by lead, mercury, cadmium, arsenic, or other metals; poisoning by carbon monoxide, hydrogen sulfide, or other gases; poisoning by benzol, carbon tetrachloride, or other organic solvents; poisoning by insecticide sprays such as parathion, lead arsenate; poisoning by other chemicals such as formaldehyde, plastics and resins.
5. *Disorders Due to Physical Agents (Other Than Toxic) Materials.* Examples: heatstroke, sunstroke, heat exhaustion and other effects of environmental heat; freezing, frostbite and effects of exposure to low temperatures; caisson disease; effects of ionizing radiation (isotopes, X-rays, radium); effects of nonionizing radiation (welding flash, ultraviolet rays, microwaves, sunburn).
6. *Disorders Due to Repeated Trauma.* Examples: noise-induced hearing loss; synovitis, tenosynovitis, and bursitis; Raynaud's phenomena; and other conditions due to repeated motion, vibration, or pressure.
7. *All Other Occupational Illnesses.* Examples: anthrax brucellosis, infectious hepatitis, malignant and benign tumors, food poisoning, histoplasmosia, coccidioidomycosis.

Recordable Cases. Any occupational injuries or illnesses that result in
1. *Death,* regardless of the time between the injury and death or the length of the illness.
2. *Nonfatal Occupational Illnesses,* other than fatalities, that result in lost workdays.
3. *Occupational Injuries* that result in transfer to another job, or require medical treatment (as defined below), or involve loss of consciousness or restriction of work or motion.

Medical Treatment. Treatment (other than first aid) administered by a physician or by registered professional personnel under the standing orders of a physician. Medical treatment does *not* include first-aid treatment (one-time treatment and subsequent observation of minor scratches, cuts, burns, splinters, and so forth that do not ordinarily require medical care) even though provided by a physician or registered professional personnel.

Establishment. A single physical location where business is conducted or where services or industrial operations are performed (for example, a factory, mill, store, hotel, restaurant, movie theatre, farm, ranch, bank, sales office, warehouse, or central administrative office). Where distinctly separate activities are performed at a single physical location (such as contract construction activities operated from the same physical location as a lumber yard), each activity shall be treated as a separate establishment. For firms engaged in activities that may be physically dispersed (such as agriculture, construction, transportation, communications, and electric, gas, and sanitary services) records may be maintained at a place to which employees report each day. Records for personnel who do not primarily report or work at a single establishment (such as traveling salesmen, technicians, engineers) shall be maintained at the location from which they are paid or the base from which personnel operate to carry out their activity.

Work Environment. The physical location, equipment, materials processed or used, and the kinds of operations performed in the course of an employee's work, whether on or off the employer's premises.

Lost Workdays. The number of days the employee would have worked but could not because of occupational injury or illness. The number of lost workdays should not include the day of injury or onset of illness or any days on which the employee would not have worked even though able to work. The number of days includes all days (consecutive or not) on which, because of the injury or illness,
1. The employee was assigned to another job temporarily
2. The employee worked at a permanent job less than full time
3. The employee worked at a permanently assigned job but could not perform all duties normally connected with it.
   For employees not having a regularly scheduled shift (such as, certain truck drivers, construction workers, farm labor, casual labor, part-time employees), it may be necessary to estimate the number of lost workdays. Estimates of lost work-days shall be based on prior work history of the employee *and* days worked by employees, not ill or injured, working in the department and/or occupation of the ill or injured employee.

The BLS recordable occupational injury and illness rate is identified as the *incidence rate,* with a base of 200,000 man-hours exposure.

Incidence rate = $N/MH \times 200,000$

where: N = Number of injuries and/or illnesses

MH = Man-hours = Total hours worked by all employees during the reference year

200,000 = Base for 100 full-time equivalent workers working 40 hours per week, 50 weeks per year

The incidence rate is calculated annually by the BLS according to the SIC (*Standard Industrial Classification*) *Manual* two-digit industry classifications and for the total United States industry. Separate incidence rates are calculated based on total recordable cases, lost workday cases, and nonfatal cases without lost workdays. Incidence rates are also calculated by size and unit categories and industry divisions.

An occupational injury or illness must be entered on the Log and Summary of Occupational Injuries and Illnesses (OSHA Form 200) within six working days after notification of the case (see Figure 2-1, p. 34). This form contains columns for entering the date of injury or onset of illness, employee's name, occupation of injured or ill employee, department to which employee was assigned, description of injury or illness, and part of body affected. Each case is also classified either as an injury or as one of seven classes of illnesses. In the case of a fatality, the death date is also entered. When cases involve one or more lost workdays (but not death), the number of days is entered; nonfatal cases without lost workdays (for example, temporary loss of consciousness) are also indicated.

In addition to the items entered on the Log and Summary, additional information must be recorded within six days on the Supplementary Record of Occupational Injuries and Illnesses (OSHA Form 101) (see Figure 2-2, pp. 36–37). Information on this form primarily concerns the accident or exposure that resulted in injury or illness.

Records must be kept at the lowest possible organization level in order to provide records for use near the operating location and to assure that the statistics accurately reflect the size and activity of the reporting unit. The system is designed to avoid the pooling of information from large numbers of small establishments, with one report presenting the combined records of such diverse activities as a central administrative office, warehouse operations, production operations, and maintenance activities. In a combined report, low-accident activities, such as administration, tend to dilute high-accident activities, such as production, and thus mask important accident trends.

Figure 2-1

Bureau of Labor Statistics Log and Summary of Occupational Injuries and Illnesses.

Page ____ of ____

For Calendar Year 19 ____

Form Approved
O.M.B. No. 44R 1453

Company Name

Establishment Name

Establishment Address

NOTE: This form is required by Public Law 91-596 and must be kept in the establishment for 5 years. Failure to maintain and post can result in the issuance of citations and assessment of penalties. *(See posting requirements on the other side of form.)*

RECORDABLE CASES: You are required to record information about every occupational death; every nonfatal occupational illness; and those nonfatal occupational injuries which involve one or more of the following: loss of consciousness, restriction of work or motion, transfer to another job, or medical treatment (other than first aid). *(See definitions on the other side of form.)*

Case or File Number	Employee's Name	Occupation	Department	Description of Injury or Illness	Extent of, and Outcome of INJURY				Type, Extent of, and Outcome of ILLNESS												
					Fatalities	Nonfatal Injuries			Type of Illness								Fatalities	Nonfatal Illnesses			
					Injury Related	Injuries With Lost Workdays		Injuries Without Lost Workdays	Occupational skin diseases or disorders	Dust diseases of the lungs	Respiratory conditions due to toxic agents	Poisoning (systemic effects of toxic materials)	Disorders due to physical agents	Disorders associated with repeated trauma	All other occupational illnesses		Illness Related	Illnesses With Lost Workdays		Illnesses Without Lost Workdays
Enter a nondupli-cating number which will facilitate comparisons with supplementary records.	Enter First name or initial, middle initial, last name.	Enter regular job title, not activity employee was performing when injured or at onset of illness. In the absence of a formal title, enter a brief description of the employee's duties.	Enter department in which the employee is regularly employed or a description of normal workplace to which employee is assigned, even though temporarily working in another department at the time of injury or illness.	Enter a brief description of the injury or illness and indicate the part or parts of body affected. Typical entries for this column might be: Amputation of 1st joint right forefinger; Strain of lower back; Contact dermatitis on both hands; Electrocution—body.	Enter DATE of death. Mo./day/yr.	Enter a CHECK if injury involves days away from work, or days of restricted work activity, or both.	Enter number of DAYS away from work.	Enter number of DAYS of restricted work activity.	Enter a CHECK if no entry was made in columns 1 or 2 but the injury is recordable as defined above.	CHECK Only One Column for Each Illness *(See other side of form for terminations or permanent transfers.)*							Enter DATE of death. Mo./day/yr.	Enter a CHECK if illness involves days away from work, or days of restricted work activity, or both.	Enter number of DAYS away from work.	Enter number of DAYS of restricted work activity.	Enter a CHECK if no entry was made in columns 8 or 9.
(A)	(B)	(C)	(D)	(E)	(F)	(1)	(2)	(3)	(4)	(5)	(6)	(a)	(b)	(c)	(d)	(e)	(f)	(g)	(7)	(8)	(9)	(10)	(11)	(12)	(13)
				PREVIOUS PAGE TOTALS	➡																				
				TOTALS (transcribe on other side of form.) ➡																					

FOLD

OSHA No. 200

Certification of Annual Summary Totals By _____ Title _____ Date _____

POST ONLY THIS PORTION OF THE LAST PAGE NO LATER THAN FEBRUARY 1.

34

The basic recordkeeping unit is the *establishment,* defined as a single physical location where business is conducted or where services or industrial operations are performed. Distinctly separate activities, such as contract construction activities at a lumber yard, are treated as separate establishments. Certain exceptions may be allowed where an employer defines "establishment" in a different manner or keeps records at a location other than the establishment.

The BLS–OSHA method of safety performance measurement differs in a number of ways from the ANSI Z-16.1 standard. The BLS definition of recordable occupational injuries and illnesses represents the most significant change. Certain specific cases are defined in the Occupational Safety and Health Act as "recordable" occupational injuries and illnesses. Statistics are required to be compiled on work-related deaths, injuries, and illnesses other than minor injuries requiring only first-aid treatment. In addition, the act specifically includes medical treatment cases, cases in which there is a restriction of work or motion and cases in which there is a transfer to another job. Recordable cases are divided into three classes: fatalities, lost workday cases, and nonfatal cases without lost workdays. Under the BLS–OSHA system lost workdays, not calendar days, are counted. Similarly, cases involving the assignment of a worker to a temporary job and cases in which an injured employee can work at his own permanent job but either cannot perform all of the functions or cannot perform them all day are included in the reporting system as recordable cases. The Z-16.1 provision that no time was lost (and thus no injury was recorded) as long as the employee could perform another established and available job has been eliminated in the BLS–OSHA system. Any change in work assignment necessitated by an occupational accident or illness is recordable.

Another significant change in the BLS–OSHA recording system is the elimination of the time charge provision. The Z-16.1 standard assigned fixed time charges for fatalities and permanent disabilities and measured time lost for other injuries in terms of calendar days. Under the BLS–OSHA system the exact number of lost workdays is recorded. Recording procedures where fatalities are involved are also changed. The Z-16.1 system arbitrarily assigned a charge of 6000 workdays per death (the estimated equivalent of 20 years of lost workdays). The BLS–OSHA system requires no time charge for work fatalities. In cases where an employee dies after returning to work as the result of a lingering illness or injury, employers are required to correct their records to reflect the additional fatality.

The BLS–OSHA base for reporting injury frequency rates is 100 full-time employees in contrast to the 1 million employee hours contained in the Z-16.1 standard. The incident rate is calculated based on 100 full-time equivalent workers working 40 hours per week, 50 weeks per year, which equals 200,000. As a replacement for the Z-16.1 severity rate, average days

Figure 2-2
Supplementary Record of Occupational Injuries and Illnesses.

OSHA No. 101
Case or File No. _____

Form approved
OMB No. 44R 1453

Supplementary Record of Occupational Injuries and Illnesses

EMPLOYER
1. Name _____
2. Mail address _____
   (No. and street)   (City or town)   (State)
3. Location, if different from mail address _____

INJURED OR ILL EMPLOYEE
4. Name _____ Social Security No. _____
   (First name)   (Middle name)   (Last name)
5. Home address _____
   (No. and street)   (City or town)   (State)
6. Age _____   7. Sex: Male_____ Female_____ (Check one)
8. Occupation _____
   (Enter regular job title, *not* the specific activity he was performing at time of injury.)
9. Department _____
   (Enter name of department or division in which the injured person is regularly employed, even
   though he may have been temporarily working in another department at the time of injury.)

THE ACCIDENT OR EXPOSURE TO OCCUPATIONAL ILLNESS
10. Place of accident or exposure _____
    (No. and street)   (City or town)   (State)
    If accident or exposure occurred on employer's premises, give address of plant or establishment in which
    it occurred. Do not indicate department or division within the plant or establishment. If accident oc-
    curred outside employer's premises at an identifiable address, give that address. If it occurred on a pub-
    lic highway or at any other place which cannot be identified by number and street, please provide place
    references locating the place of injury as accurately as possible.
11. Was place of accident or exposure on employer's premises? _____ (Yes or No)
12. What was the employee doing when injured? _____
    (Be specific. If he was using tools or equipment or handling material,
    _____
    name them and tell what he was doing with them.)
    _____
13. How did the accident occur? _____
    (Describe fully the events which resulted in the injury or occupational illness. Tell what
    _____
    happened and how it happened. Name any objects or substances involved and tell how they were involved. Give
    _____
    full details on all factors which led or contributed to the accident. Use separate sheet for additional space.)

OCCUPATIONAL INJURY OR OCCUPATIONAL ILLNESS
14. Describe the injury or illness in detail and indicate the part of body affected. _____
    (e.g.: amputation of right index finger
    _____
    at second joint; fracture of ribs; lead poisoning; dermatitis of left hand, etc.)
15. Name the object or substance which directly injured the employee. (For example, the machine or thing
    he struck against or which struck him; the vapor or poison he inhaled or swallowed; the chemical or ra-
    diation which irritated his skin; or in cases of strains, hernias, etc., the thing he was lifting, pulling, etc.)
    _____
    _____
16. Date of injury or initial diagnosis of occupational illness _____
    (Date)
17. Did employee die? _____ (Yes or No)

OTHER
18. Name and address of physician _____
19. If hospitalized, name and address of hospital _____
    _____
    Date of report _____ Prepared by _____
    Official position _____

Figure 2-2
(Continued)

SUPPLEMENTARY RECORD OF
OCCUPATIONAL INJURIES
AND ILLNESSES

To supplement the Log and Summary of Occupational Injuries and Illnesses (OSHA No. 200), each establishment must maintain a record of each recordable occupational injury or illness. Worker's compensation, insurance, or other reports are acceptable as records if they contain all facts listed below or are supplemented to do so. If no suitable report is made for other purposes, this form (OSHA No. 101) may be used or the necessary facts can be listed on a separate plain sheet of paper. These records must also be available in the establishment without delay and at reasonable times for examination by representatives of the Department of Labor and the Department of Health, Education and Welfare, and States accorded jurisdiction under the Act. The records must be maintained for a period of not less than five years following the end of the calendar year to which they relate.

Such records must contain at least the following facts:

1) *About the employer*—name, mail address, and location if different from mail address.

2) *About the injured or ill employee*—name, social security number, home address, age, sex, occupation, and department.

3) *About the accident or exposure to occupational illness*—place of accident or exposure, whether it was on employer's premises, what the employee was doing when injured, and how the accident occurred.

4) *About the occupational injury or illness*—description of the injury or illness, including part of body affected; name of the object or substance which directly injured the employee; and date of injury or diagnosis of illness.

5) *Other*—name and address of physician; if hospitalized, name and address of hospital; date of report; and name and position of person preparing the report.

SEE *DEFINITIONS* ON THE BACK OF OSHA FORM 200.

---

charged per permanent-partial disabling injury, and other units of measure, the BLS–OSHA statistics include injury and illness incidence rates, lost workday cases, and number of lost workdays.

While the BLS–OSHA method of safety performance measurement has some decided advantages over the ANSI Z-16.1 system, there remain a number of difficulties, particularly when the system is used as an index of within-plant safety performance.

SHORTCOMINGS OF THE STANDARD INDEXES

A close examination of the prevailing methods of evaluating safety performance reveals the following major weaknesses:

1. The standard methods of evaluation based on injury frequency rates, severity rates, and incidence rates are not sensitive enough to serve as an

accurate indicator of safety effectiveness. Only those accidents or illnesses resulting in actual losses are included in the rate computations.

2. The smaller the work force, the less reliable is the frequency rate, sever-
    ity rate, or incidence rate as an indicator of safety performance, partic-
    ularly when less than the base numbers of 1 million man-hours (Z-16.1)
    or 200,000 man-hours (BLS–OSHA) are worked during the period. In
    fact, when exposures below 1 million man-hours or 200,000 man-hours
    are worked during the period for which the rates are computed, the
    calculated numbers are technically hypothetical figures that, per se, have
    little meaning.

3. Lost-time accidents, recordable occupational injuries and illnesses,
    deaths, and other injuries reported according to present criteria are
    relatively rare events. Small units may go for a long period without a
    reportable accident or incident under the present systems of measure-
    ment.

4. Under the Z-16.1 reporting criteria, a single severe injury or death will
    drastically alter the severity rate, particularly in smaller organizations,
    and thus this index may not accurately reflect overall prevention ac-
    complishment. The problem of chance influences is also present in the
    frequency and incidence rate measures. In these cases chance determines
    whether or not an injury is of sufficient severity to be included in the
    reportable or recordable classifications in the first place. Under Z-16.1, if
    by chance a period of less than 24 hours lost time is involved, the injury
    does not appear in the frequency rate measure. Similarly, if the provi-
    sions of the BLS–OSHA reporting criteria are not met, the accident or
    illness is not included in the incident rate computations, although the
    BLS–OSHA system of measurement is more sensitive in terms of types
    of cases covered.

5. In the national statistics computed under both systems, comparisons are
    made among accidents occurring in various types of environments involv-
    ing nonparallel hazard categories. For example, the exposure and acci-
    dent experience of material handlers are lumped together with those of
    office workers, and similar data for milling machine operators are com-
    bined with those of stockroom clerks within rates computed by industry.

6. The measurement techniques presently in use are only remotely related
    to the behavioral and environmental changes that prevention program-
    ming activities or accident countermeasures are designed to produce.
    For example, how long does it take for a new safety training program to
    reflect itself in a reduced frequency rate or incidence rate? In most cases
    we must wait a considerable period of time to allow sufficient exposure
    to accumulate so that adequate data can be collected for a realistic fre-
    quency rate or incidence rate appraisal. There may be serious discrepan-
    cies between the problems identified by presently used measurement

systems and the direct appraisal of the behavioral malfunctions that safety programs and accident countermeasures are designed to influence.

7. Most present indexes of safety and health performance are based on an after-the-fact appraisal of injury-producing or property-damaging accidents or work-related illnesses. Some loss must be involved with a certain degree of severity, as defined by the reporting criteria, before an accident appears on a report form. In most cases accident causal information is derived solely from an examination of causal factors associated with loss-type accidents or illnesses. What is needed is a method for examining accidents at the noninjurious state where the *potential* for loss is involved but where the loss has not yet actually occurred.

8. Finally, many accidents, particularly the less severe ones, are never reported. Information valuable for analysis and control purposes is thus excluded from the presently used evaluation system. This problem may become especially acute when there is strong competition to show a reduction in the frequency or incidence rate index number as the basis for winning a contest or coming out on top in a safety award program. No doubt the most reliable index we have is the number of deaths. One simply cannot hide the body under the rug and forget about it. As the injury decreases in severity, it becomes progressively easier to ignore it or to remove it from the "reportable" or "recordable" category. For example, when pressure is put on the supervisor to cut down on his first-aid cases, he may tell his employees not to report their minor injuries to the dispensary, but to see him for some antiseptic and an adhesive bandage. In a multiplant operation we may put pressure on a particular plant to reduce its ANSI Z-16.1 computed disabling injury frequency rate, and the plant manager may decide to pick up all of his injured workers by ambulance each morning and transport them to work at so-called "regularly established jobs." Often these jobs are never filled except by an injured employee who is unable to return to the job he normally performs.

These are a few of the major difficulties associated with the currently used measures of safety performance. They are not presented with the intention of condemning all of the present safety measurement efforts. Frequency and severity rates compiled according to the ANSI Z-16.1 standard, which evolved from the early work of the U.S. Bureau of Labor Statistics, have provided a practical and uniform method of measuring disabling injury experience that is almost universally accepted throughout industrial and federal governmental organizations. These measures, dating back to 1920 when injury rate provisions first appeared in Bulletin 276 of the U.S. Bureau of Labor Statistics, have not only provided a universal language of safety effectiveness, but also have established a barometer for measuring how accident experience has risen or fallen over time. In addition, they have allowed comparisons to be made among various industries with similar hazards throughout the United States and many foreign countries.

Obviously, these measures have been effective tools or they would not have survived over the years—but like any standard limited to disabling injuries, they do not tell the whole story. One big problem is the accident that escapes detection because it appears below the fine dividing line established by the ANSI disabling injury or the BLS–OSHA recordable injury or illness definitions. In years past the larger industries in particular have accepted the disabling injury appraisal as a useful method of measuring accident prevention effectiveness. Their success has resulted in a gradual decline in these injuries until injury rates in many organizations have fallen so low that their use as a tool for evaluating the safety record has become extremely unreliable. With a downward trend in injury frequency and severity rates or incidence rates, many companies have falsely concluded that safety is not as much a problem in industry as it is, say, in off-the-job situations or where the privately owned motor vehicle is concerned. Unfortunately, presently used measures of safety effectiveness simply do not allow one to conclude that he is or is not doing a satisfactory job of controlling accident producing problems. None of the measures of safety performance commonly in use is acceptable as a valid means of identifying our internal accident problems. It is important that one is not lulled into believing that present measures are really descriptive of the actual level of safety effectiveness within an organization.

IMPROVED METHODS OF MEASUREMENT

One method of measuring safety performance is to define operationally what constitutes unsafe behavior and measure the frequency of its occurrence. Research has been conducted using the industrial engineering technique of work sampling or activity sampling to estimate the extent to which unsafe behavior exists throughout an industrial plant. This technique, called *behavior sampling* when applied to the industrial safety problem, has been experimentally tested several times and is now being used as a practical measuring tool in industry.

Behavior sampling consists of making a number of observations of behavior at random points in time. Instantaneous decisions are made as to whether the observed behavior is safe or unsafe. In the general random method of selecting times for making the observations, every point in time has an equal chance of being selected. This allows an inference to be made concerning the state of the entire population from which the sample was chosen. Thus it is possible to compute either the percent of workers involved in unsafe acts or the percent of time the observed workers are behaving unsafely. The technique of behavior sampling and its applications are described in detail in Chapter 16.

Another method for measuring unsafe behavior as well as unsafe conditions has been developed and tested experimentally in industry. This procedure, known as the *critical incident technique,* was originally applied by Flanagan (1954) and is regarded as an outgrowth of studies in the Aviation Psychology Program of the U.S. Army and Air Forces in World War II. It is an accident study method in which an interviewer questions a number of persons who have performed particular jobs and asks them to recall within a specified time period unsafe acts and/or unsafe conditions they have committed or observed. The persons questioned are selected on a stratified random sampling basis, with stratifications designated according to the type of exposure, quantity of exposure, degree of hazard present, and other criteria considered important to the representativeness of the sample. The objective is to discover causal factors that are critical, that is, that have contributed to an accident or accident potential situation. The unsafe acts and unsafe conditions identified by this method then serve as the basis for the identification of accident potential problem areas and the ultimate development of countermeasures designed to control accidents at the no-loss stage.

The technique has been tested experimentally in industry by Tarrants (1963) and found dependably to reveal causal factors in terms of errors and/or unsafe conditions that lead to accidents. Research has also shown that the technique actually reveals a greater total amount of information about accident causes than previously available methods of accident study and provides a more sensitive measure of injurious and noninjurious accident performance. The critical incident technique and its use as a measure of safety performance are described in detail in Chapter 17.

The various measures of safety performance described in this chapter are compared in Figure 2-3 (p. 42) in terms of their range of sensitivity within a total accident loss continuum. Movement to the right increases the accident severity and accident losses, while movement to the left increases relative accident frequency. Injury frequency rate, injury severity rate, and the S/F ratio include only the disabling injury and fatality classifications. The frequency of accidents within these categories is relatively low because accidents that result in injuries of sufficient severity for them to appear in these classifications are rare events in comparison with all accident occurrences. The severity of injuries appearing in these categories in terms of monetary, time, and production losses is relatively high.

The measure of number of nondisabling injuries includes those medical treatment, job transfer, work restriction, occupational illness, loss of consciousness, and so on, injuries that conform to the recordable criteria but do not result in sufficient severity to cause them to appear in the disabling injury, incidence, or fatality classifications. The frequency of accidents in this

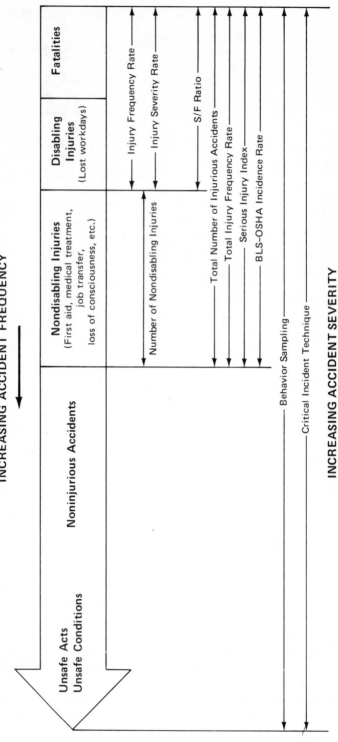

Figure 2-3
Range of Various Safety Performance Measures Within a Total Accident Continuum.

42

classification is relatively higher than the frequencies appearing in the disabling injury, incidence, and fatality classifications, while the severity, by definition, is relatively smaller. The measures of total number of injurious accidents, total injury frequency rate, the serious injury index, and the BLS–OSHA incidence rate generally include accidents resulting in injuries or recordable incidents that appear in all three categories of nondisabling injury, disabling injury, and fatality. In the case of the serious injury index certain types of injuries are excluded from the rate computation, as previously described. The incidence rate is based on recordable occupational injuries and illnesses as previously defined.

Both behavior sampling and the critical incident technique provide for the identification of accidents resulting in no injury as well as those resulting in injuries with varying severity, from nondisabling injury to fatality. Since research has shown that the causal factors involved in noninjurious accidents are the same as those involved in injurious accidents with varying severity and that noninjurious accidents occur much more frequently than injurious accidents (Tarrants, 1963), measurement techniques which identify accident problems at the noninjurious stage provide the safety professional with a greater opportunity for identifying potential injury-producing problems before losses occur. It is toward the objective of accident *loss* prevention that present safety programs can most reasonably be directed. Measurement techniques such as *behavior sampling* and the *critical incident technique* make possible the capability of identifying all accident problems regardless of the degree of severity involved, and they should ultimately permit attainment of the desirable objective of total accident loss control.

Chapter 3

A DEFINITION OF THE SAFETY MEASUREMENT PROBLEM

Present attempts to control accidents and their consequences can best be described as trial and error, chiefly because adequate measures of the effectiveness of this control do not exist in practice. Control must begin with sound measurement. The degree to which accident control is possible is a function of the adequacy of the measures used to identify the type and magnitude of potential injury-producing problems existing within a field of concern.

Because most safety measures are postmortem or after-the-fact in nature, they have little but historical value. On more pragmatic grounds our measures often are inadequate because line managers seem little inclined to believe them. Consider the paradox in the safety professional's task. The more he reduces accidents as measured by present techniques, the less support he is apt to get from management, and vice versa. Most safety professionals have experienced the sudden enthusiasm and strong support by top management in implementing a prevention programming recommendation immediately following a severe accident. (The more severe, the stronger the support, in many cases!) Unfortunately, many safety professionals have also had the subsequent experience of the pendulum rapidly swinging in the opposite direction. With injury frequency rates fairly low and no dramatically severe accidents occurring over a period of time, many managers have seen fit to *reduce* safety emphasis, even to the extent of wiping out the safety specialist's job, in a misguided attempt at "tightening the belt" in a cost-reduction effort. If the true facts were known, a sound-thinking manager might more logically *increase* his safety emphasis in times of austerity, since he would be more interested at those times in conserving the remaining personnel, equipment, and material resources available for the accomplishment of objectives. The need for increased emphasis on loss prevention at these times might appear more reasonable to line managers if they could be provided with information based on better measures of the true, *before-the-fact* loss potential of the various operations and activities within their organizations.

THE COLLECTION OF NONINJURIOUS ACCIDENT DATA

An accident causal factor identification technique is needed that will identify noninjurious accidents as well as those involving disabling injuries. This technique should also be capable of identifying unsafe conditions or defects in the environment that have an·accident-producing potential. The inclusion of *noninjurious* accidents obviates many of the difficulties associated with present measurement techniques. Since noninjurious accidents generally occur much more frequently than disabling injury or property-damaging accidents, even small organizations can collect a sizeable amount of causal data in a relatively short time. Moreover, psychological studies have shown that people are more willing to talk about "close calls" than about injurious accidents in which they were personally involved, the implication being that some special skill or ability allowed the person to avoid injury, and since no loss ensued, no blame for the accident would be forthcoming.

The general justification for the collection of noninjurious accident data is that the severity of accidents appears to be largely fortuitous. Accidents from the same causes can recur with high frequency without a resulting injury. Which accident does produce an injury appears to be determined largely by chance. (There are some exceptions, such as falling from a six-story building or falling into a vat containing 100 percent concentrate sulphuric acid.) The conclusion that the causes of noninjurious accidents can be used to identify sources of potentially injurious accidents is supported by one of the findings of the critical incident study described in Chapter 17. This suggests that the real importance of any accident is that it identifies a situation that could *potentially* result in injury or damage. Whether an accident does or does not result in injury from each occurrence is less significant.

IDENTIFICATION AND MEASUREMENT OF UNSAFE BEHAVIOR

A measurement technique is needed that will permit identification of the behavioral derivatives of noninjurious accidents. Safety professionals are almost unanimous in believing that behavior is related to accidents. If a worker goes without proper safety goggles in certain environments, foreign matter can get into his eyes. If a circular saw is operated without its guard in place, the operator can lose a finger or hand. In fact, unsafe acts have the potential for producing disabling injuries *by definition* since most acts were originally classified as unsafe because they contributed to a disabling injury, recordable injury, or fatal accident. Thus, it is the behavioral act *itself* that serves as a more sensitive criterion of safety performance, not the fortuitous, probabilistic disabling or recordable injury. When an injury does occur, the acts that the worker was engaged in at the moment of the accident are still of major importance. A method of measuring unsafe behavior directly is needed. Viewed in another light, the initiation of such accident prevention methods as training programs, safety contests, safety posters, inspections, and safety committees, is principally for the purpose of modifying behavior and indirectly to reduce injurious accidents. If one grants that the degree of change in behavior in a positive direction is a measure of the effectiveness of a safety program, then ways of detecting this change should serve as a guide for the development of the safety program itself.

IDENTIFYING THE "ACCIDENT STATE"

Most present-day safety efforts are based on after-the-fact appraisals of loss-producing causes that happen by chance to produce an accident of sufficient severity to be included within the limits of existing reporting criterion. If the ANSI Z-16.1 is used, this criterion involves more than 24 hours lost time or some permanent disablement or death. The BLS–OSHA incidence rate involves certain recordable incidents based on limited reporting criteria. Even the more frequently occurring first-aid cases are a loss-producing, after-the-fact means of identifying accident problems. These less severe injuries, unfortunately, are rarely given the attention needed to probe fully into their causes. Safety professionals and managers need to accept the necessity for modifying present methods of accident problem appraisal, with their extreme sensitivity to fortuitous severity oscillations, and to seek new measures which will improve their accident problem identification and control capability.

The contemporary safety professional concentrates most efforts on solving problems, that is, providing answers, when the emphasis should be placed on looking ahead and finding the right questions. The problems need to be measured rather than their consequences. The basis for allocating accident prevention resources must be examined in order to receive the greatest return from our efforts. At present, little is known of the effect of a particular combination of prevention efforts on the system we are concerned with controlling.

As a first step toward the development of a conceptual understanding of new safety measurement techniques, it is important that a functional definition of the term "accident" be established. If one is to identify the so-called "accident state" and use the associative information as a basis for achieving loss control objectives, a fairly clear understanding of the nature and scope of the accident phenomenon itself is needed. Acceptance of the following definition (based on a suggestion by Walter A. Cutter, formerly Director, Center for Safety Education, New York University) provides a functional framework that will broaden the scope of prevention activities and allow more effective application of usable measurement concepts to the appraisal of safety performance.

Accident. An unplanned, *not necessarily* injurious or damaging event, that interrupts the completion of an activity and is invariably preceded by an unsafe act and/or an unsafe condition or some combination of unsafe acts and/or unsafe conditions.

This definition has several implications for the problem of safety performance measurement.

1. Note that accidents do not necessarily result in injury or damage to property each time they appear, but the *potential* for future injury or damage is always present. It will later be shown that relatively few accidents actually involve injury or property damage.
2. The definition is functional in that it identifies the entire spectrum of loss-potential and loss-producing causal factors that are of interest to the safety professional.
3. The definition emphasizes the need for before-the-loss analysis of causal factors that have not yet produced injury or property damage but have the inherent capability of producing such losses in the future.
4. Note that accidents are *invariably* preceded by unsafe acts and/or unsafe conditions. A program designed to measure these causal factors directly and to use data derived from these measures as a basis for developing causal factor countermeasures will provide a sound basis for achieving the objective of long-range accident loss control.

APPRAISING THE SAFETY PROGRAM'S EFFECTIVENESS

The real problem in developing new measures of safety performance is to find a valid, reliable, and sensitive criterion of safety effectiveness. Moreover, any measure of effectiveness must be able to show the magnitude and duration of the effect, since accidents occur in the passage of time. It is also recognized that accidents are a function of environment as well as behavior. Hence, if the effects of programs designed solely to change human behavior are studied, the count of accidents might tend to hide a change in behavior unless the environmental characteristics are known. Because lost-time accidents, disabling injuries, and recordable incidents are statistically rare events and first-aid cases and serious injury frequency rates are subject to large-scale reporting inaccuracies, the safety specialist is faced with only an intuitive notion of the effectiveness of various accident prevention methods. A way to appraise the *internal effectiveness* of an accident prevention program by directly measuring its influence on an acceptable criterion of safety performance as it fluctuates over time is desirable.

WHY SAFETY PERFORMANCE MEASURES ARE NEEDED

Generally measures are needed to reveal how well one is doing—to answer the question "Did the accident prevention program pay off?" In a more specific sense, however, one must recognize that the *main function* of a measure of safety performance is to describe the *safety level* within an organization, establishment, or work unit. For this reason the argument that injurious accidents *in themselves* are adequate measures of safety quality is open to serious question. Injurious accidents are one consequence of worker behavior within specified working conditions; as such, they reveal very little about antecedent behavior and machine–environment malfunctions that are important contributors to current and future accident problems. In effect, then, measures of safety performance must help *prevent*—not *record*—accidents. They must be directional in time and space. They must describe when and where to expect trouble and must provide guidelines concerning what one should do about the problem.

A second purpose of a safety performance measure is to report continuously on the change in safety level within an organization and to evaluate the effects of accident prevention efforts as rapidly as possible. It should not be assumed that the mere *recording* of accidents brings a true picture of the existing safety level. For the most part, the *lack* of safety instead of the *presence* of safety is measured when various techniques of safety performance evaluation are applied.

New measures are needed that will enhance the ability to predict and control accident losses. The best one can hope for is a combination of instruments that will add new dimensions to the ability to identify and evaluate accident problems. At the same time, one must be careful not to overdo efforts in some premature attempt to satisfy an immediate obvious need or a particular urgent demand. A technique should be selected for its applicability to a particular situation, the relative cost involved in using it, the criticality of the component or system under study, the desired output, its compatibility with other programmed activities, and its meaningfulness to managers and those who must use it. The method of measurement should not become an end in itself. For example, there should not be a preoccupation with the mechanics of gathering, digesting, and producing numbers, rates, or indexes to the exclusion of reality. Measurement should not be perceived solely as an incidental or burdensome task wherein numbers are created on schedule at lowest cost. Nor should one continue to tolerate the old malady of purposely introducing bias, screening data, or redefining criteria to guarantee the illusory achievement of some desired safety goal.

The real importance of any accident is that it identifies a situation that could *potentially* result in future injury or damage. Whether an accident does or does not result in injury from each occurrence is less significant. If one accepts the position that the severity of accident consequences is largely a fortuitous or chance occurrence, then a measuring technique that would identify the relatively high-frequency noninjurious accident could be used to identify *loss-potential* problems at the *no-loss* stage. This information could then be used as a basis for a prevention program designed to remove or control these problems before more severe accidents occur.

Before we explore some conceptual issues relating to safety performance measurement, it will be useful to discuss "Why accident data?"—the subject of the next chapter.

Chapter 4

WHY ACCIDENT DATA?

A simple, yet far-reaching answer to the question "Why accident data?" is that the entire safety movement was born, grew in stature, and lives today because of data. Furthermore, the safety movement would most surely die if statistical data were not available to provide the nourishment, the energy, and the guiding force in support of the nation's accident prevention efforts.

Let us take a look at a few facts. Isn't it true that your acceptance, and the nation's acceptance, of the need for safety programs is based upon evidence that accident problems exist and that their existence is of some vital importance? The thoughts and actions of countless individuals and organizations are strongly influenced by accident data. They show conclusively, in continuing analysis, that accident problems exist and that their existence has a profound influence on the nation's human and material resources.

What if these data—these "statistics"—were not available? It seems quite reasonable to speculate that, in the absence of a means of identifying the accident problem, the interest of company managers, state and federal government agencies, and the general public in accident prevention would be greatly reduced, possibly nonexistent, unless someone close at hand were involved in an accident. Actually, work accidents are rare events in the life of any given individual, and few persons can form any clear understanding of their importance as a social and economic problem solely from their own experience. Statistics of the kind just described expand the scope of accident knowledge and create a realization that there is a problem in urgent need of solution.

The safety movement began with statistics developed to identify the problem, grew as statistics provided the basis for accident prevention decision-making activity, and lives today because statistics have shown the positive results of these prevention efforts and have created and stimulated the urge to action!

If one accepts the premise that the safety movement must have injury and accident statistics, the next question is, What responsibilities do the state and federal governments have for providing such statistics? The answer no doubt is obvious. Only governments have the resources to produce the comprehensive basic statistics needed to identify meaningfully the problems and measure progress. Outside of government there are, of course, many special statistical activities, some of a continuing nature, that serve particular purposes and are most useful to the safety movement. But these sources provide limited information and data are usually collected from a relatively narrow segment of the population.

For example, the National Safety Council (NSC) produces work injury statistics, including the annual publication of a very useful booklet entitled *Accident Facts*. For the most part, the NSC work injury data are collected from organizations with memberships in the Council. These organizations, as a whole, have better injury records than the total universe of organizations with similar missions, and they all have sufficient interest in safety to become a member of the council. Thus, their collective work injury record is much better than that of organizations reporting to government agencies such as the Bureau of Labor Statistics or a state safety agency.

State and federal safety agencies all have a statutory duty to foster and promote the welfare of all workers, and keeping the occupational safety and health movement alive and vigorous is of the essence in such a responsibility. Unfortunately, only a few state agencies have adequate resources to do properly the task they should be performing in this area.

A second question is, Given adequate resources, how can our statistical service objective best be achieved? Progress evolves from a restless discontent with even the best of what is currently practiced. It is not enough just to achieve the currently acceptable level of performance. If one is to achieve more effectively the objectives of accident prevention, continual appraisal of our present statistical methods with a view toward their improvement is needed. The answer to our question, "Why accident data?" takes on a new dimension when we also ask, "Where do we go from here?"

AREAS FOR IMPROVEMENT IN ACCIDENT STATISTICS

A close examination of our present activities in the accident statistics field reveals several areas where improvements might be introduced:

1. Validity and reliability of accident data
2. Problem identification and gathering of decision-making information
3. Improvement and expansion of data
4. Application of scientific procedures

The validity and reliability of accident data need improvement. Statistics have been defined as a body of methods for making wise decisions in the face of uncertainty. The quality of the decisions made from them is dependent upon the quality of the information from which they are derived. John V. Grimaldi, of the Institute for Safety and Systems Management of the University of Southern California, has stated that "there are no bad decisions, only bad information. If all of the facts are known, the decisions often suggest themselves." In the context of accident prevention applications, wrong or inadequate information about accident causes leads to wrong decisions concerning what countermeasures to recommend.

The quality of decisions is a direct function of the validity, reliability, and comprehensiveness of the information upon which the decisions are based. Accident statistics will improve as (1) better data collection methods are developed, (2) population sampling techniques are improved, (3) more precise causal classification categories are established, and (4) the system for checking the validity and reliability of the reported accident information is improved.

Additional problem identification and decision-making information along with instructions concerning its meaning and use need to be presented. Statistics serve in at least two capacities. First, they provide methods for organizing, summarizing, and communicating data. Second, they provide methods for making inferences beyond the observations actually made to statements about large classes of potential observations. The set of methods serving the first of these functions is generally called *descriptive* statistics. Most of the accident and injury statistics presented by state and federal governments, private organizations such as the National Safety Council, and individual companies are descriptive in nature. But if one wishes to generalize from the knowledge of a sample to the population of which the sample is only a fraction, a body of methods is needed that will permit conclusions extending beyond the immediate data.

The techniques that enable us to draw inferences about population from a knowledge of samples are called *inferential* statistics. The ability to make objective decisions is greatly enhanced when techniques of inferential statistics are introduced. They not only aid in determining what inferences are possible, but also how viable they are. The addition of this new decision-making information to the body of accident and injury statistics will reveal a margin of error when some population measure such as an accident frequency rate is predicted. Then hypotheses about various measured relationships within the population can be tested. Actually, the purpose of description is to prepare the way for inference. In practice, inferential statistics are perhaps more important than those of a descriptive nature. Samples are studied, but they seldom are of interest in and of themselves. Inferential statistics permit going beyond the data, and they reveal how much risk is taken in doing so. Additional information that would enhance inferential capability and greatly aid the user in making decisions include measures of dispersion (such as the standard deviation, the variance, and the range) and procedures for testing hypotheses (such as the t-ratio, the chi-square test, and analysis of variance).

Data need improvement and expansion. This point ties in closely with the presentation of additional decision-making information. In analysis work the techniques of inferential statistics need to be applied to put new meaning in the data interpretation process. Data analysis is probably the weakest link in the accident statistics development system. It is important to keep in mind that statistical data have to be interpreted—they seldom if ever "speak for themselves." Statistical data in the raw simply furnish facts for someone to reason with. They can be extremely useful when carefully collected and critically interpreted. But unless handled with care, skill, and, above all, objectivity, statistical data may seem to prove things that are not at all true. It is sometimes said that statistics are used the way a drunk uses a lamp post: for support rather than for light. Special care must be taken in analysis not to bias the results because of personal interpretations and conclusions that are unwarranted by the nature of the original data. The frequently heard comment that statistical conclusions are wrong is often supplemented by the view that when they are not wrong they are self-evident and trivial. A statistician has been defined as "a person who draws a mathematically precise line from an unwarranted assumption to a foregone conclusion" (Wallis and Roberts, 1956).

Misuses and misinterpretations in the analyses of statistics are probably as common as valid uses. Darrell Huff (1954) in his interesting book *How to Lie With Statistics* and Allen Wallis and Harry Roberts (1956) in their book *Statistics: A New Approach* have made excellent presentations of statistical fallacies and errors in data analysis. They describe errors in the use of basic definitions underlying an investigation, errors in the application of those definitions in the measurement or classification of individuals or objects, and errors in the selection of individuals for measurement. They also illustrate errors in the use of the resulting data by making comparisons improperly, by failing to allow for such indirect causes of differences as heterogeneity of groups with regard to important variables, by disregarding the variability that is usually present even under apparently constant conditions, by technical errors, and by misleading verbal or graphic presentations.

The scientific procedures necessary for making causal determinations need to be understood and applied. Ideally, data collection and analysis represent an attempt to utilize the scientific method for the purpose of identifying problems and assessing the worthwhileness of various alternative solutions or countermeasures. The more the rules of scientific methods are satisfied in designing a data collection system and conducting data analysis, the more confidence can be placed on the objectivity of the findings.

ESTABLISHING A WORKABLE DATA COLLECTION AND ANALYSIS METHODOLOGY

Researchers generally recognize five states that recur in intelligent problem solving:

1. *Planning* (including problem identification and analysis). The problem is defined and its parameters established in the planning stage. Planning involves clarifying objectives, establishing policies, selecting specific countermeasures, establishing procedures, mapping programs and campaigns, and setting up day-to-day operational schedules. It is essential that planning also include establishing procedures for evaluating program effectiveness so that cause-and-effect determinations can be made. The preferred causal analysis technique involves experimental and control groups and the random assignment of a representative sample of individuals to each of these groups. This analytical model should be "built in" at the planning stage. Planning also includes the selection of at least one criterion of effectiveness and the prescription of appropriate methods of measurement.

2. *Observation.* The data analyst observes what happens as the independent variable (the new program or countermeasure) is introduced to the experimental group and the dependent variable (criterion) changes or remains stable over time.

3. *Hypothesis.* In order to explain the facts observed, conjectures are formulated into a hypothesis or tentative proposition expressing the relationships believed to have been detected in the data. In formulating hypotheses prime consideration must be given to the criterion or measurement standard to be used in conducting the analysis.

4. *Prediction.* Based on the hypothesis or theory, deductions are made concerning the consequences of the hypothesis formulated. Here is where experience, knowledge, and perspicuity are important. Predictive reasoning can help lead to more basic problems as well as provide operational or testable implications of the original hypothesis. At this stage one anticipates or predicts what will be seen if certain observations not yet made are made.

5. *Verification.* The data analyst collects new data and tests the predictions made from the hypothesis. The essence of verification is to test the relation between the variables identified by the hypothesis. In hypothesis testing one measures the risk of incorrect interpretation objectively in terms of numerical probabilities. The procedures of statistical hypothesis testing permit determination in terms of probability, of whether the

observed difference in the measured criterion is within a range that could easily occur by chance or whether it is so large that one can be confident that the differences are actually meaningful.

What is important in this procedure for the safety program administrator is the overall fundamental idea of scientific problem solving as a controlled rational process of reflective inquiry. Data analysis methodology applied to accident control problems enables the safety program administrator to gain knowledge through the use of experiments as opposed to intuitive speculation and the application of trial-and-error prevention methods. A test of a hypothesis is a means of determining the validity of some prediction or expectation. Establishing a working hypothesis is an important step in control planning. It suggests the kinds of data needed and how they should be arranged and classified. It also establishes the basis for making programming decisions.

There are, of course, numerous practical problems associated with actual application of analytical methods in a given situation. In the evaluation of safety program effectiveness, the primary objective is to determine the extent to which a given program or countermeasure is achieving some desired result. The "success" of our data collection and analysis system is largely dependent upon its usefulness to an administrator in reducing accident losses (for example, deaths, injuries, property damage, costs, and the like). Thus, while scientific criteria may determine the degree of confidence placed on the findings of an appraisal study, administrative criteria will play an even larger role in determining the usefulness of the study once it has been done. In applied data collection and analysis activities, the potential utility of our findings must be constantly monitored. Similarly, the desired use of the data (the kinds of decisions we expect to make) should determine the type and quantity of data sought. Initially, our decision-making information should be defined, along with the kinds of data needed to make those decisions. Collecting data that are not needed for decision making is wasteful and unnecessary. "Institutionalizing" the data collection system to the point where removing obsolete information becomes difficult if not impossible should be avoided. Collecting data which repeatedly provide information we already know is a useless exercise. A rifle approach aimed directly at the target is preferable to a shot gun aimed in the general direction of the issues of concern.

Numerous experimental designs can be used in evaluating safety program effectiveness, each having its strengths and weaknesses in relation to the need for decision-making information. There is no perfect design that fits all situations. Some evaluation specialists make a distinction between the experimental and the so-called ex post facto approach to information gathering. The experiment is a scientific form of inquiry in which the evaluator manipulates and controls one or more independent variables and observes one or more dependent variables for evidence of change. For example, a public education program designed to increase driver usage of lap and shoulder safety belts can be identified as the independent variable and the change in safety belt usage as measured by behavior sampling techniques would be the dependent or criterion variable. In this experiment one attempts to change the behavior of drivers by means of the public education program focused on a specific action target; then one evaluates the effect of this program by observing the subsequent change in behavior.

In ex post facto evaluation something is done or occurs after an event with a retroactive effect on the event. In this kind of evaluation study the independent variable or variables have already occurred. The evaluator starts with observations of the criterion and retrospectively studies the independent variables to determine their possible effects on the criterion or dependent variable.

The safety specialist who investigates an accident is, in effect, conducting ex post facto evaluation. He examines certain evidence and retrospectively studies the independent variables: in this case, the behavioral, equipment, and environmental factors that he believes were causally related to the accident result. Most present-day safety program activities are based on similar after-the-fact appraisals of losses or crashes that have already occurred. It should be clear that a more desirable model for accident problem analysis is the controlled experiment since the investigator can have more confidence that the relationships he discovers are really significant in terms of their cause-and-effect implications.

One of the preferred evaluative techniques involves the use of a control group. Comparison groups are necessary for the internal validity of any evaluation program. The purpose of the control group is to rule out variables that are possible "causes" of the effects we are studying other than the variables we hypothesized to be the "causes." In other words, control of extraneous sources of influence on the criterion variable is needed. The control of extraneous variables means that the influence of independent variables extraneous to the purpose of the study are minimized or nullified. One way of controlling extraneous variables is by randomization. By randomly assigning subjects to the experimental group and the control group, the preexperimental approximate equality of these two groups is assumed in all possible independent variables.

One of the major weaknesses in much of the evaluative research conducted in the occupational safety and health field has been the lack of adequate controls. The so-called "research" evidence that is frequently encountered supports conclusions based on uncontrolled observations of certain characteristics of persons involved in accidents. For example, in an investigation into one aspect of the motor vehicle accident problem, violations of traffic regulations by drivers in fatal crashes, it was shown that nearly one-third of the drivers were exceeding the speed limit at the time of the accident. Another study concluded that in more than 50 percent of all fatal accidents occurring in a particular region the driver had been drinking. Still another study concluded that the majority of drivers involved in one-car crashes were smoking a cigarette. Such statements contribute very little to our under-standing of the problem *unless* it is known whether the cited characteristics of the accident-involved drivers are present to a different extent among those who are not involved in accidents. Such data are also frequently biased because of the tendency of those investigating accidents to conclude that the occurrence of an accident is sufficient evidence that such violations and char-acteristics were present. These conclusions are then used in a circular fashion to support preexisting biases concerning accident causes. In scientific evalua-tion it is important to compare the characteristics of accident cases with the characteristics of the corresponding populations and situations from which they are derived.

In attempting to establish a "cause-and-effect" relationship between the introduction of a countermeasure and a subsequent reduction in the safety program effectiveness criterion, the "regression-to-the-mean" phenomenon is often encountered. This concept describes the fact that when successive samples are selected from a population and certain variables measured, there is a tendency for these variables to return from the extreme to an average condition as repeated measures are taken. If 10,000 samples of workers with respect to accident record are selected, some would have a no-accident record, some would have about an average record, and some would have a bad record. If the same groups in a second time period were subsequently sampled, some of the groups with a poor accident record the first time would continue to have a poor record, some would get better, and some would get worse. In general, the samples exhibiting the extreme characteristics in the first instance would tend to exhibit less extreme characteristics in the second instance. Workers who tended to perform poorly in the first observation would tend to improve in the second, while those who performed exception-ally well in the first observation would tend to get worse in the second.

One can conclude from this that so-called "unsafe workers" identified in one time period will tend to show an apparent improved record during a second time period, regardless of whether a countermeasure or program activity was introduced or not. Thus, an apparent improvement during the period following the introduction of a countermeasure may be falsely attrib-uted to the effects of the countermeasure when, in fact, the natural tendency for sample measures to regress toward the mean may be producing the shift in criterion values. A second or "control" group that is equally as poor but is not exposed to the countermeasure may show the same improvement, even though nothing at all was done to influence this change.

The informed interpreter of evaluation results, armed with his knowledge of evaluative techniques, can readily identify these sources of bias, or at least suspect their presence when no mention is made of the frequency distributions of the same characteristics among appropriate control populations.

The following steps are essential for evaluating safety program effectiveness:

1. Formulation of the program, countermeasure, or project to be evaluated
2. Definition of the program, countermeasure, or project goals and objectives
3. Analysis of the specific problems to be addressed by the program, countermeasure, or project
4. Application of the program, countermeasure, or project within a specific problem area, with the use of control groups as appropriate
5. Measurement of the degree of change in the evaluative criterion
6. Determination of whether the observed change is due to the program, countermeasure, or project or to some other cause
7. Determination of the durability of the effects of the program, countermeasure, or project

This application of the "scientific method" to safety program appraisal provides the most promising means for determining the relationship between a specific safety program activity or countermeasure and functional safety performance objectives in terms of predetermined measureable criteria. A further discussion of hypothesis testing and the application of inferential statistics to safety performance measurement appears in Chapter 15.

Techniques for evaluating safety effectiveness are tools, and like all tools, to be effective, they must be designed for a specific function. It seems obvious that the need in safety program evaluation today is for more scientific evaluation. Greater progress in accident prevention will be made when the objectives of a particular program are examined (including underlying assumptions), measureable criteria specifically related to these objectives are developed, and a controlled situation to determine the extent to which these objectives, and any negative results, are actually achieved is created.

THE NEED FOR A BETTER UNDERSTANDING OF THE OCCUPATIONAL SAFETY AND HEALTH FIELD

The final area of improvement with regard to accident data is the need for a better understanding of the field within which our results are applied. Statistics serve as a tool that can be used in attacking problems that arise in almost every field of empirical inquiry. The statistical approach, though universal in its underlying ideas, must be tailored to fit the peculiarities of each concrete problem to which it is applied. It is dangerous to apply statistics in "cookbook style," using the same recipes over and over, without careful study of the ingredients of each new problem. Statistical methods may be highly interesting, but it is important to recognize that statistics cannot be used to full advantage in the absence of a good understanding of the subject to which they are applied. Successful statistical work depends greatly on knowledge of the subject matter. Mere manipulation of figures or preparation of standard tables and graphs is seldom fruitful unless guided by a clear conception of the subject matter and of what relations would be worth seeking. To a considerable extent, statistical application in a particular field is an art rather than a science. As with other arts, however, there are certain basic techniques whose mastery is necessary, though not sufficient, for success.

As a closing thought on data and data analysis, a quotation from an article on statistics in general by George Hagerdorn, Director of Economic Studies for the National Association of Manufacturers, is most appropriate.

> People who really understand the figures, their virtues and their limitations, are a small minority. The great majority, who do not understand the data, are divided into two classes. First there are those who have an exaggerated *respect* for published data. They assume that if only you put the right statistics together there will emerge from this compilation of *facts* the right answer to any given problem.
>
> The second group is composed of those who have an exaggerated *suspicion* of published data. In this view, since government officials frequently cite statistics in support of their partisan and questionable proposals, there must be something wrong with the statistics themselves. . . .
>
> The truth lies somewhere in between. The figures published by the old line statistical agencies are generally prepared in an honest and competent manner. But, no matter how honestly and competently the figures are prepared, it is a mistake to imagine that government statistics are facts which *speak for themselves.* They speak clearly only to those who have taken the trouble to investigate the definitions on which they are based and the techniques used in preparing them. And they give answers to economic questions only when they are combined with a good measure of common sense, knowledge of non-statistical facts, and a basic understanding of what makes our economy work.

To government and private sector statisticians, this means that they must not only produce the most valid and reliable data possible, but also must follow standard procedures in compiling the figures, apply professional competencies in the analysis of data, and, above all, fully explain and be willing to defend their data development and analysis procedures.

No one knows all the answers; all involved must realize that. When these principles are put into practice, safety professionals can feel confident that they are doing their part to provide the data base for management decision making that will keep the occupational safety and health movement alive, vigorous, and successful.

OTHER DISCIPLINES AND SAFETY PROFESSIONALS LOOK AT SAFETY PERFORMANCE MEASUREMENT

In September 1970 the second of two symposia on safety performance measurement was conducted by the Industrial Conference of the National Safety Council (NSC).

Part II contains updated and edited reprints of eight papers prepared for that symposium by outstanding practitioner/scientists. The purpose of this important second symposium was to explore various aspects of measurement and safety performance as viewed by persons outside the occupational safety field. It was hoped to bring fresh approaches to bear on the application of measurement techniques to the occupational safety field.

Each of the authors discusses measurement issues from the standpoint of his area of expertise and how they might apply to the safety performance measurement problem. Original versions of most of these papers were first published in the September 1970 issue of *Journal of Safety Research,* an NSC periodical.

Chapter 5

BEHAVIOR
AND
ACCIDENTS

James W. Altman

This chapter is based on two fundamental assumptions: (1) human behavior is an important factor in determining the frequency, severity, and incidence of accidents in a wide variety of situations, and (2) for many situations in which human behavior is an important factor in accidents, modification of that behavior is possible in such a way as to increase or decrease the likelihood that accidents will occur and influence the probable consequences of errors that may occur.

It is our purpose to suggest some of the behavioral dynamics involved in having accidents. For this purpose, we will accept Tarrants' (1965) definition of an accident:

> An unplanned, *not necessarily* injurious or damaging event, which interrupts the completion of an activity, and is invariably preceded by an unsafe act and / or an unsafe condition or some combination of unsafe acts and / or unsafe conditions.

Little or no reference will be made to general principles of human engineering, industrial design, human learning, or instructional technology. Rather, emphasis will be placed on the following six areas that have particular relevance to behavior related to accidents:

1. Analysis of errors having accident potential
2. Task engineering
3. The role of reinforcement schedules in learning to have accidents
4. Transfer of training and safe behavior
5. Conflicting reinforcements
6. Contextual comparison

ERROR ANALYSIS

If we accept the view that neither exhortations to be safe nor abstract safety training has proven to be significantly effective, how does one go about establishing a focus on behavior to be modified to improve safety? An analytical process is suggested that might be called *human error analysis.* The principal activities of error analysis and their purposes are summarized in Table 5-1.

Human error analysis simply means the attempt to relate the quality of human performance in an identifiable system to the output of that system. In the current context we are especially concerned with quality of performance as it is related to accidents. We might as readily talk about the positive side— accuracy or reliability—except that the analysis may have to cope with fragmentary data and it is easier to conceive of ways to use fragmentary error data than partial accuracy or reliability data.

The basic approach in error analysis is to identify opportunities for error by organizing and describing the tasks and activities to be performed by personnel. Then, for each activity or action, the possible errors are identified and defined. The identification of error opportunities and definition of error possibilities provide a framework to facilitate the other activities in error

Table 5-1
Principal Activities and Purposes of Human Error Analysis

Activity	Purpose
Identification of error opportunities and possible errors	To provide a framework for maximum efficiency and success of subsequent activities
Evaluation and classification of errors	To provide a basis for pooling data and establishing action priorities
Observation and recording of errors	To provide factual basis from which to develop error-reduction techniques
Determination of variables and conditions associated with errors	To facilitate the identification of behavioral characteristics having potential for improvement
Prediction of errors	To translate information from the past into the most appropriate form to guide future action
Design and evaluation of error-reducing techniques	To establish an affective mechanism for improved performance

analysis. The techniques for task identification, description, and analysis have been extensively described elsewhere (Altman, 1966b; Miller, 1965). Because of the heavy involvement of task analysis in the development of military and other systems, this is perhaps the aspect of human error analysis for which there exists the greatest current experience and capability.

The evaluation and classification of errors involve at least three stages of activity. Typically it is desirable to complete the design of an evaluation and classification framework prior to engaging in large-scale efforts to obtain error data to insure appropriate and efficient sampling. Once preliminary data have been obtained, one should verify that the initial framework is indeed optimal for the kinds of error likely to occur in the situation and to modify the evaluation-classification structure accordingly. This modification may undergo a number of approximations in the course of data gathering and analysis. In its final form one must develop a scheme and apply it to the error data in order to derive action implications and priorities.

Observation and recording of errors provide information concerning past performance—the fundamental basis for predicting errors that can be expected in the future unless some remedial action is initiated. Where observation and recording must be selective, the prior analysis of error likelihood can serve to pinpoint those aspects of performance most prone to critical error and therefore suitable for selective observation.

Tarrants (1965) has described the use of the critical incident technique and behavior or activity sampling as methods for obtaining accident-related behavior. Ferguson and Daschbach (1968) have demonstrated the use of stop-action camera techniques to record hazards encountered by a worker in the course of his job activities.

In many instances direct observation of errors will not be possible. Where there are no survivors, where dramatic events make witnesses unreliable, where observers are not familiar with proper operating routines for the worker or workers involved, and where events occur too rapidly for reliable observation, it may be necessary to infer error occurrence from its consequences and effects. For systems under design or undergoing major evolution, errors must be extrapolated from past to new contexts or error observations must be obtained from simulations or tests.

Determination of personnel, policy, and environmental variables that are correlated with error occurrence can provide one basis for classifying errors. It can also provide direct implications for remedial action to reduce error. Prediction of errors is a way of interpolating from incomplete data, filtering out freak occurrences, and extrapolating from the past to the future. Particularly in today's fast-changing environment, one does not want to be working toward the solution of yesterday's safety problems. Ultimately, of course, the payoff from error analysis will come from the design and evaluation of error-reducing techniques—in this context, techniques for promoting safe behavior and precluding or extinguishing behaviors conducive to accidents.

Let us return briefly to the evaluation and classification of errors. This is a critically important issue in the search for ways to help workers learn safe rather than accident-facilitating behavior, for without effective evaluation and classification it will be impossible to concentrate one's efforts on the behaviors likely to have a crucial impact on safety.

Figure 5-1 presents what we find to be the major bases for error classification and evaluation. Each of these bases has been developed in detail in other writings. Our discussion in this chapter will be limited to only one way of

Figure 5-1
Bases for Error Classification

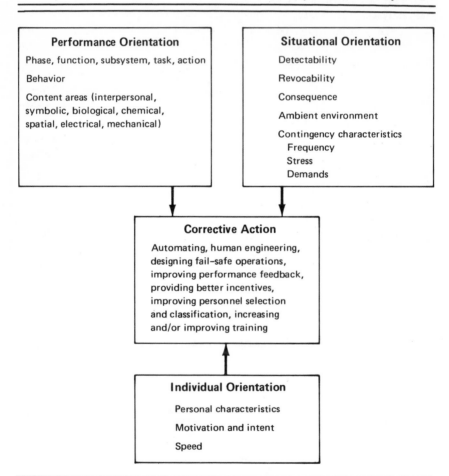

looking at selected situational factors and at behavioral classification of errors. (Error analysis is discussed more fully in Altman [1965] and error classification is more fully elaborated on in Altman [1966a].)

In Figure 5-2 we see the convergence of three situational factors to define errors critical to safety. *Detectability* includes both the probability that an error will be detected and the time-space remoteness of detection from the error occurrence. *Revocability* is concerned with the extent to which the possible consequences of an error can be alleviated if it is detected at specific points in time. *Consequences* is concerned with the effects of error on the performance or cost of the system of which the worker's performance is a part.

Detectability can have a direct impact on revocability, which, in turn, can have a direct impact on the nature and extent of error consequence. At the lower left of Figure 5-2 we find "measureable error tendency but no actual error." This suggests immediate detection, total revocability, and no significant consequence. As we move to the upper left, we encounter progressively more critical and dangerous situations, with the consequences of error

Figure 5-2
Detectability-Revocability-Consequence Matrix.

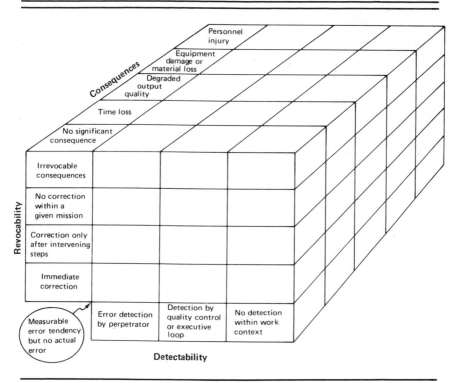

becoming increasingly irrevocable. These situations are of special concern to persons whose prime concern is safety. This is not to say that the other kinds of errors are necessarily irrelevant to safety—a minor error may mean a near accident and today's rapid correction may foreshadow tomorrow's disaster. However, the need for intervention to avoid harmful consequences becomes more urgent as revocability decreases.

In Table 5-2, the correspondence between the hierarchy of behavioral complexity and classes of possible error is shown. A behaviorally defined

Table 5-2
Classes of Error for Different Behavioral Levels

Behavior Level	Classes of Error	
	Omission	*Commission*
Problem solving	Failure to use information that would facilitate derivation of a solution Giving up before a solution is reached	Formulating erroneous rules or guiding principles Acceptance of an inadequate solution as final Perseverance on a problem known to be insoluble
Logical manipulation, rule using and decision making	Failure to identify or apply appropriate information, rules, or alternatives Delaying a required decision	Incorrect value weighting Application of a fallacious or inappropriate rule Making an unnecessary or premature decision
Estimating with discrete or continuous responding (tracking)	Failure to respond to a superthreshold target change Late response	Responding to a subthreshold target change Premature response Inadequate or excessive response Wrong direction of control action
Chaining or rote sequencing	Omitting a procedural step	Making a below-standard response Inserting an unnecessary step Misordering steps
Sensing, detecting, identifying, coding, and classifying	Failure to monitor the field Failure to record or report a target or signal change	Recording or reporting a target or signal change when none has occurred Assignment of a target or signal change to the wrong class

error classification scheme has particular relevance to learning and accidents since it has been well established (Gagne, 1965) that the course of learning complex tasks is definitely dependent upon error-free performance of its less complex behavioral components.

Having emphasized the importance of discriminating analysis of errors in defining safety training requirements, we will now violate our own precepts by considering learning and behavior without consistent distinction among varieties of error. This is done solely for efficient communication and not because such distinctions are unimportant.

The critical question that remains is how the results of error analysis can be gainfully applied for improved safety. In the next section on task engineering we explore some of the ways in which these results can be applied to the design of performance requirements for reduced error and error consequence. The three subsequent sections deal with different ways in which the results of error analysis can be used to enhance the learning of safe behavior. After these we will examine ways to use the results of error analysis in comparing accident behavior across different contexts.

TASK ENGINEERING

Task engineering involves a number of related means for reducing accident potential, including the following:

1. Eliminating or reducing hazards
2. Eliminating error-likely operations by automating functions
3. Utilizing human engineering techniques (equipment, tools, instructions, procedures, the work environment, or the organizational structure)
4. Improving the monitoring of operations to eliminate errors or their consequences
5. Improving feedback concerning errors and their consequences

Hazards

Strictly speaking, reduction of hazards is not legitimately a part of task engineering. However, analysis and design of tasks can be a powerful aid to focusing on those hazards most in need of elimination or reduction. Understanding of the relationships between tasks and hazards can also provide useful insights concerning effective ways of reducing hazards at minimum cost. Mechanical, electrical, structural, and bioengineering factors in hazard reduction are not within the scope of this chapter. The intent here is simply to point out the importance of the interface between such factors and task engineering.

Automation

Identification of error-likely operations is the first step in rational selective automation. Even if full automation of operations is not feasible, it may be possible to reduce the complexity of error-producing behaviors by using machines to aid worker performance. Where the unburdening of performance through "machine aiding" may not be feasible, machine interlocks to prevent error and/or to detect error if it occurs may be feasible. Feedback of error information to the worker automatically may be an effective way of enhancing safety. Buffering of error to prevent its full consequences may be possible where positive interlock to prevent error and correction by the worker prior to consequences are not viable alternatives. Simple automatic alerting devices may remind workers of errors of omission.

Cost, consequences of human error, and probability and consequences of failure in alternative machine operations all must be traded off for optimum design. Complete automation would, of course, eliminate any real need for engineering worker tasks—for, indeed, there would be no workers or tasks. Such complete automation, however, is unlikely to be a credible alternative in situations of interest to the safety professional. Consequently, the application of automation techniques for purposes of increased safety must usually be highly selective. Task and error information can be an important aid to selection of areas for maximum safety payoff per dollar. Specific automation techniques and mechanisms are outside the scope of task engineering and this chapter.

Human Engineering

Designing the work environment for compatibility with worker capabilities and limitation—human engineering—can influence any aspect or component potentially involved in accidents. Equipment, tools, informational performance aids and instructions, procedures, work space, and the ambient environment can all have either a direct or indirect effect on accidents. The fundamental data for human engineering relate to the nature of interaction between worker and design features. Thus, effective human engineering design is sharply influenced by the nature of behavior to be supported, especially potential error behavior.

A major potential contribution of error analysis is to help guide human engineering activities selectively to concentrate on those behaviors having greatest error and accident potential and to design features that have promise of reducing such accident potential.

Monitoring

At least three purposes directly relevant to safety can be served through the monitoring of human error. First, monitoring can often intercept and overcome the effects of error before serious accident potential and consequence develop. Second, it can provide a source of data that can be used to guide modification of design, procedures, manning, and training. Third, monitoring can provide the alerting and shaping information for feedback to the worker that will aid his learning.

However, monitoring has its price, and not just in cost to support the activity. If it is excessively obtrusive or perceived by the worker as aimed at the wrong goals, it can be quite disruptive to desired performance. The monitor can sometimes mistakenly perceive correct performance as being in error. If charged with correcting errors, the monitor may inject additional errors of his own.

Feedback

In a later section we discuss the importance of conflicting reinforcement as a detriment to safe performance. Here we are concerned simply with feedback of performance information as a mechanism for increasing sensitivity of the worker to error-generating work habits—particularly those errors with accident potential.

Five characteristics of feedback are important:

1. *Speed.* The more promptly feedback is provided after error commission, the greater the opportunity for operational rectification and worker learning.
2. *Specificity.* The more narrowly feedback focuses on the specific error, the more effective it is likely to be.
3. *Accuracy.* Any imprecision or error in feedback is likely to encourage performance that is contrary to intentions.
4. *Content.* The information content and media should be appropriate to the desired behavior. For maintaining alertness and guiding simple stimulus response behavior, bells, lights, and other signals common to conditioning may be appropriate. Complex behavior, such as decision making and problem solving, may require a more elaborate information feedback.
5. *Amplitude.* To be effective in modifying behavior, feedback obviously must be sufficiently differentiated from the rest of the stimulus field to gain the worker's attention and demand perception. However, feedback that is excessively salient in terms of stimulus load or emotional implication may actually be disruptive to desired performance.

REINFORCEMENT SCHEDULES

Suppose that we have completed a rather precise error analysis. What is our natural tendency with respect to the tasks we find to be critical? There are two things we are likely to do that have direct relevance to accidents (Altman,

1968). First, we are likely to emphasize training on those tasks with critical safety implications, watching trainee performance with special care and providing appropriate feedback on whether or not each performance is acceptable. Second, in the operational situation, we are inclined to bring a great deal of redundant capability to bear on tasks having critical safety implications.

What could possibly be wrong with these obvious "good things"—training with special care and insuring more than ample manpower for safety-critical tasks? Before we can answer that question directly, it will be necessary to take a short side trip.

First, we must define the four archetypal, or highly "purified" types of workers.

Critical worker. Feedback (reinforcement) is always a direct function of his performance.

Series worker. Positive feedback will occur only if he and at least one other worker perform correctly. If any worker in the series performs incorrectly, negative feedback will be provided.

Parallel worker. If any parallel worker performs correctly, positive feedback occurs. Only if all parallel workers perform incorrectly will negative feedback be provided.

Redundant worker. Feedback is not a function of his performance in any way.

The critical worker is one whose performance, on the task in question, is always rewarded with appropriate information about the adequacy of his performance. Let us take as an example a worker whose job includes installation of missile engine igniters just prior to launch. In either simulated or real launch he is given realistic feedback as to whether or not he correctly installed the igniters, unclouded by whether or not he was correct or incorrect on some other step or whether some other worker did or did not do his job properly.

The series worker, in contrast, receives only indirect information about the adequacy of his performance. If anyone in a critical chain of performances is incorrect, the feedback is negative. Only if everyone in the chain is correct will the feedback be positive. Imagine an operation where a number of workers participate in the assembly of an item of equipment, but the specific contribution of each individual worker cannot be pinpointed. Final inspection and test may be able to identify every misassembly, but the only feedback possible is that everyone did everything right or did what was done wrong—not who was responsible.

Unlike the series worker who receives positive feedback only if he and every other worker involved performs correctly, the parallel worker may receive positive feedback even when he is wrong. Take another assembly situation. Suppose the item being assembled undergoes a number of checks and, if needed, adjustments of its critical circuitry. It does not much matter who of a number of persons working on the assembly makes the adjustment as long as it is made prior to the final acceptance check. If anyone makes the adjustment, the feedback is that performance is correct, even though workers who should have made the check did not. Only if everyone "goofs" will there be negative feedback.

The redundant worker's performance is irrelevant to the feedback he receives. Let us go back to the missile igniter installer. Suppose, because of the criticality of his job, he is given an assistant. The assistant may be permitted to tinker with the missile, but our critical worker always checks thoroughly what his assistant has done and unobtrusively adjusts things to his own liking. The assistant becomes superfluous and redundant. The feedback is entirely a function of the critical worker's performance.

In Table 5-3 we have reduced the foregoing discussion to a series of probability statements concerning reinforcement received by our different archetypical workers. Note that in all of these cases we are assuming that the performance monitoring is perfect and feedback is complete and infallible. That is, at the point where performance is checked, all error effects are detected and the results of the error monitoring are always accurately reported back to the responsible workers.

Even though such perfection of quality control over human performance is rarely if ever achieved in practical operations, it is commonly accepted that workers will tend toward error to the extent that knowledge of results is lacking or inaccurate. It is assumed that perfection in performance monitoring is a goal toward which all safety professionals strive. The prin-

Table 5-3
Reinforcement Probabilities for Different Types of Workers

Type of Worker	Correct Reinforcement		Incorrect Reinforcement	
	Positive	*Negative*	*Positive*	*Negative*
Critical	p_1	q_1	o	o
Series worker	$p_1 p_2$	$q_1 q_2$	o	$p_1 q_2 + q_1 p_2$
Parallel worker	$p_1 [p_1 + p_2 - (p_1 p_2)]$	$1 - [p_1 + p_2 - (p_1 p_2)]$	$q_1 [p_1 + p_2 - (p_1 p_2)]$	o
Redundant worker	$p_1 p_2$	$q_1 q_2$	$q_1 p_2$	$p_1 q_2$

p_1 = the probability of a correct response by the worker of interest
p_2 = the probability of a correct response by another worker
q_1 = the probability of an incorrect response by the worker of interest—or $(1-p_1)$
q_2 = the probability of an incorrect response by another worker

ciple illustrated by Table 5-3 is different. It demonstrates that the feedback to the individual is a function of the performance role of the worker with respect to the monitoring and feedback loop, even when that loop is comprehensive, reliable, and accurate.

Except for the critical worker situation, where the number of workers is not relevant, Table 5-3 illustrates only two-worker situations for the sake of simplicity. These situations can, of course, be readily expanded to more workers.

With this background, then, we return to the questions posed earlier. What can be wrong with concentrated monitoring and feedback of safety-critical task learning and insuring more than adequate manpower resources for their accomplishment in an operational setting? Precisely this. Unless we are precise about the nature and schedule of knowledge of performance results, we may expect to see some of the unfortunate circumstances illustrated in Figure 5-3.

Figure 5-3
Theoretical Learning Curves for Different Types of Workers.

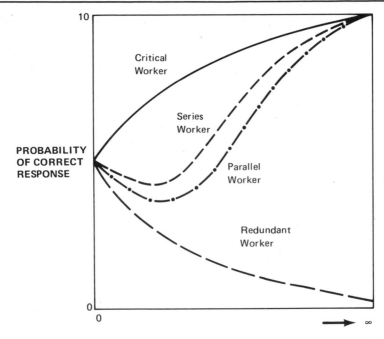

EXPOSURE TO WORK SITUATION

True, the curves in Figure 5-3 are simply illustrated extrapolations from a series of laboratory studies (Klaus and Glaser, 1968; Short, Cotton, and Klaus, 1968). But the situations posed here have a very real bearing on learning to have accidents. It is consistently observed that performance degrades temporarily when a worker goes from the close monitoring and 100 percent reinforcement likely to characterize his initial orientation to critical tasks to the more occasional reinforcement of the series or parallel worker. This dip in performance is likely to be even more pronounced, of course, when total feedback is less perfect than we have assumed in setting up the idealized series and parallel situations.

In any case, this period of degraded performance can be a time of real danger. The solution is not so simple as having a supervisor support the worker through this transition with enhanced feedback. The available evidence (Klaus and Glaser, 1968) suggests that this will merely postpone, not reduce, the performance dip.

The message comes through loud and clear. Safe performance demands learning under reinforcement conditions like those in the operational situation—with adequate safeguards for the novice while he learns but with a rigid demand that he achieve sustained safe performance *under the operational conditions of reinforcement.*

If one can achieve a permanent, more favorable feedback situation in the operational environment, that is, of course, desirable. The learning principle, however, remains the same: operational performance standards can be assured only when the worker has achieved and sustained a satisfactory level under the conditions of reinforcement that obtain in the real work environment. To depend upon performance achieved under more favorable conditions for rapid learning and reliable performance is to build a house of safety on quicksand.

The problems of the redundant worker are even more severe. His level of performance, after some initial period of learning under conditions of maximum feedback, can be expected to erode regularly under conditions of irrelevant reinforcement. True, if he remains wholly unnecessary and redundant, no great harm is done. But if his shield of critical workers is removed in an unguarded moment, an accident becomes highly probable.

Take as an example a person learning to drive an automobile. He receives excellent initial instruction. He receives verbal feedback from his instructor that helps him stay in his lane through curve and turn with nary a stray over the line. He passes his test with flying colors. Driving now on his own, he begins to crowd the center line. He never meets a policeman when over the line; instead, he meets only accommodating drivers coming in the opposite direction who are too busy avoiding a collision even to shake their fist. He continues in this mode until one day he meets, head on, a driver with inclinations like his own. An accident becomes unavoidable at this stage.

TRANSFER OF TRAINING

Earlier, we explored briefly the need in error analysis to allow for changing conditions. The rapidly changing requirements and conditions of modern industry have implications for learning and accidents. Indeed, training for safety might sometimes be almost easy were it not for contingencies and change.

What can we expect of the worker as he moves from one set of performance requirements to another that may be similar or different to varying degrees and along various dimensions? This presents the problem of transfer of training. We cannot cover this complex field with even minimal adequacy here. Rather, we will touch on a few of the more salient issues in transfer of training as they relate to safety.

Let us consider the following theoretical transfer of training functions:

Response Repertory Transfer $(RR) = p_r^2$

Response Association Transfer $(RA) = p_r^2 [1 - 2(q_{rs} + q_s)]$

Stimulus Association Transfer $(SA) = p_s^2 [1 - 2(q_{sr} + q_r)]$

Net Transfer $= RR + RA + SA$

where:

p_r = the probability that a response element in the transfer task will be in the original learning task

q_{rs} = the probability that a response element is to be found in both the original learning and transfer task, bonded to a stimulus also found in both tasks, but bonded with a different stimulus in the two tasks

q_r = the probability that a response element in the transfer task will not be in the original learning task, $q_r = 1 - p_r$

p_s = the probability that a stimulus in the transfer task will be in the original learning task

q_{sr} = the probability that a stimulus element is to be found in both the original learning and transfer task, bonded to a response also found in both tasks, but bonded with a different response in the two tasks

q_s = the probability that a stimulus in the transfer task will not be in the original learning task, $q_s = 1 - p_s$

Several qualifications are in order. Although consistent with a great variety of transfer of training experiments, these formulas are disproportionately based on association of verbal elements (Osgood, 1949). The set of curves implied by these formulas can only be illustrative since we would expect different kinds of stimulus and response content, different levels of original learning, and different reinforcement schedules, each generating its own set of functions (Martin, 1965).

Despite these qualifications, there is sufficient consistency in transfer of training results to suggest some generalizations. First, there is not one but at least three kinds of transfer to be taken into account. Response repertory or response availability is a matter of transferring the worker's familiarity with responses in one situation to responses in another situation. Transfer of response associations is a matter of the worker's carrying stimulus expectations associated with a response in one situation to another situation in which the response element may appear, whether or not the associated stimuli present in the original situation are also present in the transfer situation.

Transfer of stimulus associations is analogous to transfer of response associations, except that in the transfer of stimulus associations it is a matter of the response expectations for stimulus elements carried from the original learning to the transfer situation.

What are the expectations for these three types of transfer? Transfer of the response repertory is always expected to be zero or positive. That is, the familiarity with or ability to produce one response will not interfere with the learning of a new response. This is not to say that some responses cannot have generalized negative effects. Placing objects too close to the edge can be bad in a myriad of tasks and circumstances. It is unlikely, though, that this peculiar ability will interfere with learning other object handling or mis-handling behavior.

Another aspect of response repertory transfer, which it shares with other kinds of transfer, is the nonlinearity of transfer as one moves from small to great similarity from original learning to transfer. For little similarity, the transfer is disproportionately small. Only when similarity is very high does the transfer begin to approach a proportional effect.

Response association transfer is a function of not just response relatedness between original and transfer situations; it is also a function of the stimulus similarity between the original learning and transfer tasks and of the correspondence between the way in which response elements in both tasks are bonded to stimulus elements also to be found in both tasks.

Stimulus association transfer is symmetrical with response association transfer, but in this case it is response similarity and correspondence in the bonding of stimuli that modify the dominant stimulus similarity function. Both stimulus and response association can result in either positive or negative transfer.

Net transfer is shown as a simple additive function of the three fundamental kinds of transfer. Of course, in particular situations the relative weights assignable to the different types of transfer should be differential, but we will not be concerned with the nature of such refinements.

Space does not permit exploring even a small proportion of the curves that can be generated from the simple formulas just presented. However, we can profitably examine some of the boundary contours that they generate to see what implications they might suggest for safe behavior.

Figure 5-4 presents two curves that can be directly derived from the formulas on page 78. Recall that transfer of response repertory includes no stimulus term. It is a function only of response similarity between original and

Figure 5-4
Sample Theoretical Transfer Curves.

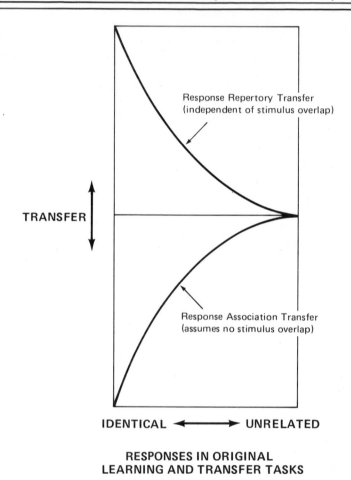

TRANSFER

Response Repertory Transfer
(independent of stimulus overlap)

Response Association Transfer
(assumes no stimulus overlap)

IDENTICAL ◄─────► UNRELATED

**RESPONSES IN ORIGINAL
LEARNING AND TRANSFER TASKS**

transfer tasks. The curve for response association, which is a function of stimulus characteristics, is based on the assumption of no recognizable stimulus relatedness from original to transfer task. When the two curves are combined across the same spectrum of response similarity, they cancel out to a zero transfer situation. In other words, when stimuli in original learning and transfer tasks are unrelated, transfer effects will be nil regardless of the similarity of responses in the two situations.

In contrast, if the stimuli are identical in original learning and transfer situations (and stimulus-response bonds are not altered), transfer will be positive as long as response similarity is above some median value. However, as responses get less similar, negative transfer increases.

We may note that rearrangement of stimulus-response connections can have the same impact on associative transfer as can stimulus or response dissimilarity from original to transfer learning. However, since stimuli and responses not present in both situations cannot be rearranged, rearrangement and similarity effects must be inversely proportional for any given transfer situation.

The maximum positive transfer can be expected when stimuli, responses, and stimulus-response bonds are all identical from original to transfer tasks; in other words, the two tasks are identical. Zero transfer will tend to occur whenever stimuli are unrelated from original to transfer situations, regardless of response similarity. Maximum negative transfer will occur when stimuli remain the same but required responses are completely different.

Consider one example of maximum potential for negative transfer in an important area of safety. Assume we have a boy aged 11 or 12 who has had years of practice with very realistic toy guns: loading, sighting, firing them at frequent intervals. He now comes into direct contact with real guns for the first time. He has been told many times how dangerous guns are but has never practiced procedures in their safe handling. It hardly seems necessary to predict what may happen if he has unsupervised access to real guns, particularly in the presence of some continued old stimuli—his buddies whom he has been "shooting" for years.

CONFLICTING REINFORCEMENT

All of our discussion about reinforcement schedules and transfer for training has been based on the assumption that the definition of performance requirements is not ambiguous nor is the worker's motive ambiguous. Unfortunately, real-life situations are seldom that simple, and we are constantly confronted by conflicting motives and reinforcement.

Three factors make conflicting reinforcements and safety a difficult problem. First, the real world of industry is seldom consistent. Yes, industrial management recognizes that safety is important. But there is that product to be turned out, and speed can conflict with accurate and safe performance. There is generally a curvilinear relationship between motivation and performance (Vroom, 1964). Too great an urgency for speed can be self-defeating in its own right by actually impeding rapid performance. In most situations, however, the standards for accurate and safe performance will be destroyed long before urgency for speed has become self-defeating.

The second factor intensifies the first. It is not bad enough that the industrial establishment has a difficult safety trade-off which it cannot always make consistently. It really does not matter what the industrial establishment chooses to reinforce in the way of safe performance except insofar as workers perceive this same intent. The policy of top-level management gives way to the immediacy of the demands of one's supervisor. The approval of one's peers may be more important than some remote safety objective.

The third factor derives from one of man's essential motives. Release of tension is extremely motivating to the human organism. Humor, sexual gratification, amusement park rides, dramatic plots, competitive skilled sports, and highway chicken probably all derive much of their popularity from the pleasures resulting from release of tension. So it is hardly surprising that work groups can engender reinforcement and escalation of unsafe practices, especially where there is an element of competition or a need for peer-group approval.

Is the outlook for the future entirely bleak? Will conflicting reinforcements make learning unsafe behavior a phenomenon beyond our control? Not necessarily. Certainly, precise analysis and attention to the learning imperatives we have discussed thus far can contribute to creative ways of reducing conflict with reinforcement of safe performance.

There are two tendencies that work to our favor. If the worker can be made to consistently exhibit it under a realistic reinforcement schedule, safe behavior will tend to be self-protecting. That is, the worker will increasingly tend to reject those performance alternatives that he does not follow over a period of time. Second, a consistent and compelling rationale and reinforcement schedule for safe performance will tend to make conflicting performance alternatives less salient (Brehm and Cohen, 1962). That is, one *can* successfully up the ante for safety and thus reduce the competition from negative forms of behavior.

CONTEXTUAL COMPARISON

Even a cursory review of literature in the safety analysis and engineering field demands the conclusion that comparison of safety from one context to another is neither a simple nor an easy matter. Even when the purpose is simply to describe differences in accident rates from one context to another, the prescription of data gathering and measurement techniques that will permit comparisons is fraught with problems. It is much more complex and difficult to provide measures of accidents that will permit the kinds of comparison across contexts that will facilitate insight into underlying dynamics and support improved design. Certainly among the more complex of these dynamics are the ones concerned with causative behaviors.

Our brief review of error analysis and task engineering and learning has touched on only a few of the complex factors involved in behavior and accidents. This suggests that we should not expect to find the problem of defining a diagnostic framework for comparing accident-related behavior in one context with another to be an easy one.

Perhaps this anticipated complexity should cause us to pause and ask what it is about such comparison across contexts that might make it worth the trouble. Fundamentally, our understanding of behavioral processes is such that we can make useful, but by no means complete or entirely accurate, estimates of accident-producing error likelihood and means of reducing that likelihood. But a potentially powerful ally to the direct analysis of error probability and consequence is the ability to use real-world or complex simulation operations as a source of error data that will permit comparison from one situation to another. Such a comparison can "partial out" the effect of behaviorally relevant variables or error likelihood and consequence. The identification of these variables and accurate assessment of the nature and magnitude of their effects can contribute in important ways to the engineering of tasks and learning for safety.

The search for variables that significantly affect accident-related behavior will long continue. Indeed, the identification of such variables will be a source of continuing payoff for safety engineering and training. Without some relatively stable frame of reference, however, this search (and the more sophisticated search for the nature and magnitude of effects) is likely to bog down in a morass of uninterpretable differences from one situation to another. Only in a partial and tentative way can we identify the elements of such a frame of reference. The following, however, seem to be essential elements to consider in comparing accident-related behavior in different situations:

1. *Roles.* The function(s) or purpose(s) fulfilled by the worker whose accident-related behavior is being considered (for example, management, planning, programming, operation, maintenance)
2. *Behavioral Aspects.* The content, levels, and variability of stimulus; psychological process; response and feedback required of the worker
3. *Performance Aspects.* The dimensions of performance (for example, speed, quantity of production, quality of production)
4. *Task Aspects.* The number of tasks assigned to a given worker, frequency

of task repetition, variations in task requirements from one performance to another, speed stress

5. *Design Compatibility.* The nature of match between performance requirements and design components (for example, displays, controls, consoles, auxiliary instruments and tools, equipment, informational performance aids, work space, physical environment, organizational environment)

6. *Learning Characteristics.* Includes reinforcement schedules, transfer considerations, and nature of reinforcement

7. *Personnel Characteristics.* Includes physical, social, intellectual, and experiential traits

The availability and utility of cross-context information will be dependent, of course, upon properties of observational and measurement techniques. Since Tarrants has described the ideal characteristics for such techniques in Chapter One of this book, we will not discuss them further here (see also Tarrants 1965, 1966).

CONCLUSIONS

We have been forced to skirt a number of complex and highly technical issues in this chapter, but the central message is simple and clear. Vague generalities and gross data will not contribute greatly to improved behavior as it relates to accidents.

Techniques holding promise for improving the measurement and control of accident-related behavior include:

1. Identification of specific error-likely behavior that can result in accidents
2. Task engineering aimed at reducing errors having accident potential
3. Schedules of reinforcement that will consistently preclude the occurrence of critical errors
4. Assurance that change does not fallaciously depend upon proficiencies that will not readily transfer or that will permit negative transfer to occur in critical operational situations
5. Detection of and specific action against the effects of reinforcements that conflict with safe performance
6. Exploiting the insight that can be gained from multivariate comparison of accident-related behavior in different contexts

In short, detailed analysis of performance requirements and learning conditions is the imperative precursor to effective safety engineering and training. Identifying actual performance errors as deviations from standards of safe performance is a most appropriate and useful approach to safety performance measurement.

References

Altman, J. W. *Human Error Analysis.* Paper presented to the Quality Control Conference, Pittsburgh, October 1965.

Altman, J. W. *Classification of Human Error.* Paper presented to the American Psychological Association Convention, New York, 1966. (a)

Altman, J. W. *Research on General Vocational Capabilities (Skills and Knowledges).* Pittsburgh: American Institutes for Research, 1966. (b)

Altman, J. W. *Learning to Have Accidents.* Paper presented to the National Safety Congress and Exposition, Chicago, October 1968.

Brehm, J. W., and A. R. Cohen. *Explorations in Cognitive Dissonance.* New York: Wiley, 1962.

Ferguson, E. J., and J. M. Daschbach. *Research in Prediction and Prevention of Industrial Accidents.* Paper presented to the National Safety Congress and Exposition, Chicago, October 1968.

Gagne, R. M. *The Conditions of Learning.* New York: Holt, Rinehart and Winston, 1965.

Klaus, D. J., and R. Glaser. *Increasing Team Proficiency Through Training.* Pittsburgh: American Institutes for Research, 1968.

Martin, E. Transfer of verbal paired associates. *Psychological Review, 72*:327–343, 1965.

Miller, R. B. Task description and analysis. In R. M. Gagne (Ed.), *Psychological Principals in System Development.* New York: Holt, Rinehart and Winston, 1965.

Osgood, C. E. The similarity paradox in human learning: A resolution. *Psychological Review, 56*:132–143.

Short, J. G., T. S. Cotton, and D. J. Klaus. *Increasing Team Proficiency Through Training.* Pittsburgh: American Institutes for Research, May 1968.

Tarrants, W. E. Applying measurement concepts to the appraisal of safety performance. *Journal of the American Society of Safety Engineers, 10*(5):15–22, 1965.

Tarrants, W. E. *Research.* A report based on the Proceedings of the Research Workshop of the Symposium on Measurement of Industrial Safety Performance, sponsored by the Industrial Conference of the National Safety Council, 1966.

Vroom, V. H. *Work and Motivation.* New York: Wiley, 1964.

Chapter 6

AN ALTERNATIVE APPROACH TO THE MEASUREMENT OF INDUSTRIAL SAFETY PERFORMANCE BASED ON A STRUCTURAL CONCEPT OF ACCIDENT CAUSATION

Murray Blumenthal

This chapter attempts to increase the usefulness of safety statistics by tying them directly to the management decisions that are needed to prevent or reduce injuries. Traditional concepts of "accident" and "causation" are replaced by a causal structure consisting of four levels. At the top level, accidents are defined as symptomatic of malfunctions or failures in the man-technological system that comprises the second level. These failures reflect the inability of people and materials to cope with the demands placed upon them by the system. The inappropriate demands, in turn, reflect inadequate management decisions at the third level. The inadequacy of management follows from the absence of adequate knowledge and/or the valuing of profits, efficiency, and so forth, over safety.

An approach to the "measurement" of safety performance, based on the independent work of Gibson (1961) and Haddon (1966), relating accidental injuries and energy sources correlates symptoms, system malfunction, and management decisions. Specifically, the approach includes in one overall table: industry, bodily location of injury, injury severity, relative frequency, energy type, suggested type of management countermeasure, and judged adequacy of emergency medical care and transportation.

Other measures are suggested for the measurement of *management's* occupational safety performance. These are the percentage of total overhead invested in safety research and the *slopes* of their cost-effectiveness ratios.

Finally, it is suggested that by going beyond a "market" conception of the worth of human life, justification could be provided for increased expenditures for the prevention and amelioration of accidental injuries. It is the purpose of this chapter to examine new approaches and techniques that will enhance accidental measurement, control, and prediction capabilities.

In the occupational context the level of safety achieved is a combined result of philosophic, economic, legal, technological, managerial, and motivational influences among others. The measurement of occupational safety performance also is subject to these influences, and a critical examination of safety measurement will have to take them into account. For example, the earliest safety measurements can be traced to the emerging industrial safety legislation in the late nineteenth and early twentieth centuries that required factory inspections, the guarding of dangerous machinery, and the assumption of financial liability by employers (National Safety Council, 1974).

WHAT IS AN "ACCIDENT"?

Safety measurement generally involves the evaluation of the event (or its consequences) that has been labeled "accident." There are many conceptions of the nature of an accident, and the conception that is selected will, in turn, influence the measurement procedure employed. For example, at one extreme, the fatalistic view of accidents ascribes their causation to forces outside of direct human control. From this point of view, there is little to be gained from their measurement. The phrase "act of God" is still to be found in the legal literature, and it has even been proposed as a category in an accident classification system (McGlade and Laws, 1962).

Paralleling the fatalistic view, in one sense, is the belief that so-called accidents are random events, again, with little to be gained from their tabulation.

Still another view ascribes accident causation to invariant human characteristics, identified as "accident proneness" (Greenwood and Woods, 1919). Here, the measurement is of selected personality characteristics of people.

It is the point of view of this chapter that "the term 'accident' is a prescientific concept that emphasizes a lack of intention and the unexpectedness of an event" (Blumenthal and Wuerdemann, 1968) and that emphasizing these features is unlikely to contribute to the effectiveness of safety programs any more than doing so would increase the effectiveness of disease prevention programs. The occurrence of disease is also generally unexpected and unintended.

Further, the "term [accident] has its roots in a time when events that deviated from the expected aroused superstitious fears and were often explained in demonological terms" and that "accidents shared this characteristic with mental and physical illness, meteorological and planetary phenomena, etc. With the development of science, these previously mysterious phenomena yielded to . . . naturalistic . . . explanations" (Blumenthal and Wuerdemann, 1968).

WHAT "CAUSED" THE ACCIDENT?

Another concept having a direct impact on occupational safety performance measurement is "causation." Here again, the concept has been variously interpreted. A simplistic definition of causation holds that every effect has its single cause and that the aim of accident investigation is to locate that cause, while the aim of safety measurement is the tabulation of the various causes uncovered by investigations.

Writing for the investigators of traffic mishaps, Baker (1975) states:

> It has been customary to name a single factor as the "cause of the accident." Even people who recognize that nearly all accidents have more than one factor may urge us to report only the "most important." . . . Naming a single factor as most important would be like trying to pick out the most important link in a chain or the most important leg on a chair. . . . What we mean when we do this is usually that the "most important" factor is either the easiest factor to discover or the easiest to control (p. 29).

Sometimes proposed as a more adequate treatment of causation is the concept of "contributing conditions." However, the *Traffic Engineering Handbook* (Baerwald, 1965) reports that when contributing conditions "appear in state and city accident summaries, they are often *erroneously* accepted as accident 'causes'" (italics added).

One study has concluded that "the request for 'contributing conditions' ignores the same logical and procedural difficulties as the request for 'the cause' or 'causes' of an accident. The investigator can identify only those factors that are available to relatively superficial observation and that are close to the accident in time and space" (Blumenthal and Wuerdemann, 1968). In the National Safety Council's *Accident Prevention Manual* (1974) requests for "accident causes" are found under "hazardous conditions" and "unsafe acts."

A PROPOSED APPROACH TO CAUSAL ANALYSIS

In contrast with "single-point" or "link-chain" concepts of causality, it is proposed that there be established a causal *structure,* one made up of successive levels or layers starting at the top with a symptomatic level. At this level accidents are regarded as symptomatic events indicating a malfunction of a system at the level below, which is made up of human and technological components.

A malfunction, or system failure, occurs at the second level, when human and/or material capabilities are exceeded. For example, human beings are incapable of continuous, uninterrupted attention to a task. If the operation of the system and safety of an operator depends upon his meeting such an unrealistic expectation, then, inevitably, an unsafe condition will occur as part of the breakdown of the system. Similarly, the expectation that a given machine part under considerable stress can function indefinitely without fatigue will just as inevitably result in a system failure.

The analysis of causation need not stop at the operating system level; it can usefully continue to the next underlying level, that of *management.* It is management that makes (or fails to make) the assessment of human and material capabilities and oversees the conception, design, implementation, operation, and modification of the occupational system.

The causal analysis can continue into a level below management that includes both *knowledge* and *values.* Failures in the system may occur because management does not "value" safety above other values, such as the economic, despite the assertion, in 1948, of the president of Jones and Laughlin Steel Corporation that "*if we can't afford safety, we can't afford to be in business*" (as quoted in National Safety Council, 1974).

The trade-off between safety and other values—ultimately economic—is evident in the statement of the president of the U.S. Steel Corporation in 1906, that the corporation "expects its subsidiary companies to make every effort *practicable* to prevent injury to its employees." (italics added—as quoted in National Safety Council, 1974).

The level of safety that management is able to achieve is also rooted in the state of knowledge available to decision makers.

This causal analysis (Blumenthal, 1968) has extended through three underlying layers of the causal structure and is summarized in Table 6-1.

Given the conception of an "accident" and "causation" outlined above, what are the implications for occupational safety performance measurement?

First, without clarification, the term "measurement" is ambiguous. Measurement, mensuration, tabulation, and investigation are used almost interchangeably in this chapter. An examination of *Accident Facts* (National Safety Council, 1976) indicates that occupational safety data reported in this source are limited exclusively to counting or rates and then they are ranked

Table 6-1
Causal layers Comprising an "Accident"

Symptomatic	"Accidents" as symptoms of a system failure
Operating System	Human and/or material capabilities exceeded
Management	Failure in conception, design implementation, operation or modification of the operating system
Value and Knowledge	Inadequate knowledge of human or material characteristics Other system outcomes Values over safety

in order of magnitude for comparable categories. For example, the number of accident-related injuries involving the head, eyes, arms, hands, thumbs and fingers, legs, feet, toes, and trunk are counted and ranked. Or the frequency and severity rates for industries reporting to the NSC are reported and ranked. Total and average cost of injuries associated with an ambiguous category "source of injury" are reported, and so on. More analysis and interpretation are needed.

It is apparent that, in Stevens' (1951) terms, nominal and ordinal scales are used exclusively, and principally to describe events at the symptomatic level in the preceding table.

ALTERNATIVES TO PRESENT MEASUREMENT APPROACHES

It is likely that presently collected statistics could be made more useful if a study of their present and potential uses by decision makers were made. For example, who presently uses the tables appearing in *Accident Facts,* for what specific decisions?

Would modifications in collection, tabulation, or reporting procedures enhance their use? There are very few cross-tabulations, so that questions such as "What kinds of injuries are associated with different industries?" or "Do younger workers incur different kinds of injuries than older workers?" cannot be answered by presently available tabulations.

Another alternative is suggested by Gibson's (1961) conclusion that "injuries to a living organism can be produced only by some energy interchange." Gibson suggests a "classification of injury according to the forms of energy involved." He identifies these energy forms as mechanical, thermal, radiant, chemical, or electrical.

Haddon (1966) amplifies Gibson's approach with an indication of the primary injury produced by the "delivery to the body of energy in excess of local or whole body injury thresholds," as shown in Table 6-2.

Haddon suggests, in order of priority, the following actions that can be directed against the destructive effects of the energies discussed in these tables:

1. Prevent the marshalling of hazardous energy
2. Prevent or modify its release
3. Separate it and the susceptible structure in time or space
4. Interpose a barrier which blocks or attenuates its action
5. Raise injury thresholds so that damage is prevented or substantially reduced
6. Provide as rapidly as possible the optimum in emergency care and transportation, and/or
7. Provide later the clinical, corrective, and rehabilitative services necessary to reduce to the maximum possible extent the damage already produced

Table 6-2

Illustrations of Class I Injuries: Those Due to the Delivery of Energy in Excess of Local or Whole Body Injury Thresholds

Type of Energy Delivered	Primary Injury Produced	Examples and Comments
Mechanical	Displacement, tearing, breaking, and crushing, predominantly at tissue	Injuries resulting from the impact of moving objects such as bullets, hypodermic needles, knives and falling objects; and from the impact of the moving body with relatively stationary structures as in falls, plane and auto crashes. The specific result depends on the location and manner in which the resultant forces are exerted. The majority of injuries are in this group.
Thermal	Inflammation, coagulation, charring, and incineration at all levels of body organization	First-, second-, and third-degree burns. The specific result depends on the location and manner in which the energy is dissipated.
Electrical	Interference with neuro-muscular function, and coagulation, charring and in-	Electrocution, burns. Interference with neural function as in electroshock therapy.

Haddon extends this approach in Table 6-3 (p. 94) to include injuries "due to interference with normal or whole body energy exchange."

In applying these countermeasures to Class II injuries, "it is the source of the interference with normal function which is blocked."

Gibson's and Haddon's conceptions suggest an approach to occupational safety performance that integrates the upper three levels of the "causal structure" outlined earlier. This approach simultaneously relates

Occupation
Bodily location of injury
Injury severity
Relative frequency
Energy type
Suggested type of management intervention
Judged adequacy of emergency medical care and transportation

Table 6-2
(Continued)

Type of Energy Delivered	Primary Injury Produced	Examples and Comments
	cineration, at all levels of body organization	The specific result depends on the location and manner in which the energy is dissipated.
Ionizing radiation	Disruption of cellular and subcellular components and function	Reactor accidents, therapeutic and diagnostic irradiation, misuse of isotopes, effects of fallout. The specific result depends on the location and manner in which the energy is dissipated.
Chemical	Generally specific for each substance or group	Includes injuries due to animal and plant toxins, chemical burns, as from KOH, Br_2, F_2, and H_2SO_4 and the less gross and highly varied injuries produced by most elements and compounds when given in sufficient dose.

Source: W. Haddon, Jr., "The Prevention of Accidents," in D. W. Clark and B. MacMahon (Eds.), *Textbook of Preventive Medicine*. Boston: Little, Brown, 1966, Ch. 33.

Table 6-3
Illustration of Class II Injuries: Those Due to Interference With
Normal Local or Whole Body Energy Exchange

Type of Energy Exchange Interfered With	Types of Injury or Derangement Produced	Examples and Comments
Oxygen utilization	Physiologic impairment, tissue or whole body death	Whole body—suffocation by mechanical or chemical means (for example, by drowning, strangulation, CO and HCN poisoning) Local—"vascular accidents"
Thermal	Physiologic impairment tissue or whole body death	Injuries resulting from failure of body thermoregulation, frostbite, death by freezing

Source: W. Haddon, Jr., "The Prevention of Accidents," in D. W. Clark and B. MacMahon (Eds.), *Textbook of Preventive Medicine.* Boston: Little, Brown, 1966, Ch. 33.

The segment of a hypothetical table given in Table 6-4 illustrates the way that the data could be organized. The same type of tabulation and analysis (using the same categories) could be done for the other energy forms.

In Table 6-4 there are actually a total of 336 injuries recorded. The frequencies in category 6 (emergency medical care or transportation) also appear in the other categories and thus are not counted as part of the total. By converting frequencies to percentages, Table 6-4 illustrates that in Industry A approximately 5 percent of the mechanical injuries to the head were fatal, 30 percent resulted in permanent disabilities, and 65 percent in temporary disabilities. It is judged that the circumstances in 47 percent of the head injuries were deficient in Haddon's countermeasure 1 (prevention or the marshalling of hazardous energy) and 42 percent were judged deficient in countermeasure 3 (separating the energies and the susceptible structures in time or space).

It can also be noted that under countermeasure 6, emergency medical care was judged to be inadequate in 56 percent of the fatalities, 22 percent of the permanent disability injuries, and 23 percent of the temporarily disabling injuries.

Under the "?" the frequencies indicate that a judgment about the required countermeasure could not be made in 17 percent of the fatalities and 15 percent of the permanent and 8 percent of the temporarily disabling injuries.

Table 6-4
Proposed Tabulation and Analysis of Accidental Occupational Injuries
for Industry A

	Mechanical Required Countermeasures							Thermal Required Countermeasures							Etc.
	1	*2*	*3*	*4*	*5*	*6*	*?*	*1*	*2*	*3*	*4*	*5*	*6*	*?*	
Head															
Fatalities	8		7			10	3								
Perm. disabled	25		60			22	15								
Temp. disabled	125		75			50	18								
Eyes															
Fatalities															
Perm. disabled															
Temp. disabled															
Etc.															

With a completed table, the relative severity of injuries occurring to the different parts of the body and under the various energy categories could be compared, as well as the relative judged deficiency of the various types of countermeasures. The components of a complete table could be totaled by rows or by columns for further comparisons.

This approach is intended to supplement, rather than replace, other measurement methods.

Such a compilation could provide at least a partial basis for legislative, administrative, and research priorities. For example, if it were found that in Industry A 70 percent of the instances of fatal injuries involving the head and mechanical energy were lacking measure 6 (relating to "emergency care and transportation), then a required course of action would be indicated.

A similar approach, using energy types and countermeasure priorities, could be used (1) as a check on the adequacy of work-place conceptualization and design, (2) for the diagnosis or monitoring of hazards in an existing work place, and (3) for evaluating the adequacy of present safety standards.

Safety professionals will note that educational and motivational approaches are not included in Haddon's list of countermeasures. However, his approach is in accord with Stonex's (1963) summary of the occupational safety principle: "Anticipate every type of accident which may occur because of machine or human failure and then establish safeguards to eliminate the hazard or minimize the injury when failure occurs."

A "public health" approach to community health similarly does not depend upon the continuing, motivated and skilled use of health measures by the population at risk. Instead, priority health strategies seek measures that are effective without depending upon individual cooperation, such as in community-wide water and sewage systems as a more effective approach than individual families boiling their drinking water or disposing of their sewage.

Both Stonex's and the public health approach place the responsibility for safety clearly in the hands of the management level of the causal structure. Therefore, the next possible application of occupational safety performance measurement concerns the extent to which management effectively carries out its responsibilities.

For example, would it be useful to compare similar industrial organizations as to

1. The percentage of their total overhead invested in safety research
2. The *slopes* of their accident and accident-severity rates
3. The *slopes* of their cost-effectiveness ratios

Management, as discussed earlier, is constrained by the trade-offs that have to be made by economic and other considerations. The value of human life, defined by the National Safety Council (1976) as "the present value of all future earnings lost," reflects only economic criteria. Surely a human life is worth more than the "present value of all future earnings"—both to the holder of that life and to others. An informal pilot effort by Joksch (noted in a communication to the author) suggested that if individuals could be guaranteed their freedom from fatal accidents, they would be willing to pay "premiums" ordinarily required for a life insurance policy of approximately $1 million. By finding a rational basis for increasing the economic worth of a human life, justification could be provided for increased expenditures for prevention and amelioration of accidental injuries.

References

Baerwald, J. E. *Traffic Engineering Handbook.* Institute of Traffic Engineers, 1965.
Baker, J. S. *Traffic Accident Investigator's Manual for Police,* 2nd ed. Evanston, Ill.: Traffic Institute, Northwestern University, 1975.
Blumenthal, M. Dimensions of the traffic safety problem. *Traffic Safety Research Review, 12:*7–12, 1968.

Blumenthal, M., and H. Wuerdemann. *A State Accident Investigation Program, Final Report,* Vol. 1. Project 24, RFP 162, Contract FH-11-6688 with the National Highway Safety Bureau of the U.S. Department of Transportation. Hartford, Conn.: The Travelers Research Center, 1968.

Gibson, J. J. The contribution of experimental psychology to the formulation of the problem of safety—A brief for basic research. In *Behavioral Approaches to Accident Research.* New York: Association for the Aid of Crippled Children, 1961.

Greenwood, M., and H. M. Woods. The incident of industrial accidents with special reference to multiple accidents. *Bulletin No. 4. Industrial Fatigue Research Board* (London), 1919.

Haddon, W., Jr. The prevention of accidents. In D. W. Clark and B. MacMahon (Eds.), *Textbook of Preventive Medicine.* Boston: Little, Brown, 1966.

McGlade, F., and D. F. Laws. Classifying accidents: A theoretical viewpoint. *Traffic Safety Research Review,* 6:2-8, 1962.

Mine safety, 78 mute witnesses for reform, *The New York Times,* December 1, 1968.

National Safety Council. *Accident Prevention Manual for Industrial Operations,* 7th ed. Chicago: The Council, 1974.

National Safety Council. *Accident Facts,* 1976 ed. Chicago: The Council, 1976.

Stevens, S. S. Mathematics, measurement, and psychophysics, in *Handbook of Experimental Psychology.* New York: Wiley, 1951.

Stonex, K. A. Relation of cross-section design and highway safety. *Traffic Safety Research Review,* 7:1963.

Chapter 7

THE ERROR-PROVOCATIVE SITUATION: A CENTRAL MEASUREMENT PROBLEM IN HUMAN FACTORS ENGINEERING

Alphonse Chapanis

In March 1962 a shocked nation read that six infants had died in the maternity ward of the Binghamton, New York, General Hospital because they had been fed formulas prepared with salt instead of sugar. The error was traced to a practical nurse who had inadvertently filled a sugar container with salt from one of two identical, shiny, 20-gallon containers standing side by side, under a low shelf in dim light, in the hospital's main kitchen. A small paper tag pasted to the lid of one container bore the word "Sugar" in plain handwriting. The tag on the other lid was torn, but one could make out the letters "S..lt" on the fragments that remained. As one hospital board member put it, "Maybe that girl did mistake salt for sugar, but if so, we set her up for it just as surely as if we'd set a trap" (Chapanis, 1965b, p. 6).

RECIPE FOR AN ACCIDENT

In order to discuss measurement, it is important that we fully understand the phenomenon we intend to measure. Let us first examine the concepts of "accident" and "accident prevention" in the man–machine–environment setting. It seems clear that there is no single common factor that provides the key to accident prevention. In fact, the reason behind this point of view is now so generally accepted (see Jacobs, 1960; Old, 1966) that it merits the status of an axiom. We shall state it as Axiom 1.

99

AXIOM 1: *Accidents are multiply determined. Any particular accident can be characterized by the combined existence or coincidence of a number of events and circumstances.*

What are the basic ingredients of an accident? For the great majority of significant accidents there are four—people, objects, physical environment, and social environment.

PEOPLE.   The kinds of accidents that we are primarily concerned with involve people in one way or another. This means that we shall eliminate from consideration, for example, a great variety of natural accidents, catastrophes, and disasters in which people were only minimally involved.

One of the most basic things we can say about people is that they are different. Indeed, this observation is so elementary and so obvious that we often overlook it, especially when we attempt to make models of people or of systems involving people. Not only do people differ, but the differences between people are very large. An octogenarian is not at all like an astronaut; a 10-year-old slum child is not at all like the president of a large corporation; and Farrah Fawcett-Majors is not at all like Jimmy Carter.

People differ for three obvious and very important reasons. First, some differences between people are inherited. Sex, skin color, height, reaction time, and intelligence—all these and many more traits, are, for the most part, determined biologically. Superimposed on, and interacting with, our biologically determined characteristics are the experiences, knowledges, skills, and attitudes that we accumulate throughout our lifetimes. The skills of a ballet dancer, a concert pianist, a nuclear physicist, or an airline pilot are largely learned skills, skills acquired through long hours of training and education, both formal and informal.

In making these statements we are not concerned with how much of any particular trait is biologically determined and how much is determined by education and training. This is not really relevant here. The primary point is that we are what we are partly because we were born that way and partly because we learned, or were taught, how to be that way.

The third reason people differ is that their behavior is constantly being warped, molded, and changed by a variety of more or less transient influences: foods, beverages, drugs, fatigue, and diseases being some of them.

OBJECTS.   The second ingredient of most accidents is some kind of an inanimate object: a tool, utensil, appliance, machine, or piece of equipment. These are the things with which we get involved when we have accidents. In listing these agents mentally you may let your imagination range freely because the objects that are involved in accidents are even more diverse and varied than their human partners.

PHYSICAL ENVIRONMENT.   Accidents do not happen in a vacuum; they always occur in an environment of some kind. These environments are the third ingredient of our recipe. Environments range from hot to cold, from dry to moist, from bright to dark, from clean to dirty, and along many other dimensions that have been catalogued and measured by industrial hygienists and other scientists.

SOCIAL ENVIRONMENT.   Environments may, of course, be defined narrowly or broadly. The way in which environments have been addressed above implies a fairly narrow definition of the term. But there is more to an environment than just such physical things as temperature, humidity, and airborne contaminants. A number of social factors contribute to safety or to its converse, accidents, just as surely as do the things already mentioned. These can be identified as the social environment, or social climate, in which accidents occur.

Included in the social environment are the formal rules, regulations, and laws that influence, or are supposed to influence, human behavior. These formal regulations and laws are entrusted to individuals, organizations, or agencies that make up any one of several administrative superstructures under which people normally live and work. These, too, form part of our social environment. Also in the social environment, and just as important as the formal rules and regulations, are the informal conventions, habits, and attitudes that help to shape and mold human behavior. For example, people who live in small communities just naturally seem to drive differently in the areas where they live than they do in the center of a large city, even though the traffic laws may be identical in both locations. To continue, the stereotype of the "Sunday driver" identifies a set of driving habits that seems to have some validity. As one more example, certain segments of our society seem to be characterized by deliberate and cautious behavior while others are conspicuous for their bravado and recklessness.

It is not of very great importance whether one considers the social environment as a separate ingredient of accidents, distinct from the physical environment, as has been addressed here, or whether one thinks of the social environment as simply another part of the environment as a whole. The important thing is to recognize that the social environment is there and that it has to be considered in the total accident picture. One last point: The fact that it may be extremely difficult to measure the effects of regulatory laws, or of social conventions, on our behavior does not in any way diminish their reality.

All that has been said above about these basic ingredients of an accident is neither startling nor new. The chief value of addressing the problem in this way is that it reminds us of the complexities and the difficulties of doing anything novel or revolutionary in the field of accident prevention. Indeed, one of the thorny problems in this field is how one can even make general statements with any degree of validity in the face of the almost infinite variability of people and things. One thing is certain: accidents cannot be prevented by considering any one ingredient in isolation. An effective attack on accidents has to take into account all of these ingredients and all of their complex interrelations.

It is important to note that the ingredients that contribute to accidents are, for all practical purposes, the very same elements that enter into all man-machine systems. Moreover, when we say that these ingredients have to be considered together, we are, in fact, talking almost the same language as the systems engineer. Therefore, let us look at accidents through the eyes of a systems engineer.

THE DESIGN OF MAN-MACHINE SYSTEMS: BASIC PHILOSOPHY

We live in an age of man-machine systems, many of them huge, awesome, amazing, exciting, productive, and fearsome. Whatever adjectives we might apply to one or another of our man-machine systems, however, at least one adjective applies to all of them—man-made. All man-machine systems are designed and constructed by people.

The Elements of Systems Design

Another interesting characteristic of many systems is that they have impressive reliabilities. Man himself is fallible. Yet somehow he has been able to construct man-machine systems that function almost perfectly. The elements of systems design are, moreover, the same ones that we have mentioned above as being the basic ingredients of accidents. How does the systems engineer get reliable performance out of the very same components that produce those accidents? One part of the answer is that the systems designer takes into account all of the constituents of his system—in whatever amounts, sizes, or shapes he has to take them—and then designs a system around them. Let us review the elements of systems design once again with this in mind.

1. *Personnel (People)*. Systems are designed for particular kinds of people. A space vehicle is designed around a highly select group of astronauts. A telephone, by contrast, is designed to accommodate a wide spectrum of people. When a systems designer starts his work, he has to have a clear idea of the kinds of people who will most likely be using his system.

2. *Training.* Personnel must be trained so that they will have the amount and kinds of information and skills that are necessary to operate the system. In this connection, note that training is not left to chance. It is a deliberate part of systems design. Good systems design includes the design of a training program.

3. *Equipment Design.* In designing equipment the systems designer has to take into account a great many purely physical factors: materials, components, inputs, outputs, weights, power requirements, costs, and so on. Also among the factors to be considered are the ways in which the design interacts with the people in the system, in the environment where man and machine will be working.

4. *Operating Procedures.* The instructions, procedures, and rules that set forth the duties of each operator in a system comprise the operating procedures. Some rules may specify how the system is supposed to behave with regard to other systems, or in the space through which it operates. Navigation rules, for example, establish which of two vessels have priority when their paths intersect. Similar rules of the road apply to vehicular traffic. Although many operating procedures and rules are arbitrary, a good set of them contributes greatly to safe and orderly operations.

5. *The Environment.* Systems, like people, operate in particular environments and they often create environments of their own. The internal and external environments of a system have to be known and specified carefully before systems design can begin.

The five elements identified above are often called components of the *personnel subsystem* as distinguished from the hardware subsystem. Human factors engineering makes sure that these elements of the personnel subsystem are fully considered in systems design.

Interrelations among the Elements of Systems Design

Perhaps the most important characteristic of a system is that everything interacts with everything else. The situation is something like the illustration shown in Figure 7-1 (p. 104). First, the selection of personnel interacts with the kind and amount of training the personnel will have to receive before they can operate any particular system. If personnel selection procedures are set up to pick out operators with high levels of appropriate abilities and skills, then these operators will ordinarily not require much additional training before they are ready to go to work. On the other hand, if unskilled operators, or if operators with low levels of relevant ability, are selected, they will more likely need a considerable amount of training before they can become useful and productive.

Figure 7-1
Interrelations Among the Elements of Man-Machine Systems Design.

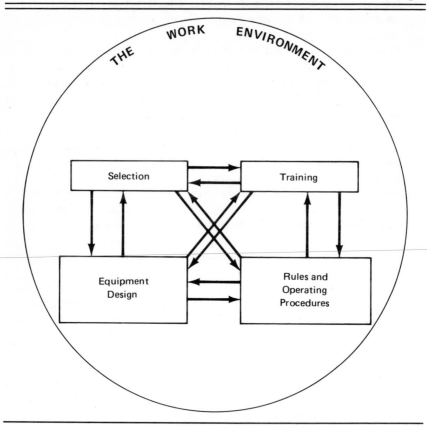

THE INTERACTION BETWEEN EQUIPMENT DESIGN AND PERSONNEL SELECTION. The way in which equipment is designed also interacts with personnel selection. For example, about eight percent of otherwise normal males are color-blind. Many of these men cannot reliably distinguish red from green. How can we cope with this fact in designing a man-machine system, for example, for a system of transportation? One way of dealing with it might be to select only color-normal men as operators of our system. By eliminating all color-blind men from our population of operators, we are free to use colors and color codes for signaling, identification, or other purposes.

An alternative way of dealing with the fact of color blindness is to redesign the system so that color-blind operators will not be handicapped in their operation of the system. This kind of redesign can be done in several ways. One is to use certain colors that almost all people, both color-normal and color-blind, can easily distinguish. For example, although many color-blind people have difficulty in distinguishing conventional red and green traffic lights, few of them would have trouble with traffic lights if the green were made bluer and the red made yellower. Other ways are to use codes that do not depend on color, such as shape, brightness, and position codes, and so do not handicap the color-blind operator.

The purpose of this example is to show the kind of trade-off, or balancing of factors, that is usually possible between personnel selection and equipment design. Perhaps an even better way of illustrating this idea is to remind you that machines can be designed for use even by seriously handicapped persons. Telephone switchboards, typewriters, and production machinery, for example, have been designed for use by totally blind persons, by amputees, and by multiply handicapped persons. Moreover, when such workers use machines that are properly matched to their limitations, they often have impressive safety records. It is not really necessary to do very much screening of personnel if someone is willing to spend the time and effort to do a good design job.

THE INTERACTION BETWEEN EQUIPMENT DESIGN AND TRAINING. A third set of interrelations in Figure 7-1 is that between equipment design and training.

Figure 7-2
A Micrometer of Conventional Design. (From A. Chapanis,
Man-Machine Engineering. *Belmont, Calif.: Wadsworth, 1965.)*

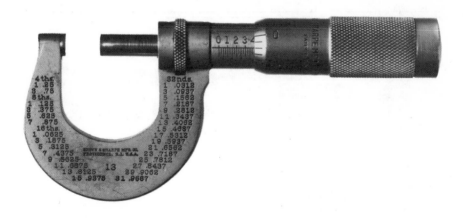

As a simple example, Figure 7-2 shows a micrometer of conventional design. Can you read the setting? First, the numerals 0, 1, 2, 3, and so forth, on the frame of the micrometer represent tenths of inches. Second, notice that between any pair of adjacent numbers there are four marked divisions. This means that each marked interval corresponds to 0.1 ÷ 4, or 0.025 inches. In addition, the thimble (the section at the end of the micrometer) makes one complete revolution per marked interval on the frame, and around the circumference of the thimble are 25 small marked intervals. Hence, each of these marked intervals corresponds to 0.025 ÷ 25, or 0.001 inch. Now, with this information, can you read the micrometer? You will undoubtedly agree that a novice would have to spend at least a little time in training before he could cope with this instrument.

Next, look at the micrometer in Figure 7-3. This one substitutes a simple, countertype indicator for the complicated series of scales in Figure 7-2. The improvement is so obvious, that further words of explanation are hardly needed. Apprentice machinists need a considerable amount of practice before they can read a micrometer like that in Figure 7-2. The one in Figure 7-3 can be read with no training at all.

The example given here is simple and obvious. It is, nonetheless, important because we know that even skilled machinists occasionally make errors in reading conventional micrometers. A more sophisticated example of the interrelation between equipment design and training is given by Gibbs

Figure 7-3
A Micrometer Designed According to Good Display Principles.
(From A. Chapanis, **Man-Machine Engineering.** *Belmont, Calif.:*
Wadsworth, 1965.)

(1952), who redesigned the scales on some conventional machine tools (lathes and jig boring machines). Tests showed that with the new devices unskilled apprentices were able to do certain jobs faster and better than skilled machinists could with conventionally designed machine tools.

OPERATING PROCEDURES. Figure 7-1 shows that operating rules and procedures also interact with all the other elements of systems design. The common denominator of most operating rules and procedures is, of course, language. People who work with machines and the hardware of systems often fail to realize how important language is to the proper use and operation of these machines. The fact is, however, that the words used in instructions, operating manuals, and maintenance manuals often help speed a system to needless disaster. This point is so important that it is worth some elaboration.

Almost everyone has had at least one experience with instructions that accompany most assemble-it-yourself items. Some of these instructions are so complicated that even people with college degrees have difficulty understanding them. A less homely example is the large device sketched in Figure 7-4. When I say large, I mean that a man, if he wanted to, could easily crawl

Figure 7-4
*A Schematic Illustration of a Large Device That Has to Be Sprayed with Protective Finishes. (From A. Chapanis, "Words, Words, Words," *Human Factors, 7:1–17, 1965.)*

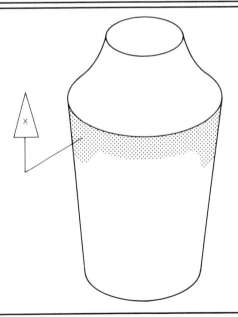

into this shell. It so happens that parts of this thing need to be carefully shielded against some severe environmental stresses such as extreme heat. Sprays are applied to protect this gadget, and these sprays are called *finishes*. This thing was designed to be transported to a paint shop with some instructions—a part of which appear in Figure 7-5. Let us look at one of these instructions: "Finish l98 all over but may have 684B on areas designated . . . ." What exactly does this instruction mean? Exactly what is the man in the paint shop supposed to do? Four interpretations that I received in a small survey are shown in Figure 7-6. Perhaps you will understand why this set of instructions caused some confusion.

Some of my favorite examples, however, come from military equipment. I have read a number of instruction manuals written for various pieces of electronic equipment. Most of these were, in my opinion, well beyond the audience for which they were intended. The soldiers' solution to this difficulty was sometimes simple and effective. They translated the instructions to meet their needs. To illustrate, the instructions for one item contained this sentence:

> WARNING: The batteries in the AN/MSQ-55 could be a lethal source of electrical power under certain conditions.

Figure 7-5
The Designer's Instructions for the Device Illustrated in Figure 7-4.
(From A. Chapanis, "Words, Words, Words," Human Factors, 7:1–17, 1965.)

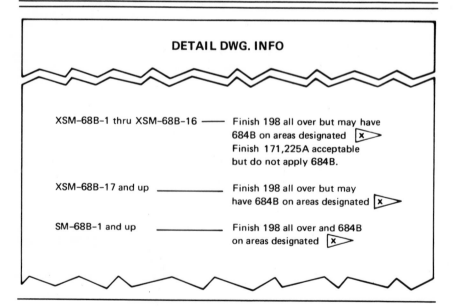

Figure 7-6
Interpretations of One of the Instructions Illustrated in Figure 7-5.
(From A. Chapanis, "Words, Words, Words," Human Factors, 7:1–17, 1965.)

WHAT IT SAYS:

Finish 198 all over but may have 684B on areas designated ⟦x⟩ .

WHAT IT COULD MEAN:

1. Finish 198 or 684B on areas marked ⟦x⟩ ; 198 everywhere else.
2. Finish 198 all over even if 684B was applied first on areas marked ⟦x⟩ .
3. Finish 198 all over first. 684B optional afterwards on areas marked ⟦x⟩ .
4. Finish 684B first on areas marked ⟦x⟩ . Then 198 all over.

On the equipment itself, however, someone had placed next to the terminals a slip of paper on which he had printed in large red letters:

LOOK OUT! THIS CAN KILL YOU!

What a marvelous job of translation this is! Some unknown soldier had cut through the turgid statement that "the batteries . . . could be a lethal source of electrical power under certain conditions" and had extracted the heart of the idea: "This can kill you." He used just four words, each a single syllable, short, blunt, clear, and to the point.

Instructions do not have to contain many words to be confusing, of course. I am sure you will agree that the displays in Figure 7-7 (p. 110) could actually lead to confusion, errors, and accidents.

A special problem of operating rules and procedures is of fairly recent origin. It has arisen because of the way communications, commerce, and transportation have shrunk our world. The problem is associated with difficulties in preparing instructions for people who do not have English as their native tongue. Numerous stories have been recorded of airplanes that have crashed (see Ruffel Smith, 1968), trucks that were ruined, and tractors that were never used because, in part, Americans could not communicate effectively with the people who were using, or trying to use, the machines that were produced (Chapanis, 1965a). When you recall that American long-range jet aircraft are being used by almost every major airline of the non-Communist world, you will appreciate that we do indeed have here a language problem of major proportions.

Figure 7-7
Signs on One of Our New Interstate Highways

To conclude this section on operating rules and procedures the following observations are offered. Operating procedures interact with all the other elements of systems design. Complex instructions require highly selected and highly trained personnel. When equipment is complicated, instructions and operating rules tend to be complicated as well. Simple instructions, on the other hand, can be used by less highly selected and less highly trained operators. When the design of equipment is simplified, operating rules and procedures can also be simplified, even though this may not always occur in practice.

THE WORK ENVIRONMENT.   Finally, Figure 7-1 does not show, but should show, that there are interrelations between the work environment, on the one hand, and personnel selection, training, equipment design, and operating procedures, on the other. These are so obvious that a simple example may suffice. In designing a system for use in torrid regions, one can select people who are better able to withstand hot temperatures, one can train them through acclimatization to withstand these temperatures better, or one can design the system with air conditioning so that ordinary people can operate the system. The physical environment also interacts directly with operating procedures because the latter usually specify such things as the duration of work periods and of rest periods and the pace or rate at which work is done. In these ways operating procedures may be used to compensate for otherwise intolerable temperatures or other conditions.

To summarize, the systems designer and the human factors engineer view a system as a complicated set of interrelationships between people, training, equipment design, operating procedures, and the environment. All these elements are variables in their equations and all can be manipulated in the final product. Moreover, when a system fails it does not fail for any one reason. It usually fails because the *kinds of people* who are trying to operate the system, with *the amount of training* they have had, are not able to cope with *the way the system is designed, following procedures* they are supposed to follow, *in the environment* in which the system has to operate. The elimination of such failures may be brought about by varying any one or more of these five basic elements.

THE HUMAN ENGINEERING DESIGN OF EQUIPMENT

As should now be clear, the human factors engineer is usually the specialist who is concerned with the personnel subsystem, that is, with the integration of man into a man-machine system. The basic approach of the human factors engineer is that of manipulating personnel selection, training procedures, design of equipment, operating procedures, and the environment to produce the best system performance possible.

Error-Provocative Situations

In their work human factors engineers find that many man-machine systems are error-provocative. Other people have sometimes used the terms "error-inducing" or "error-producing." Whatever we call it, an error-provocative situation is one that almost literally invites people to commit errors. Let us consider a few examples.

Some years ago, a former student and I reported the results of a study on medication errors made in a large hospital (Safren and Chapanis, 1960). This is, of course, a problem that affects all of us and that interests most of us. Over a period of seven months we were able to collect data on 178 errors in the administration of medicines in this hospital. The type of error that had the highest frequency was one in which a patient received or almost received a wrong dosage of a drug. These, incidentally, were not small errors, but errors in which a patient might receive as much as 24 times the required dosage of a drug!

q. h.   *q. n.*

Let us examine one reason why errors like this could happen. The abbreviation on the left in Figure 7-8 is *q.h.*, which stands for *quaque hora*, "once every hour." The abbreviation on the right in Figure 7-8 is *q.n.*, which stands for *quaque nocte*, "once every night." Whether a patient receives a particular medication once a day or 24 times a day depends on the height of the stem of the *h* or *n*. Considering the way in which most people write, this is most certainly an error-provocative situation, a situation that almost literally invites people to make mistakes.

Take another example from the study. Figure 7-9 shows a small pharmacy bottle. It is sealed at the top and when a pharmacist, nurse, or physician wants to draw off some medicine he (or she) inserts the needle or a hypodermic syringe through the rubber top that seals the bottle and draws off the required

Figure 7-9
The Label of a Small Pharmacy Bottle

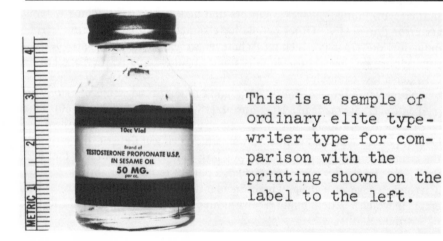

This is a sample of ordinary elite type-writer type for comparison with the printing shown on the label to the left.

amount. Now look at the label on this bottle. One can easily see at the top that the bottle contains 10 cc. One can also see that the medicine has a concentration of 50 mg. But can you see the critical words "per cc" at the very bottom? These letters are so small that they can easily be, and sometimes are, overlooked even by people with normal vision. As a result, a nurse or a pharmacist may compute a dosage thinking that the medicine is packaged in a concentration of 50 mg per 10 cc instead of 50 mg per cc. This would yield a dosage that is 10 times the required amount—not a trivial error. Once again, it is apparent that this is an error-provocative design.

Consider one more example of a different sort. Studies show that most people expect a toggle switch to move up or to the right to turn something ON and down or to the left to turn something OFF. Figure 7-10 shows a col-

Figure 7-10
Four Toggle Switches From the Laboratories at Johns Hopkins University.
(From A. Chapanis,* Man-Machine Engineering, *Belmont, Calif.:
Wadsworth, 1965.)

lection of four toggle switches in the laboratories at John Hopkins University. They show that all four possible directions of movement are used to turn something on. The consequences of an error in moving a toggle switch in the wrong direction in this laboratory are probably not very serious. At most, such an action might result in a temporary inconvenience to someone. But we know for certain that accidents have occurred in aircraft because pilots could leave one aircraft and get into the cockpit of another aircraft of the same type, made by the same manufacturer, and find switches that moved in different directions. The same kind of accident has been documented in large passenger buses. Once again, it is obvious that these are error-provocative designs.

Errors are so numerous in many man-machine systems that examples could be cited almost endlessly. Indeed, their frequency is matched only by their importance. At the very least, errors are disruptive of normal routines and result in inefficiency. At the worst, errors may result in accidents and fatalities. The reduction and elimination of errors is one of the main objectives of human factors engineering. It is for this reason that in the title of this chapter error-provocative situations are referred to as a "central" problem in human factors engineering.

Designing for Human Limitation

Once we recognize the existence of error-provocative situations, it is an easy step to address this issue in human engineering design. This can be stated in the form of another axiom.

AXIOM 2: *Given a population of human beings with known characteristics, it is possible to design tools, appliances, and equipment that best match their capacities, limitations, weaknesses.*

The gains that can be realized by concentrating on the design of equipment are often very great, so much so that another axiom can be stated as follows:

AXIOM 3: *The improvement in system performance that can be realized from the redesign of equipment is usually greater than the gains that can be realized from the selection and training of personnel.*

These two axioms together account for the relative emphasis that human factors engineers place on the various elements of systems design. It is necessary to stress the importance of equipment design in the control of errors because in the past too much attention in accident research has been focused on the human being. No one denies that accidents occur because of human acts—someone making a wrong decision, or a wrong movement, or failing to respond to a signal. One can ascribe such failures to carelessness, to inattention, to impulsiveness, or to some other characteristic of the human operator. But this approach often leads up a blind alley. Everyone is inattentive, everyone is careless, and everyone is impulsive at some time or another. To say that an operator was inattentive, careless, or impulsive is merely to say that he is human. It tells us nothing about how one could prevent such accidents from occurring in the future.

The human factors, or the systems, approach looks at other parts of the system. The evidence is clear that people make more errors with some devices than they do with others. Some highway intersections earn the reputation, justifiably, of being "death traps," whereas others are accident-free. A good systems engineer can usually build a nearly infallible system out of components that individually may be no more reliable than a human being. The human factors engineer believes that with sufficient ingenuity nearly infallible systems can be built even if one of the components *is* a human being.

ERRORS VERSUS ACCIDENTS

Thus far in this chapter we have discussed mostly errors. What do errors have to do with accidents?

The Essential Identity Between Error- and Accident-Provocative Situations

Human factors engineers usually do not worry about the distinction between errors and accidents and for a good reason.

> AXIOM 4: *For purposes of man-machine systems design there is no essential difference between an error and an accident. The important thing is that both an error and an accident identify a troublesome situation.*

Axiom 4 is such an important point that it requires some elaboration. Consider the following true case study.

> The home economics classroom in a modern city high school has rows of gas ranges on which students learn how to concoct various culinary specialties. One day Janice leaned over a stove to look into a pot on one of the back burners. At the same time, she reached for a control to increase the heat under the pot. By mistake, she grasped the wrong control and ignited the gas in one of the front burners. Janice exclaimed, "Oh, darn," and corrected her error.

Barbara, on the next stove, went through exactly the same sequence of actions. Barbara, however, has long hair and when she turned up the flame in the front burner, the flame ignited her hair.

What is the essential difference between these two occurrences? The difference is a trivial one: Janice had short hair, whereas Barbara had long hair. Because of this trivial difference, we say that Janice made an error but that Barbara had an accident.

From the human engineering standpoint, both incidents are equivalent. They both identify an error- or an accident-provocative situation. Indeed, investigation reveals that some linkages between controls and burners are error-provocative. Figure 7-11 shows four of many different control-burner arrangements that one can find on stoves. Chapanis and Lindenbaum (1959) found that in tests involving 60 people, each of whom had 80 trials on one of these stoves, no one ever made a mistake on the arrangement labeled 1. Out of 1200 trials, 76 errors were made on arrangement 2. Arrangements 3 and 4 were the poorest, and both were about equally bad. To repeat, people tend to make many errors on some arrangements of stoves; other arrangements are easy to use and are almost completely error-free. Whether you have an acci-

Figure 7-11
Four Control-Burner Arrangements
Tested by Chapanis and Lindenbaum (1959)

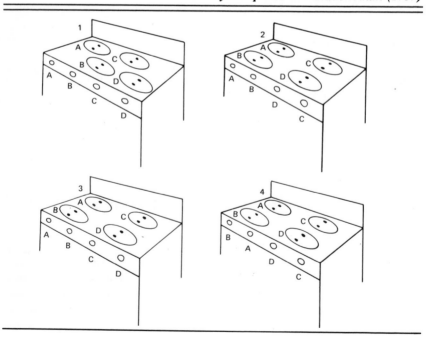

dent or merely make an error with one of these stoves is, however, largely a matter of chance.

Let us consider one more example. In one of the classic studies in the field of human engineering, Fitts and Jones (1961) showed that pilots, even experienced pilots, occasionally misread a three-pointer altimeter by 1000 feet. Whether this kind of misreading results in a simple error or in an accident, depends on a number of chance, and largely irrelevant, factors, such as the height of the terrain or obstacles in the environment, the locations of other aircraft in the vicinity, and so on. The really important part of that study was not that pilots did, or did not, have accidents because they misread their altimeters. The really important part was the fact that pilots could, and did, misread their altimeters. Something about the design of these altimeters was error-provocative and potentially dangerous. As a result of these early studies, we now have altimeters that are much easier to read—or, conversely, that are much harder to misread.

The Advantages of Analyzing Error-Provocative Situations

Working with error-provocative situations has several advantages over working with accidents per se. Some of these have already been ably stated by other people (Tarrants, 1965, for example). We list only five here for consideration:

1. It is easier to collect data on errors and near-accidents than on accidents.
2. Errors and near-accidents occur much more frequently than do accidents. This means, in short, that more data are available.
3. Even more important than the first two points is that error-provocative situations provide one with clues about what one can do to prevent errors, or accidents, before they occur. Too many measures of accidents are after-the-fact measures. It is all very well, for example, to use *total dollar cost* as a way of assessing the severity of an accident problem, and there are many people to whom this kind of measure is undoubtedly useful. But the total dollar cost of an accident provides absolutely no information about how one can prevent accidents like this from happening in the future. Essentially the same criticism can be made about several other indexes that have been used for measuring accidents.
4. The study of errors and near-accidents usually reveals all those situations that result in accidents plus many situations that could potentially result in accidents but that have not yet done so (Tarrants, 1965). In short, by studying error-provocative situations, we can uncover dangerous or unsafe designs even before an accident has had a chance to occur. This, in fact, is one of the keys to designing safety into a system before it has even been built.

5. This last point is a rather technical, but important, one. If we accept that the essential difference between an error and an accident is largely a matter of chance, it follows that any measure based on accidents alone, such as number of disabling injuries, injury frequency rates, injury severity rates, number of first-aid cases, and so on, is contaminated by a large proportion of pure error variability. In statistical terms, the reliability of any measure is inversely related to the amount of random, or pure error, variance that contributes to it. It is likely that the reason so many studies of accident causation show such marginally low relationships is the unstable, or unreliable, nature of the accident measure itself. Statisticians tell us that the correlation between any two measures X and Y has an upper limit. That upper limit is set by the reliability of X or Y, whichever is lower. If Y is some sort of a measure based on accident occurrences, it cannot by its very nature be a very reliable measure (see Lybrand et al., 1968, pp. 33–38; Teel and DuBois, 1954). On the other hand, there is evidence to suggest that near-accident data, or at least "accident behaviors," constitute a more reliable measure of potentially dangerous situations (see Whitlock et al., 1963).

THE PREDICTION OF ERROR-PROVOCATIVE SITUATIONS

Because human factors engineering is more generally concerned with the design of new products and systems rather than with the redesign of systems that are already in operation, it emphasizes the identification of error-provocative designs before they are actually built. This is an extremely difficult and challenging task, and we do not have, by any means, the information and the techniques to do this infallibly. We do have, however, some general guidelines and some methods that help us in our work, which we shall discuss next.

Avoiding Design Characteristics That Increase the Probability of Error

Although we do not have as much data as we would like to have about human performance in specific systems, we know enough about errors in general to specify some of the kinds of things that should, and should not, be done. We know, for example, that operators are likely to make errors of judgment if they are required to interpret instructions for themselves. On the other hand, we know that errors are often avoided or are quickly corrected if an operator can be given immediate feedback about the consequences of what he has done. We know, too, that a powerful way of increasing the reliability of systems, and of decreasing errors, is to use parallel, or redundant, links (see Chapanis, 1960, pp. 549–553; Fogel, 1963, pp. 716–733).

In general, the probability of error increases when the job, the system, or the situation does the following:

1. Violates operator expectations
2. Requires performance beyond what an operator can deliver
3. Induces fatigue
4. Provides inadequate facilities or information for the operator
5. Is unnecessarily difficult or unpleasant
6. Is unnecessarily dangerous (Weislogel, 1960)

OPERATOR EXPECTATIONS.   We have already alluded to the fact that people generally expect controls and displays to move in certain ways. For example, most people expect that an increase in a system function, such as an increase in speed or an increase in temperature, will be displayed by the movement of a pointer from left to right on a horizontal scale or from down to up on a vertical scale. Conversely, people expect pointers to move in the reverse direction to show a decrease in a function. When consistencies of this kind can be found in a large percentage of people, the consistencies are often called "population stereotypes." A great many such stereotypes, or natural expectancies, have been studied and tabulated in human engineering textbooks and guides (see Chapanis, 1965b; Morgan et al., 1963). We can say, based on a great deal of research, that a system should never require an operator to do things that are unnatural or unexpected.

OVERLOADING THE OPERATOR.   Systems that demand more than an operator can deliver are often said to overload the operator. This overloading can occur in many ways. First, a system may require the operator to do what no ordinary human being can do. For instance, it may require the worker to respond to signals embedded in so much noise that no ordinary person could be expected to hear them, or it might require the operator to exert forces that are beyond what any reasonable person might be expected to exert. Second, a system may overload an operator because it is designed for someone who is more highly selected, or highly trained, than the person who is actually trying to use the system. Third, a system might overload the operator because it requires him to respond to too many inputs, or to make too many control movements, or to make responses too rapidly.

Whatever the source of the overloading, the evidence seems unmistakable: overloading an operator increases the probability that he will commit errors.

FATIGUE.   In the annals of science fatigue has proved to be as elusive as the study of accident causation. Nonetheless, the evidence seems clear that people commit more errors when they are tired than when they are not. Fatigue, like overloading, may arise from several sources. It may be physical or it may be mental. A longshoreman becomes fatigued from the sheer amount of physical exertion that he indulges in during a day. An air traffic controller, on the other hand, may not exercise his muscles very much but may become just as exhausted as the longshoreman. The fatigue of the air traffic controller is mental fatigue, exhaustion brought about by the barrage of stimuli that bombard him, by the necessity of making correct and rapid decisions and by the knowledge that a single incorrect decision might easily result in disaster. Fatigue may also come from awkward postures that have to be maintained for long periods of time, seats that are uncomfortable, insufficient lighting, excessive noise, and many more conditions too numerous to catalog here. An important aim of human factors engineering is to design against such fatigue-inducing, error-producing situations.

INADEQUATE FACILITIES AND INFORMATION.   People are much more likely to make errors when they have to work under makeshift conditions or are provided with inadequate information. Numerous errors have been made in attempts to repair something without the proper tools or under improper work conditions. We can all recall with feelings of suppressed frustration and rage a number of assemble-it-yourself items that we have ruined or damaged because the instructions that came with the kit were obscure, incomplete, or even incorrect. To do a job well, the operator has to have a well-designed work place, a good working environment, the proper tools and auxiliary equipment, and instructions that are simple, clear, and complete. All of these requirements are an important concern of the human factors specialist.

UNNECESSARILY DIFFICULT OR UNPLEASANT TASKS.   Many people seem to regard work as a necessary evil, as something to be "gotten through" so that they can enjoy the better things of life. When jobs are made unnecessarily complicated or when the work environment is unpleasant, this tends to increase levels of frustration and to make workers careless and slipshod in their performance. This is a motivational problem, not directly related to the abilities of an operator per se. Although the evidence is not as conclusive as we would like, it appears that streamlining procedures and making a work situation more pleasant and comfortable have the additional benefit that they help workers perform more safely.

UNNECESSARILY DANGEROUS TASKS.   Another motivational problem, related in some ways to the point above, involves tasks that are unnecessarily dangerous. If, for example, a technician has to reach up among high voltage lines without proper equipment or safeguards, he is much more likely to become "rattled" and do his work less carefully and less thoroughly than if the danger did not exist. Safe designs do not expose operators to needless risks.

Measurement Techniques and Devices

The guidelines suggested above are useful, but they are so general that they are often not of much help when one comes to the practical business of designing particular pieces of equipment, work places, or jobs. Some of the measurement techniques that help the human factors specialist anticipate and evaluate error-provocative designs are given below.

CHECKLISTS AND EVALUATION GUIDES. Through extensive research and experience human factors specialists have accumulated considerable evidence on kinds of designs that are more likely to produce errors and those that are not. Some suggestions about the contents of such items are contained in the examples presented. The following are a few more checklist items to illustrate how they are often structured:

1. Are emergency controls located where they are easily accessible, regardless of the momentary position of the operator?
2. Are the sizes of markings, numerals, and labels on visual displays large enough to be seen by the operator from his normal working position?
3. Are displays located so that an operator can see them easily from his normal work position?
4. Are controls that require fine, precise adjustment assigned to the hands, those that require the application of large forces to the feet?

Evaluation guides are very much like checklists, and both perform essentially the same function. Evaluation guides usually contain detailed lists of design criteria that can be used for assessing various items of equipment. Both checklists and evaluation guides are usually arranged under headings that are related to specific kinds of equipment, as, for example, Test Points, Cathode-Ray Tubes, Control Devices.

The amount of information of this kind that is available is considerable and it can be very useful. By studying blueprints, preliminary designs, and mockups with a measurement checklist or evaluation guide in hand, the human factors specialist can discover many error-provocative designs before they are ever built.

If automobile designers were to make better use of such evaluation checklists and guides, they could easily eliminate some of the error-provocative designs that characterize modern vehicles. As an example, on one of our family cars the brake pedal is so situated that if my foot is not exactly on the center of the pedal when I depress it, my foot strikes a protuberance at the base of the steering column. This protuberance prevents the brake pedal from being depressed fully. On several occasions this fault in design has caused near-accidents. Fortunately, it has always been possible to recover in time and brake correctly. So far, that is. On a second family car the rear-view mirror is so located that when the seat is adjusted to a comfortable driving position, there is no way in which the driver can get a clear, unobstructed view in the mirror through the rear window of the car. These and other design errors in these automobiles are unnecessary. We know how to anticipate such faulty designs. All that is necessary to correct them is for someone to *use* the techniques that are available.

SIMULATORS.   Simulators and functional mockups are also extremely useful tools for the human factors specialist. Accidents and near-accidents, the effects of various stresses, overloading of the operator, and alternative design configurations have all been studied in complete safety with simulators of modern jet aircraft. Driving simulators, simulated air traffic control systems, submarine simulators, and space vehicle simulators are other examples of devices that have been used successfully in discovering error-provocative situations.

One of the most impressive examples of this kind of study is a series of experiments undertaken by Milton Grodsky and his staff at the Martin Company for the National Aeronautics and Space Administration. The simulator filled an enormous room and included, among other things, a three-man Apollo Command Module, a two-man Lunar Excursion Module, a star field for navigation, and an elaborate computer control and data collection system. Three-man crews lived in the simulator for seven days and engaged in realistic missions simulating all aspects of a flight to the moon and return to earth. During the course of these missions investigators collected millions of items of data on 40 different measures of performance. The analysis of these data then allowed the investigators to estimate human reliabilities for the various kinds of performance and to see when and where errors were most likely to occur.

Other important advantages of simulators are that they can be fully instrumented to measure and record all of the kinds of data that the scientist or design engineer needs for his purposes. In addition, a simulator is usually much cheaper than the device it mimics. The cost of a jet simulator is only a fraction of the cost of a jet aircraft. Finally, simulators can be used as multipurpose devices. They can be used for training operators, for example, as well as for studying the system.

TASK ANALYSES. A task analysis is a procedure for systematically identifying exactly what an operator needs to know, what he needs to do, and what he needs to have in order to do his job properly. A task analysis also lists the characteristic human errors that are likely to result at each step of the operation. A good task analysis is a highly complicated instrument and to describe it completely would require far more space than is available here. For our purpose of examining measurement techniques, it is sufficient to state that in a good task analysis potential errors are carefully listed and analyzed under specific headings. As part of the error analysis, the human factors engineer also tries to suggest ways in which each source of error might be obviated in the final design (see Pickrel and McDonald, 1964).

THE STUDY OF SYSTEMS IN EXISTENCE. Sometimes the human factors specialist can obtain a great deal of useful information about error-provocative situations by studying systems that are already in existence and that resemble the one being designed. The possibilities here are several. Activity analyses, collecting operator opinions, and the critical incident techniques are ways of studying systems in operation to discover troublesome design features, operations, and procedures. Used properly, these measurement techniques can be extremely useful (see Chapanis, 1959; Tarrants, 1965).

HUMAN RELIABILITY DATA. Just as the engineer is able to give reliability figures for the various components he uses in his work, so human factors engineers have started to collect data on human reliabilities. At the present time reliability data have been tabulated for a great variety of detailed human actions, such as reading a label that contains one or two words, a label that contains three to five words, a label that contains six to eleven words, punching one to five numbers into a keyboard, flipping various numbers of toggle switches, and so on. The eventual aim of this technique is to compile a detailed and valid set of reliability figures from which one can synthesize various human operations on paper and predict what the error rates are likely to be for the entire operation. Presumably, therefore, an engineer could make paper-and-pencil analyses of various alternative designs and predict which of the alternatives would yield the best performance.

The difficulties, both practical and conceptual, in collecting human reliability data of this kind are many. It is still too soon to decide how useful and productive this approach will be. About all we can say at the present time is that it is available, that it is still under active development, and that it bears watching (see Altman, 1964; Meister, 1964).

LABORATORY EXPERIMENTATION. Well-designed laboratory experiments provide the ultimate in control of variables, precision of data, and elegance of statistical analysis. Used properly, they can be very helpful in the identification and analysis of error-provocative situations. The study of stoves already referred to (Chapanis and Lindenbaum, 1959) was a laboratory experiment. One sequel to the discovery that pilots often misread altimeters in common use in the 1940s (Fitts and Jones, 1961) was a laboratory experiment (Grether, 1949). (The sequence of dates is confusing here because the Fitts and Jones study was prepared in 1947 as an air force memorandum report with limited circulation. It was reprinted in 1961 in a more readily available form.) In his study Grether was not only able to confirm what Fitts and Jones had reported about the difficulty of reading the conventional altimeter, but he was also able to demonstrate that a number of other kinds of altimeters could be read with far greater accuracy. On the basis of Grether's findings, and the findings of other studies related to his, pilots now have altimeters that are virtually foolproof.

For all of their potential usefulness, the results of laboratory experiments have to be applied to practical situations with some caution. In the process of abstracting variables for study, laboratory experiments often change basic variables so that they no longer resemble those in real life. In addition, the methods used to test conditions in the laboratory are often artificial and unrealistic. These and other limitations of laboratory experimentation (see Chapanis, 1967) mean that the method must be applied judiciously. Still, with all its limitations, the laboratory experiment is a powerful tool in the arsenal of the human factors engineer.

SOME LIMITATIONS OF STUDYING ERROR-PROVOCATIVE SITUATIONS

Every scientific method of measurement and evaluation has its limitations and the general approach described in this chapter is no exception.

Errors and Accidents Are Not Symmetrical

All accidents start out as errors, but not all errors end up as accidents. This means that, in their noninjurious forms, all accidents are simply errors. This point was discussed earlier in this chapter, and the data collected by Tarrants (1965) were cited to support it. However, the reverse is not true. Many errors, in their most extreme forms, never result in accidents. It is difficult to imagine, for example, how simple typing errors, errors in using slot machines, or errors in using a dial telephone, just to name a few, could ever be called accidents. Errors of this kind might be annoying, distracting, and even costly, but they would never by themselves be referred to as accidents.

In studying error-provocative situations, therefore, the accident investigator is certain to uncover potentially dangerous situations, but he might also uncover situations that are only error-provocative and no more.

The Identification of Error-Provocative Situations Does Not Necessarily Reveal Cures

Knowing that a certain situation is error-provocative does not necessarily tell you how to eliminate it. Most of the examples in this chapter are success stories, and a few of the solutions almost literally fell out of the data. But solutions are not always simple, neat, and tidy. Identifying situations where errors are most likely to occur is half the battle. Nonetheless, eliminating sources of error may still require a great deal of scientific sleuthing. As has already been stated, when a human factors engineer uncovers an error-provocative situation, there are several major options available to him. He may select personnel more carefully, he may train personnel better, he may change the environment, he may redesign the situation, or he may follow two or more of these routes simultaneously. It is not always clear which route, or combination of routes, will be most productive. Sometimes one or another potential route may even be excluded on other grounds.

An example will illustrate this point. In the study of hospital medication errors to which we have already referred (Safren and Chapanis, 1960) it was found that one major class of errors was that the medication was given to the wrong patient. Most commonly, it appeared to us, this occurred because a nurse failed to check the name on the bedside tag against the name on the medicine ticket. What is the best way to eliminate errors of this kind? The solution is not immediately obvious. One option, selecting nurses who would be less likely to make errors of this kind, was ruled out at the start. Nurses are scarce enough as it is, and no one was willing to tamper with the selection procedures in any way that would reduce the supply. Under the circumstances the only available solution choices were additional specialized training and changing the situation. Attention to these alternatives produced a number of suggestions concerning ways of improving procedures and ways of identifying patients. In any case, the ultimate proof of these suggested measures lay in further tests to be sure that they really did result in improvements.

Error-Provocative Situations Are Not the Ultimate Measure of Accidents

However useful it might be to identify error-provocative situations, doing so does not provide the universal measure that some safety professionals and accident investigators seem to be searching for (see Elsby, 1964). It would be ideal if we had some kind of yardstick that would enable us to measure safety performance in much the same way that we can measure height, weight, visual acuity, reaction time, or intelligence. The techniques that have been discussed in this chapter make no pretense of providing that kind of a precise measure.

Indeed, the measures of accident-provocative situations that have been described are subject to many of the same criticisms that Tarrants (1965) has voiced about other common measures of safety performance. This does not in any way detract from the usefulness of the measurement techniques noted in this chapter. It means only that we must clearly understand what the study of error-provocative situations will and will not do and concentrate on applying these techniques to situations where their use is most appropriate.

SUMMARY

This chapter has presented a measurement approach that the human factors engineer uses in tackling error-provocative situations. The essential ingredients of man-machine systems design are personnel selection, personnel training, equipment design, operating procedures, and the work environment. These ingredients all interact and are variables that can be manipulated individually or in combination.

The human factors engineer is primarily interested in predicting and eliminating error-producing designs before systems are even built. To this end he has formulated a number of guiding principles and devised a number of techniques of his own, or adopted them from other disciplines, to help him in his job of error prediction. The techniques include checklists and evaluation guides, simulators, task analyses, studies of systems resembling the one under consideration, human reliability analyses, and laboratory experimentation. The human factors engineer is also interested in identifying errors occurring within existing man-machine systems. One measurement approach that aids in error identification is the critical incident technique described in detail in Chapter 17. This is a pragmatic approach aimed not so much at measurement for the sake of measurement but rather at measurement for the sake of practical control. The goal of the human factors engineer is to design man-machine systems that are truly safe, efficient, comfortable, and convenient. The goals of the safety professional and the human factors engineer are, in part at least, common goals. This description of the human factors engineer's philosophy and measurement methods is presented as a contribution to our reaching those common goals.

References

Altman, J. W. Improvements needed in a central store of human performance data. *Human Factors, 6*:681–686, 1964.

Chapanis, A. *Research Techniques in Human Engineering*. Baltimore: The Johns Hopkins Press, 1959.

Chapanis, A. Human engineering. In C. D. Flagle, W. H. Huggins, and R. H. Roy (Eds.), *Operations Research and Systems Engineering*. Baltimore: The Johns Hopkins Press, 1960.

Chapanis, A. "Words, Words, Words," *Human Factors, 7*(1):1–17, 1965. (a)

Chapanis, A. *Man-Machine Engineering*. Belmont, Calif.: Wadsworth, 1965. (b)

Chapanis, A. The relevance of laboratory studies to practical situations. *Ergonomics, 10*:557–577, 1967.

Chapanis, A., and L. E. Lindenbaum. A reaction time study of four control-display linkages. *Human Factors, 1*(4):1–7, 1959.

Elsby, C. H. Measurement of industrial safety performance: The search is on. In *National Safety Congress Transactions,* Vol. 12. Chicago: National Safety Council, 1966, pp. 112–113.

Fitts, P. M., and R. E. Jones. Psychological aspects of instrument display. I: Analysis of 270 "pilot-error" experiences in reading and interpreting aircraft instruments. In H. W. Sinaiko (Ed.), *Selected Papers on Human Factors in the Design and Use of Control Systems.* New York: Dover, 1961, pp. 359–396.

Fogel, L. J. *Biotechnology: Concepts and Applications.* Englewood Cliffs, N.J.: Prentice-Hall, 1963.

Gibbs, C. B. A new indicator of machine tool travel. *Occupational Psychology, 26*:234–242, 1952.

Grether, W. F. Instrument reading. I. The design of long-scale indicators for speed and accuracy of quantitative readings. *Journal of Applied Psychology, 33*:363–372, 1949.

Jacobs, H. H. Conceptual and methodological problems in accident research. In H. H. Jacobs, E. A. Suchman, B. H. Fox, et al., *Behavioral Approaches to Accident Research.* New York: Association for the Aid of Crippled Children, 1961, pp. 3–25.

Lybrand, W. A., G. H. Carlson, P. A. Cleary, and B. H. Bauer. A study on evaluation of driver education. Unnumbered report of the American University, Development Education and Training Research Institute (5185 MacArthur Boulevard, N.W., Washington, D.C. 20016), July 31, 1968.

Meister, D. Methods of predicting human reliability in man-machine systems. *Human Factors, 6*:621–646, 1964.

Morgan, C. T., J. S. Cook, III, A. Chapanis, and M. W. Lund (Eds.). *Human Engineering Guide to Equipment Design.* New York: McGraw-Hill, 1963.

Old, B. S. (Project Supervisor). *Summary Report, The State of the Art of Traffic Safety: A Critical Review and Analysis of the Technical Information on Factors Affecting Traffic Safety.* Cambridge, Mass.: Arthur D. Little, 1966.

Pickrel, E. W., and T. A. McDonald. Quantification of human performance in large, complex systems. *Human Factors, 6*:647–662, 1964.

Ruffell Smith, H. P. Some human factors of aircraft accidents involving collision with high ground. *Journal of the Institute of Navigation, 21*:354–363, 1968.

Safren, M. A., and A. Chapanis. A critical incident study of hospital medication errors. *Hospitals, 34*(9):32–34 *et passim, 34*(10):53 *et passim,* 1960.

Tarrants, W. E. Applying measurement concepts to the appraisal of safety performance. *Journal of American Society of Safety Engineers, 10*(5):15–22, 1965.

Teel, K. S., and P. H. DuBois. Psychological research on accidents; Some methodological considerations. *Journal of Applied Psychology, 38*:397–400, 1954.

Weislogel, R. L. Detection of error-producing designs. In J. D. Folley, Jr. (Ed.), *Human Factors Methods for Systems Design.* Pittsburgh, Pa.: Document AIR-290-60-FR-225. The American Institute for Research, 1960, pp. 63–78.

Whitlock, G. H., R. J. Clouse, and W. F. Spencer. Predicting accident proneness. *Occupational Psychology, 16*:35–44, 1963.

Chapter 8

SUGGESTIVE, PREDICTIVE, DECISIVE, AND SYSTEMIC MEASUREMENTS

C. West Churchman

In this chapter the concern is not so much with the technical characteristics of measurements as with their use, with "safety" as a central illustration. Measurement is a special type of information that is applicable in many different contexts ("broad") and permits a high degree of differentiation ("deep"). This thesis has been explored in general in an earlier essay (Churchman, 1961, Chap. 5). Here we will take the basic idea further by classifying information into four categories: *suggestive, predictive, decisive, and systemic.* To get their flavor before exploring their meaning in depth, consider the following pieces of information about Factory A as they apply to these categories:

1. *Suggestive.* The fatality rate per man year in Factory A has been 0.01.
2. *Predictive.* Of all serious accidents in Factory A, 20 percent were caused by failure of the workers to wear safety equipment, such as specially designed masks.
3. *Decisive.* The special masks when used in Factory A would cost $20,000 per year, and the benefit per year in accident reduction would be $100,000.
4. *Systemic.* Reduction in accidents within Factory A is the third most important project of the system.

In considering these categories, it will be worthwhile placing this effort within the general setting of the literature on measurement. One large section of this literature is logical and attempts to classify various types of measurement scales (see Stevens, 1959). A second part is technical and concerns itself with specific calibration techniques, for example, the reports of the various sections of the American Society for Testing Materials. In neither branch of the literature is there a specific interest in the way in which measurement is used, except to state that it is "quantitative" and to imply that this is a "good thing."

Merely to assign numbers to events or objects in the world is not necessarily a good thing at all; it may, in fact, be as dreary a process as one can imagine, as anyone not interested in baseball can attest after listening to two fans discussing baseball statistics. A better idea of what is involved in measurement can be obtained by recalling the "Maine mile"; if you ask a Maine farmer how far it is to Grand Lake, he might well reply "about a mile." What he means is "some distance" or "not around the next bend." A more literal-minded measurer would want to distinguish between "about a mile" and "about 2 miles." Measurement, as we interpret it here, is above all as literal as it can be, and the ambition of the measurer is to create finer and finer distinctions and broader and broader usage.

To illustrate breadth and depth, consider the measurement of the tides. From certain basic properties of the earth and moon, one can generate tables that tell what the depth of tide will be at various places at various times, with precision, in feet, out to the first or second decimal place. A contrast is some statement about the number of pedestrians killed in a city during a month; little useful information is to be culled from this statistic alone.

One might wish to say that statistical data, such as the pedestrian death rate above, that describe the number, or rate, or relative frequency of events may be *suggestive* in other contexts. "Suggestive" means "evocative, presented partially rather than comprehensively." (All definitions are drawn from *The Random House Dictionary* [1967].) To make this suggestive definition more precise, we need to depict, or model, the thought process of the information receiver. We assume that he has one or more purposes (goals, ends, objectives) and that the receipt of information in some way affects his capability of attaining these. The information is absorbed into the receiver's "relevant picture of the world," which is modified in one way or another by the information.

For example, a traveler consults an airline guide to determine when planes leave from San Francisco to Washington. The information in the guide is, from the point of view of the user, suggestive. Of course the guide is not suggesting that a plane might conceivably leave at 9:15 A.M.; rather it is suggesting times of departure that might or might not fit into the traveler's plans. Indeed, the airline guide assumes that the traveler has a fairly well worked-out plan of what he wants to do. The item in the guide helps him to fill out one unknown of this almost completed plan. It is not the responsibility of the guide publishers to question the advisability of a traveler's taking a specific trip.

In these terms we can now describe the four types of information listed above. Rather than speak of bits of information, for example, numbers or descriptors, we need to consider an information system and its relationship to a potential user. Suggestive information systems make only very weak assumptions about what the user wants and how he should get it. Predictive information systems go a step further and tell the user what would happen if he were to do so-and-so. Decisive information systems go further still and model the user as a decision maker within a bounded system, where the boundaries are given (for example, by the user). Finally, systemic information systems attempt to relate the bounded decision-making to other "higher level" considerations.

Thus, in the examples previously given an information system that generates statistics about fatalities in a factory makes only a weak assumption about what a user should do; it is entirely up to him whether he closes down the plant, puts on a safety campaign, or does nothing at all. Of course, the suggestive information system makes *some* assumption about the user because it vaguely assumes that its information is relevant to the user's decision making. An example of a predictive information system is one that tells the user that if the workers were to wear the available safety equipment religiously, the accidents would go down 20 percent. This information does not necessarily imply any action on the part of the user, but if the information is relevant, then clearly a class of possible actions are implicitly assumed by the information system. For example, the predictive information seems to be saying that the user might seriously consider a safety rule enforcement campaign to require the wearing of safety equipment.

Decisive information systems go much further toward depicting the user's problem. They recognize that safety by itself is only a part of the picture; reducing accidents or accident potential, in the decisive system, must be coupled with other objectives, such as production or, in the case of highways, "throughput." The coupling consists of finding a common dimension for these partially conflicting objectives so that the user is aware of the relative merit of alternative plans or programs. Thus, a cost-benefit description of an accident reduction program aids the user in deciding whether to implement the program. The broader systemic information system enables the user to place safety programs in context with other "efficiency" or "cost-reduction" activities.

Although the main concern of this chapter is to assess the potentials of decisive and systemic information in the area of safety, it is not intended to denigrate either suggestive or predictive information. Most information in today's culture is suggestive—the information to be found in newspapers, books, TV, "data banks," libraries, and so on. Up to a short time ago, most occupational safety information was suggestive. This is well borne out in the *Proceedings* of the National Safety Congress (NSC) during recent years. The argument will later be presented that all indexes are essentially suggestive rather than decisive or systemic. There has been a great deal of interest in the NSC in the development of indexes. On the other hand, the attention paid by the NSC to the "critical incident technique" (see Tarrants, 1965) indicates an ambition to go beyond correlations, which tend to be suggestive, to cause-effect models, which are predictive. Also, the interest of the NSC in costs of accidents and measuring the results of accident reduction programs is a step toward obtaining decisive information.

The point to be made about suggestive information is that it is cheap, and, perhaps more important to safety programs, the information system can be quite limited in what it assumes about the real world. In areas that deal primarily with rare events whose consequences are so difficult to evaluate, this restricted assumption making can have its advantages. Of course, someone eventually has to make a decision, and whoever he is, he will have to make the assumptions.

DECISIVE INFORMATION SYSTEMS AND SAFETY MEASURES

"Decisive" means "determining, . . . putting an end to a controversy, . . . resolute, determined." The difference between suggestive and decisive information can best be illustrated by one of the classical puzzlers of safety measurement—the ratio or index. For example, in transportation there has long been a debate as to whether fatality per passenger mile is a better or worse index of performance than, say, fatality per trip. All ratios of these types are suggestive; the decision maker has to fill in the gap by what he assumes about the world. The relevant question for the decision maker is how a mile or a trip contributes to his values. It is probably the case that manned space travel is one of the safest modes of travel based on fatality per passenger mile, but no one would view it as being especially "safe."

Ratios are essentially suggestive types of information; the numerator represents one kind of value and the denominator another, partially conflicting value. If we measure accidents per man-hour in a plant, we are comparing two opposing values: the value of safety and the value of production. But the ratio does not tell us which is the more important value. In the case of an army battalion in a war, we need not be shocked if this ratio is relatively high, unless we feel the war is a mistake. Thus, the meaning of a ratio depends on some value system of the decision maker.

Decisive information is rarely expressed as an index. Instead, all goals are described along one, unifying value scale, for example, dollars or utility.

In the example given at the beginning of this chapter, the cost and benefit of a piece of safety equipment are both expressed in dollar terms. The decision maker, in this bounded productive system, can then evaluate what a certain policy (enforcing the wearing of masks) would be for him. The "controversy" has been ended, so to speak. This does not mean that all doubt is removed, of course, because all measurement is subject to error. Indeed, some estimate of the reasonable bounds of benefit minus cost should be given by the decisive information system.

Returning to the transportation illustration, we see that neither fatality per mile nor per trip is decisive. To become decisive, the information system must say something definite about the value return to a passenger when the transportation system produces one unit for him. Whether this unit is a mile or a trip or something else depends on the purposes of the decision maker. If he is a cab driver and is paid in terms of the number of miles he drives, then "mile" may be appropriate. If he is a consultant and is paid in terms of the number of trips he makes, then "trip" may be appropriate. I say "may be" because value analysis is always complicated and simple scales are rarely appropriate.

I am suggesting, of course, that safety professionals should be trying to establish decisive measures, but I also point out the extreme difficulties and pitfalls of this attempt. For one thing, if a program starts to move into this area, it gives up its own identification. Specifically, in one decisive mode we cannot really speak of safety as such, as a separate and identifiable aspect of a system. Instead, we have to think of safety as an inseparable element of a more comprehensive measure. We have to recognize that people do not want just to be safe, as their use of freeways and jets attest. Indeed, very great concern with one's own bodily safety may be a sign of neurotic or even pathological behavior.

To some extent, the attempt to find decisive measures of safety performance may seem inappropriate because safety seems so fraught with intangibles. Many people are shocked by the attempt to express the loss of a limb or a life in dollar terms. Such dollar measures are often obtained by calculating the lost income that an accident produces. But lost income may not at all reflect the real pain, shock, or grief. Based on this method of accounting, an old man turns out to be worthless, or even costly, so that a few induced accidents among the aged might seem to be a good thing.

However, such a superficial economic benefit-cost analysis is really no more than suggestive. The point to be made is that none of the goals of the decision maker are really "intangible," but they may very well be hidden. Various techniques have been suggested for revealing them (see Ackoff, 1961, Chap. 2). All of these techniques are based on the reflection that when a decision maker finally decides in a rational manner, he is directly comparing various goals. Indeed, if one of his goals is dollar income, then *in prin-*

ciple all of the other goals can be expressed in dollars if his choice is rational because behavior choices indicate how much of dollar income he is willing to sacrifice for each of the goals. In other words, if an information system classifies a goal as "intangible," what is meant is that the system does not intend to touch it, not that it is "untouchable."

If occupational safety programs move into decisive information, I should expect to see for each industry the emergence of manuals that indicate the probable benefit of a specific safety program, minus the probable cost, where benefit and cost are estimated in terms of the value systems of the clients of the industry (customers, stockholders, workers, managers, public). "Probable" implies that the manual will report sources of error as well as error estimates.

If such manuals do come into being, their introductory material should emphasize how strong a role judgment has played in shaping the data. Judgment is a central aspect of decisive information, just because so much needs to be assumed about the user.

MOVING TOWARD SYSTEMIC INFORMATION SYSTEMS

A movement in the direction of decisive information systems for safety is a realistic program for the next decade. If such a movement occurs, its success will be greatly enhanced by a consideration of what will follow. One of the most challenging of all stories about the future, be it 1984 or 2000, is to estimate what the future will take its future to be. A guess, in the area of information, is movement toward systemic information systems. Today these are bound to appear unrealistic, and even idealistic, but taking them seriously may be the most important aspect of designing realistic systems.

Systemic information systems have no "data," if by "data" one means "what is given." But otherwise, the ambition of a systemic system is to remove the necessity of externally induced instructions; obviously, there will be varying degrees of success in attaining this ambition.

A first step away from decisive toward systemic information will be a self-conscious one, in which the cost of measuring is included as an aspect of the information system. That is, the question "Why measure?" will be meaningful. This is a mere beginning, however, as is the attempt to relate benefit-minus-cost to the "next higher level" of the whole system. In the latter case one is trying to evaluate a safety program, say, against other cost reduction programs of a company. The systemic information system is interested in the next higher level (for example, as a basis for budgeting), but this is only a step toward its more comprehensive ambition, which is information about the "whole system."

The history of measurement in astronomy makes a reasonably good model of the progress from suggestive to systemic measurement. Early navigators could plot their course by the stars, but they had no sound predictive method. The development of Ptolemaic, Copernican, Keplerian, and Newtonian theories provided the predictive measurements of the solar system. Then the fixed stars became unfixed and no longer just plain stars, but a whole genus and species of matter in motion. The whole system is still way out there in the mysterious vastness of an expanding universe.

The lesson of this bit of history is that the systemic information system, unlike the decisive, is never closed and final. Proponents of decisive measures, such as cost-benefit analysis and program budgeting, often refer to their method as the "systems approach," but, of course, it is not so in any comprehensive sense. The aim of the decisive information is to bring controversy to an end, to bring action into being, to supply enough information to enable the decision maker to be resolute. It accomplishes this aim by arbitrarily setting "feasible" boundaries on the system. The aim of systemic information systems, on the other hand, is to keep controversy alive, to use action as experiment, to create problems whose study will increase the scope of understanding. Measurement is no longer the servant of decision making; it is the raison d'être of decision making. One decides in order to measure the better. This no doubt seems fantastic to an age so caught up in the practical, but even today one might say that some social decisions (for example, those of the U.S. Supreme Court) had experiment in mind as much as welfare: to understand what people want and how they behave, one needs to decide things in a certain way.

AN IDEAL SYSTEMIC INFORMATION SYSTEM

If your "fancy" is caught by this "advertisement," I would like to close this chapter by depicting the eternal life of a systemic information system. If your "fancy" has not been "hooked" you can stop here and spend the saved time contemplating the possibility of a practical decisive information system.

The systemic information system is no different from any other in its collecting the rudiments of its information; it counts and questions, listens and smells, using whatever instruments seem appropriate. But it never regards these rudiments as givens. Instead, they have a tentative existence. Each set of rudimentary information is interpreted by means of a model of reality, to be a description of some aspect of reality. For any aspect of reality there must exist significantly different ways of describing it. For example, if we are describing the safety of a piece of machinery, we can interview workers, take past records, use the critical incident technique, or conduct an engineering analysis. Each one of these methods provides the information system with rudimentary information that is then interpreted as describing the specific aspect of reality.

The systemic information system next compares these "adjusted" pieces of information. If they differ significantly, the system tries to alter or enrich its model of reality to explain the differences and arrive at a set of information that is consistent. If the information is all exactly alike, the system refines the rudimentary set, for example, by inventing instruments that read out to one more decimal place. If the information differs but not significantly, the system seeks to find another model of reality that also produces a consistent description of the aspect of reality. That is, the system seeks to find an equally plausible explanation of the "facts," the facts themselves being created out of the model and the rudimentary information. Once an alternative explanatory model is found, the original model and its newly created "deadly enemy" fight out their battle in a series of "crucial experiments," until one or both are defeated, and the system looks for another model.

Part of the central strategy of the system is the selection of those "aspects of reality" that provide it with the best measure of its progress. The guideline is the one suggested at the beginning of this chapter: choose those aspects of reality that have the largest inferential power to the user in different contexts and times.

Such an eternally restless system might remind us of a neurotic compulsive—or else a God: "Behold, he that keepth Israel shall neither slumber nor sleep" (Psalm 121:4).

References

Ackoff, R. L. (Ed.), *Progress in Operations Research,* Vol. 1. New York: Wiley, 1961, Chap. 2.

Churchman, C. W. *Prediction and Optimal Decision.* Englewood Cliffs, N.J.: Prentice-Hall, 1961.

The Random House Dictionary. New York: Random House, 1967.

Stevens, S. S. Measurement, psychophysics, and utility. In C. W. Churchman and P. Ratoosh (Eds.), *Measurement: Definitions and Theories.* New York: Wiley, 1959.

Tarrants, W. E. Applying measurement concepts to the appraisal of safety performance. *Journal of the American Society of Safety Engineers, 10*(5):15–22, 1965.

Chapter 9

THE MEASUREMENT OF SAFETY ENGINEERING PERFORMANCE

John V. Grimaldi

For centuries puzzling problems have been yielding to the power of facts scientifically marshalled, but among the many enigmas that remain, some, though important, will not be solved unless their merits are realized more widely. How to implement safety most effectively is one puzzle whose virtues have not yet commanded enough interest to enable its mastering. And the absence of a reliable safety performance measure may be as responsible for the lack of interest in a solution as it is the result of our inability to answer the question satisfactorily.

CHARACTERISTICS OF THE PROBLEM

Although there is reason to believe that society has been seriously concerned about safety since at least the beginning of recorded history, there does not seem to have been much realistic effort to achieve it. At most, safety has known sporadic popular attention, and then only with respect to incidental issues that surfaced out of the mass of inadequacies.

Safety Is a Function of Emergencies

Practically without exception a dreadful event (for example, the 1968 Mannington coal mine catastrophe) or an increasingly intolerable injury and property damage experience (such as the familiar automobile accident totals) has been the initiator of every new safety consideration. Each is dealt with as if it were a distinctive disorder to be corrected, when, in fact, these unwanted events are most likely symptomatic of a larger systemic deficiency that must be corrected in toto or the incidental failures will persist—as indeed they have.

In the absence of practically any understanding that the complex safety problem cannot be solved by the simplistic approaches consistently used to attack it and because the inciting events demand quick attention, hasty antidotes usually are proposed. But when time has pushed the initiating event into the past, the improbability of an immediate recurrence encourages acceptance of the expediency as "good enough." It is permitted to stand, usually without challenges, until a new intolerable experience occurs. Then another expeditious remedy is likely to follow. Thus safety has developed relatively haphazardly. Surely, reacting to emergencies generally does not allow the time to be taken that normally is needed for studying a problem. Under such circumstances a solution is likely to be inadequate. Often it is little more than a fixing of the obvious or assumed failure points. Indeed, the practice of safety seems to be essentially a patchwork of expediencies.

Prescriptive Correctives Dominate

As one safety step after another developed, the substantive as well as the expedient accumulated, until now there is a massive store of regulations, codes, rules, standard operating procedures, and safety program techniques. An observer can easily conclude that the answers to accident problems must be given either in the existing safety recipes or in some which eventually will be devised. Under the circumstances continued reliance on established corrective prescriptions is not surprising. Their acceptance is so strongly fixed and their application is so widespread that it seems almost impossible to de-emphasize reliance on them.

In the presence of almost endless lists of patent hazard correctives, most accidents it would appear should be preventable. That they are not has been evident frequently. The resulting frustrations probably account substantially for the popular emphasis on safety awareness programs, since errant behavior may be the most visible explanation for accident prevention failures. The rationale seems to be plain; given a resource for countering established accident causes, only ignorance of its applicability or benefits will limit its use. This uncomplicated notion suggests the need for safety observance promotions where it is seen that established accident prevention procedures are not employed diligently. The implication seems clear that when the promotional programs are implemented fully, accidents can be stopped. However, such simplism, which familiar safety efforts frequently exemplify, overlooks the complex nature of the hazard control problem and thereby disables the development of reliable safety performance measures. Safety achievement also is handicapped.

Indeed, safety awareness may not be a significant day-to-day deficiency at all—at least not in the United States where safety education often begins in the grade school, if not at mother's knee, and safety reminders abound all around. The benefits of safety and the substance of its common recipes surely are known. What is unknown—or perhaps more accurately—what is difficult to relate are the advantages *and the disadvantages* of the safety option in an inherently hazardous situation where some attraction exists and where the likelihood of immediate or significant harm does not seem too great. Rewards often require some exposure to risk. But how much? What are the tolerances for the harm to be expected from an unwanted reaction, if it occurs? Can an event be controlled to an acceptable level? These are some of the questions that knowingly or intuitively in some degree seem to occur in decisions involving the application of safety precepts. Therefore, it would appear that safety achievement is best not measured simply in terms of the occurrence of unqualified events, with an absolute null point as the performance objective. Reality seems to deny such a premise.

The Safety-Risk Paradox

There is a normal predilection, it seems, for taking risks if the possible consequences appear more attractive than the apparent probability and degree of harm that exposure to them may present. When the choice goes awry, plainly a bad judgment was made. But the ambivalent attraction between safety and risk apparently is a deeply ingrained human bent, which often disables wise decisions. And it does not seem to be alterable easily, even though current and historical disasters clearly teach the need for cautious risk selection and control. Caution, therefore, remains more an admonition than a precept. Yet this may not be altogether unfavorable, although in the view of conventional safety objectives it may be unpopular to think so.

Daring and impatience may have enabled man's progress as much as they have occasioned countless costly interruptions in his pursuits. It seems that there is, in general, an intuitive recognition of the need for venturesomeness that opposes the obvious desirability of safety. Society, like the individual, apparently responds to day-to-day perils, in accordance with the dictates of the most influential culture and needs at the moment, by accepting a level of safety that it finds tolerable. Thus a zero accident incidence, although ideally desirable, is not the universal standard for safety effectiveness. And administrators, on whom safety achievement depends, probably cannot be convinced that they should be held accountable for such ideal effectiveness.

The safety-risk paradox may seem incidental among significant puzzles, but for the manager and scientist who will bother to be interested in its implications a fascinating consideration appears—one that should be worthy of the best and most intensive effort that can be marshalled to weigh it, since it seems basic to the whole problem of achieving desired goals through organized effort.

The inclination (and sometime need) for chance-taking appears to be fundamental in daily living. Venturesomeness, indeed, may be as responsible for human progress as the quality of the intelligence that has employed it. Its importance to society seems to be understood implicitly as evidenced by the familiar toleration of ventures that end adversely. Usually it is only when undertakings involve perils that are thought to be harmful almost immediately and in an intolerable degree that the prospects or risks are considered unacceptable. Frequently it appears that the level of objectionable consequences is largely and exclusively a function of the expected gain from the venture. Risk versus gain trade-offs seem to be assessed instinctively (but sometimes analytically) according to such criteria as the threat to the mission, the probability of a harmful consequence, the degree of physical loss and suffering that could occur as well as the sensitivity of legal and ethical relationships that are inherent to the venture and that are balanced more or less against the rewards expected from taking the risk.

SAFETY PERFORMANCE MEASUREMENT CONSIDERATIONS

It seems that the determination of what is safe (or, from a risk-taking point of view, appropriate or accepted) is based on a fluid set of values. People appear to want protection principally from the kinds of harm they cannot tolerate, but the tolerances are not uniform. They seem to vary from individual to individual and within the individual according to the several discretionary factors one sees.

Safety Decision Variables

There are four judgment factors that, although not usually calculated rigorously, may intuitively govern safety decisions:

1. The likelihood (probability) of an unwanted consequence(s) occurring
2. The maximum degree of harm that could result from the consequence
3. The social sensitivity of the issues associated with the possible consequence (that is, their legality, morality, or ethicalness)
4. The magnitude of the gain expected from the action taken

Possibly each factor is related instinctively in some degree when a decision is required.

The immediacy, harmfulness, and visibility of a possible consequence ordinarily would govern the thoroughness with which a safety decision is reached; and motivational factors, which derive from the individual's heredity and culture, probably influence any interpretation of the results and the action taken. Thus, while it may be possible to regularize the process whereby a safety decision is made, it may be unlikely that any result would be weighed identically, very often. Therefore, disparate actions, with preventable accidental events following, probably are to be expected. It appears that a performance measure that uses accidental events in some scalar fashion to quantify safety effectiveness cannot reliably and validly reflect the quality of the safety process. However, when it is considered that perhaps the more common safety decision necessarily is not always to avoid hazardous situations but rather to enter them when it is believed an unwanted consequence (if it should occur) is limited to an acceptable level, then safety performance is seen to be at least as much a function of the success in limiting consequences as in preventing them. A meaningful safety performance measure necessarily, therefore, would seem to be concerned with appraising the effectiveness with which events are controlled to levels that cause no greater harm than the individual (or the system) is prepared to tolerate. The variances in objectives, which influence safety decisions and consequences, are seen frequently in conflicting intentions at the administrator-operator interface; consequently, they deplete conventional admonitory safety programs.

Safety Performance Measures Are Important to Achievement

When striving for accomplishment it appears that objectives, and the variations that occur in their interpretation, are not the only obstacle to achievement. Drucker (1954) points out that "the real difficulty lies indeed not in determining what objectives we need, but in deciding how to set them. There is only one fruitful way to make this decision: by determining what shall be measured in each area and what the yardstick of measurement should be. For the measurement used determines what one pays attention to" (p. 64).

Safety achievement relies heavily on management effectiveness. In our society we organize the efforts of individuals to achieve common goals. This requires leadership at each level of the hierarchy. A manager cannot prevail, however, unless his objectives are clear, attainable, and measurable. Then he is able to plan to meet them, assign appropriate responsibilities to his subordinates, and determine accountability for their performance. Otherwise direction becomes intuitive, as so often seems to be the case in safety.

When the performance measure is relevant and reliable—at least to the point where its margin of error is acknowledged—it can feed back information that is useful for correcting systemic deviations. This is the expected purpose of gauging devices generally. It is a value that is sought particularly in measurement applications for management systems. It should have critical importance, therefore, for optimizing safety since, as in the case of many functions which may not be reliably carried out unless there is *management* of the performance effort, occupational safety achievement relies to a large extent on such directional influences according to how they are expressed downward through the organization's echelons. (The work of *managing* is considered here to be: persuading others to work—usually with respect to standards that are established for individual jobs.) Indeed, the industrial hierarchy is an outcome of the need for distributing and channeling the responsibility, authority, and accountability for fulfilling the corporate objectives. And safety is no more likely to be accomplished than any of industry's responsibilities, if the hierarchical chain does not activate it as well. For this reason it appears especially important for the safety performance measure to be responsive to managerial measurement requirements. This suggests that the measurement device should assist, in its area of application, the fulfillment of the steps that generally comprise the work of managing. Thus, in the diagram of inputs to managerial work in Figure 9-1 it is seen that the manager PLANS the accomplishment of his *objectives,* ORGANIZES to achieve his plans, COORDINATES the workings of his organization, and then MEASURES its progress, providing thereby a loop that enables optimization according to the sensitivity, clarity, reliability, and validity of the measure's feedback. It is "the means of self-control" (Drucker, 1954), and its implications for safety achievement seem clear.

In its design or selection there is the artful influence on achievement that a performance measure exerts that must be considered or it may in fact misdirect the effort. Drucker (1954) reminds us that "the measurement used determines what one pays attention to." It appears, therefore, that the measure should deal clearly with what it is that is essential to the performance objective. If not, something other than the wanted results may obtain. Perhaps to a significant extent this likelihood has inhibited safety achievement.

Measures of safety performance customarily have been "accident rate" yardsticks, probably on the supposition that numerical differences in accident tabulations, over periods of time, mirror the relative intensity of safety's programs. However, the types of events that are measured, semantic differences in definitions as well as frequently dubious relationships between a zero (or even changing) incidence rate and the quality of the safety effort are inclined to disqualify accident rates as meaningful indicators of safety performance. Common experience indicates that when given goals that are unrealistic, people generally choose performance levels that suit their convenience, knowing that they cannot be held accountable for failing to reach an ideal. Moreover, accident indexes, by focusing on the zero rate as the implicit objective, tend to direct the safety effort toward risk avoidance (that is, "accident prevention") when the common inclination (or capability) may

Figure 9-1
Flow Diagram of Managerial Work Inputs and Measurement Feedback.

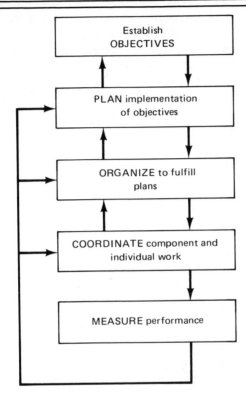

not generally permit the achievement of absolute safety perfection. The nominal interest instead seems to be concerned with taking on risks, when they hold some attraction and when the consequences will not be intolerably harmful. In such a circumstance the accident prevention ideal—as a constant objective—may be unrealistic and is likely to be qualified by performers according to what they consider to be their practical needs.

Safety in practice, therefore, often becomes a complex problem involving establishing the merits of a risk as a function of its inherent hazards as well as determining the acceptability of available safety measures to control it. Indeed, safety is not a simple matter of avoiding inherently threatening activities or devoting one's self to proper but restrictive precepts. Consequently, irrelevancies are apt to occur between accident rate expressions and real day-to-day hazard situations that may falsify the usefulness of accident data for feedback on the course of safety performance.

Course of Action Determined by What Is Measured

In order for a measure to be reliable, it is of paramount importance that its units be employable, with reasonably duplicative accuracy, by all who are trained to apply it. Then the information obtained can be interpreted reliably among analysts, and intelligent inferential dialogue becomes possible. The accident unit of popular safety measures, however, does not permit such transference of interpretations. Indeed, it may not even allow intelligent explanations of their observations by individual measurers. This limitation is due partly to the absence of a standard definition for "accident."

Dictionaries, for example, indicate the vagueness of the term by giving the differences in its usage. Even the most complete reference apparently cannot cope with the variety of shadings that apply. A rose may be "a rose is a rose," as Gertrude Stein says, but an *accident* is not always "an accident," speaking either colloquially or technically. Popularly it may be an "injury," although there does not appear to be any lexicographic support for this synonymity. On the other hand, "accident" rarely is used as a synonym for "fire" or "infection" even though either may be harmful and sudden and—like injuries—may be reduced by studiously applying established diagnostic techniques and safeguarding principles. Furthermore, an unexpected, unwanted event that happily produces no injurious consequence rarely is classified as "accidental," although the dictionary would seem to permit that label.

A most distracting variant is the practice of identifying as "accidents" the injurious events that were predictable and whose consequences could have been controlled satisfactorily. In the writer's experience the larger proportion by far of the occupational cases seen are so identified. Such events do not completely satisfy the condition of unexpectedness that the word "accident" connotes, but by implying uncontrollability and unpredictability the accident label in these cases falsely suggests the cause is at least in part attributable to circumstances beyond the influence of those responsible for its elimination or control. The insidiously destructive effects of a partial excuse for the occurrences ("It could not be helped. It was an accident") and the resultant structure imposed on the application of accountability in managed systems ("A man cannot be answerable for what he cannot help") erodes the base of the safety performance effort. Besides, the differences, though they seem small, can affect significantly the interpreting of measures that are based on such data.

Making the accident label more understandable without clarifying what is wanted or meant by safety performance, however, would not settle the issues of the safety performance measurement problem. As Drucker (1954) pointed out, the *measure* determines what is attended to. Presently the safety experience indicators focus attention on classes of events—principally fatalities and disabling injuries, but implicitly on all injuries as well. The thrust of the interest in safety, therefore, is in the direction of reducing injuries (or synonymously "accidents") to the zero point. The approaches followed, perhaps logically and appropriately under the circumstances, employ guarding and safe practice steps that hope to neutralize injury-producing exposures. However, it is the opinion here that the control of such events involves concepts and methods that are far more complicated and encompassing than those conventionally concerned with reducing injury totals and that devotion to injury reduction distracts from the central need for developing a body of effective hazard control knowledge. This necessity may be seen most clearly, perhaps, when the nature of "safety engineering" work is considered.

Safety engineering presently appears to be without definition. Almost any form of "accident" reduction program may be graced by the title. Semantically, however, it would seem that safety engineering should be concerned primarily, if not wholly, with the application of science to the properties of matter and sources of energy in the pursuit of man's *safety.* If this were its practice, it would approximate engineering work generally. It would be differentiable, nevertheless, by virtue of its special engagement in the identification and practical control of failure points in operating systems. It would be separable as well from *safety management, safety education,* and other functional safety titles that might seem to have special roles to play but are involved usually in similar accident reduction objectives, often without any remarkable difference between their approaches. However, there is a question as to whether an engineering specialty in safety has a role to play. It is of interest here since the measurement of safety engineering performance, which is the concern of this chapter, would be only an academic proposition if the work to be evaluated had little or no substance.

Engineers generally in their professional practice are well aware of the importance of safety and are attentive to it. Therefore, unless the safety engineering function truly has a special purpose, it can be a hollow occupation and may be filled easily by almost any type of safety approach. But when the complexities of the requirements for controlling unwanted events are considered, *safety engineering,* it seems, may have significant value. The implications of its practice suggest it may provide the key to solving the troublesome day-to-day problem of controlling *hazards* practically in operating systems.

(In this discussion *safety engineering* is considered to be essentially the scientific application of engineering and related knowledge to the problem of identifying systemic failure points and evaluating their risk with respect to the effect of possible countermeasures. The overall objective is to develop information that will enable the designing of systems so that their inherent hazards will be known, assessed, and eliminated or controlled practically to an acceptable level of risk. Here, *hazard* is differentiated arbitrarily from *risk.* A *hazard* in this context is considered to be any threat, peril, or possible cause of harm, while a *risk* is viewed as a hazard whose potential for producing harm (and rewards) has been evaluated and is known. Thus a *risk,* as conceptualized in this chapter, is a hazard whose degree of harmfulness, probability of causing harm, social sensitivity, and merits, if any, have been determined as reliably as may be possible or necessary.)

The need, and the capability for satisfying it, to which a properly constituted engineering specialist in safety could respond may be evident when it is considered that engineers generally treat trouble spots—which seem to have a small possibility for causing harm—as if they had no probability of occurring. The major determinant usually is the degree of consequence that is thought to be possible if a failure should occur. The greater the expected severity, the more the show of concern. This is a necessary and often proper posture if designs are to be developed practically. However, when the information upon which such judgments are based is deficient, an incorrect decision will follow. In this view there are no bad safety decisions—just bad information.

Inasmuch as an implicit objective, if not the specific duty, of responsible leadership is to safeguard against serious deterrents to its mission and particularly to limit the risks that may be inherent in its ventures, then definitive information on the presence of perils, their criticality, and their means of control may be said to be imperative. Such intelligence, however, is not usually readily at hand. Its development generally requires aggressive inquiry that is conducted systematically from an informed vantage point. It is believed here that this requirement is not met normally and that it demands a special occupation if the complex information that often is needed is to be provided. This specialty, it seems, occurs in the practice of safety engineering, as it is regarded in this chapter. And when the performance of *safety engineering* is considered, its effectiveness would appear to be a function of how well management is enabled to reduce to the lowest practical level the risks that occur in its operations. Although safety ordinarily is concerned principally with injury-accident reductions, this interest may not be de-emphasized, but in fact implemented more practically, by the characteristics of a performance measure as implied in the above consideration.

Engineers (and managers) are concerned usually with completing assignments according to certain goals: function, performance, schedules, and cost. Influences on each are attended to normally according to the magnitude of the effect that is experienced or sensed. The severity of a possible consequence, particularly when recognized before an event has occurred, therefore, is a powerful inciter to action. It is a sensitive subject to which alert administrators respond—and which may provide a base for measuring safety engineering performance—since it is meaningful, relates to a critical area of managerial interest, and, therefore, has some good expectation of being attended. It will be seen later in this chapter that an investigation has indicated there may be a substantial relationship between administrative effectiveness (represented by the ability to limit the unwanted increases in cost that are caused by unexpected contingencies) and the organization's ability to limit severe injury occurrences. The price of progress is risk. But when it is considered that there are implicit (and often explicit) delta limits for the severity of a consequence—if a risk is taken—as well as for the practicality of its controls, the engineer (and other decision makers) must have knowledge that will enable the best balanced decision within the necessary tolerances. The complexities of reaching a balanced decision and the need for information that will enable it are illustrated by Figure 9-2. A frequent complication in the implementation of safety is the fiscal relationship between prevention and the possible cost of unwanted events. While it may be argued that accident prevention reduces expense this may not always be so. Many times the cost of control may be considered to exceed the estimated damage. Certainly with an increased investment in control functions, damages can be expected to reduce. But at some point a practical fiscal optimum occurs. Usually this is not calculated. The relationship between damage and control costs is evident.

Often the information required for a judicious safety decision is not immediately at hand, particularly when the risk-cost-gain consideration involves a number of variables, as well it might in many operations. There is a distinct need, therefore, for a specialized and systematic function to assemble, analyze, and distill, data that can persuade correct decisions. Such

Figure 9-2
The Relationship Between Damage and Control Costs.

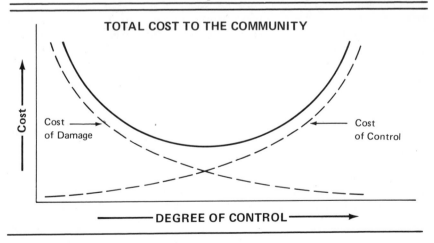

work appears to be uniquely that of safety engineering (even though it may seem to be intrinsic to the responsibilities of engineers generally) since experience indicates that engineering designers are inclined to view their projects broadly with respect to their missions, frequently missing thereby incidental failure points in the system. Unidentified hazards and others that though recognized were not pressed, on the assumption that they were insignificant, often have triggered unwanted consequences that have affected severely the public as well as employees. Safety engineering performance, therefore, is seen to be unmeasurable by standard[1] means inasmuch as it reflects simply the rate of disabling work injury occurrence.

"Accident" Injury Data Base Measures

The most common methods of quantifying safety performance are ratios of unwanted events (generally a specific classification of injury) and the exposure usually stated as a function of time, for example, man-hours in measures of work injury experience.

Frequency and *severity* ratios have been the accepted standard measures by which a company appraises its work safety effort. Generally, these terms express the incidence of major injuries and the severeness of the major injury experience, each with relation to the man-hours of work exposure during the period that is measured.

The Z-16.1 standard of the American National Standards Institute (1967, r. 1973) defines *work injury* as "an injury suffered by a person which arises out of and in the course of employment." *Injury* is construed to "include also occupational disease and work-connected disability."

Only *disabling* work injuries are used in determining the frequency and severity rates. However, since time loss is the value factor to which the standard measures relate, the injuries tabulated are described frequently by the sometimes misleading term "lost-time cases." The ANSI standard describes four types of disabling (or lost-time) injuries, which are paraphrased here:

1. *Death.* Tabulated irrespective of how long after the "work injury" occurred, if attributable to it.
2. *Permanent Total Disability.* Any injury that (by the ANSI definition) irreversibly incapacitates the individual for employment in a gainful occupation.
3. *Permanent Partial Disability.* An injury where irreversible loss of use of a body member occurs, as defined in the ANSI Z-16.1. (A combination of complete loss of use of two or more body members is considered permanently totally disabling.)

4. *Temporary Total Disability.* Injuries that do not result in death or permanent impairment but render the injured person unable for one or more days to perform effectively throughout a full shift, the essential functions of a regularly established job that is open and available to him. Decisions as to whether an injured worker is capable of performing a regularly scheduled job are made by the company's physician. The majority of the injuries that count in the standard measure are qualified by this fourth classification.

It may be observed that interpretive variations can occur readily in the application of the temporary total disability classification. Also it is seen that the four classifications are highly selective, since the injuries and their classifications necessarily are arbitrarily defined. For example "inguinal hernia" work injuries, until their repair, are considered by the standard to be partial disabilities; after repair they are expected to be reclassified and counted according to the rule for temporary total cases. Physician treatment cases and first-aid cases are not counted. Obviously, neither are noninjury-producing unwanted events, which are of interest to safety effectiveness determinations. Furthermore, cases that are considered indemnifiable by the workers' compensation laws of the states may not be included in the standard rates since the laws of the states varyingly define compensable occupational injuries; the states' definitions, therefore, often are not compatible with the Z-16.1's disabling injury qualification.

There are sound reasons generally for the limitations and the disparities that mark the use of the ANSI standard measure. It is difficult to explain, however, the confidence given to frequency rates as a meaningful indicator of the hazard control effort. The ANSI measure, clearly, by title and design, is concerned with "recording and measuring [disabling] work injury experience" (American National Standards Institute, 1967, r. 1973.) It may appear logical to conclude that the rate of work injury occurrence is a direct function of the safety effort, but anyone who has noticed that disabling injuries do not necessarily follow risk-taking ventures (even when hazardous acts almost certainly could be expected to produce harm) will realize that zero and low frequency rates may occur in spite of a poor safety approach. The converse also can be observed. Probably, there is a relationship between accident incidence and safety effort, but the appropriateness of appraising *safety* performance solely by the incidence of a single class of events may be as questionable as evaluating the strength of a bank by its rate of withdrawals. There are external factors, in either case, which should be known before a performance estimate is formulated.

The Z-16.1 describes three "measures of injury experience" (American National Standards Institute, 1967, r. 1973.):

1. *Disabling Injury Frequency Rate* (the most popular expression of industrial safety performance). Its formula:

$$F = \frac{\text{Number of Disabling Injuries} \times 10^6}{\text{Employee Hours of Exposure}}$$

Example: Assuming an establishment experiencing 12 employee disabling injuries, during a 1-month measurement period, and a total employee hours of exposure amounting to 2,189,243, then, substituting in the frequency formula,

$$F = \frac{12 \times 10^6}{2,189,243} = 5.48 \text{ disabling injuries/million}$$
employee hours of exposure

2. *Disabling Injury Severity Rate* (essentially a weighted frequency rate). This measure expresses the days actually lost due to temporary total disabilities and the days charged (arbitrarily by an ANSI schedule) for the fatal and permanently disabling cases. The formula, where *total days charged* equals temporary total days lost plus schedule charges for permanent disabilities, follows:

$$S = \frac{\text{Total Days Charged} \times 10^6}{\text{Employee Hours of Exposure}}$$

Example: Assuming for the above example that the establishment experienced 10 temporary total cases that collectively totaled the days lost due to disability and one death for which the schedule charge is 6000 days, plus one loss of an eye for which the schedule charge is 1800 days, the combined total days lost and charged would be 7872 days. Substituting in the equation,

$$S = \frac{7872 \times 10^6}{2,189,243} = 3596 \text{ days lost and charged per}$$
million employee hours of exposure

3. *Average Days Charged per Disabling Injury.* This measure is the ratio of severity to frequency rates. It may also be calculated as the ratio of the total days lost and charged to the total of disabling injuries.

$$\frac{S}{F} = \frac{\text{Total Days Charged}}{\text{No. of Disabling Injuries}}$$

Example: Using the data for the assumed establishment in the above examples and employing the given ratios,

$$\text{Average Days Charged per Disability Injury} = \frac{3596}{5.48} \text{ or } \frac{7872}{12} = 656$$

The Z-16.1 standard refers also to two "nonstandard measures," which it offers as supplements to the standard frequency and severity rates. One is the *serious injury frequency rate*. This rate includes all the disabling injury cases counted in the standard ratios and adds several categories of nondisabling injuries as well. A ratio is then computed. Like the standard frequency rate, the serious injury frequency rate is an expression of the total injuries *for its classification* as a function of 1 million hours of employee exposure.

The other nonstandard measure is the *disabling injury index,* which multiplies the standard frequency rate by the severity rate (then divides by 1000 to obtain a more manageable number) and is included in the standard "as an aid to those companies which want a measure of their combined frequency and severity experience." It is intended "primarily for ranking different establishments, organizations, or industries," but one can question the need inasmuch as comparisons between the index's computed values may be essentially the same as those for the severity rate multipliers. The examples of the frequency and severity rate calculations given above illustrate this. The frequency rate ordinarily is small numerically (in the order of 10) while the severity rate generally may be significantly larger (in the order of 1000). Comparisons between the products of the severity frequency multiplicands, therefore, will tend to have differentials that proportionally approximate those between the severity rates.

An inclination to employ the severity rate when appraising occupational safety performance is of some interest. As noted earlier, the frequency rate popularly is the indicator of relative safety effectiveness. The ANSI Z-16.1, for example, makes the point, with respect to its introduction of the "nonstandard" serious injury frequency rate, that "for internal purposes such a rate may measure safety program effectiveness more adequately than the disabling injury frequency rate" and contributes thereby its authoritative approval of frequency rates for the measurement of safety performance. Similar recognition is not given severity ratios, however, but the deference shown them by accommodating the "companies which want a measure of their combined frequency and severity experience" is noteworthy. It indicates some practitioners see (or sense) the value of appraising safety effectiveness as a function of the ability to control serious consequences, rather than simply with respect to the rise or fall of injury rates.

The safety literature has reported criticisms of the Z-16.1 frequency rate (Attaway, 1966; Grimaldi, 1962). The point of contention is the temporary total disability classification. This classification often impairs cross-operations frequency comparisons since identical injury cases are not necessarily classified by definition. A distinction occurs with respect to whether the injured person returns to work without loss of time.[2]

Because the Z-16.1 standard selects one class of work injury (the disabling) and because of the classification differences permitted by the temporary total disability definitions, some practitioners have turned to incidence rates using a wider spectrum of work injuries for frequency rate performance comparisons. De Reamer (1958), for example, has suggested total injury frequency rates as a means of indicating "average safety performance." The proposal is that frequency ratios be computed for all of the work injuries reported to the plant dispensary. It is alleged that this rate "provides a sensitive barometer of accident-prevention performance" and that, "such a measurement is needed if supervisors are to be held accountable for safety." However, total injury frequency rates, or the somewhat similar in purpose serious injury frequency rate (referred to as a "nonstandard" measure in Z-16.1) do not appear to enjoy wide popularity, possibly because practitioners are doubtful that injury data accurately reflect the quality of an operation's safety effort. The attitude of the medical treatment staff, for example, can be a significant but unquantifiable influence on the number of cases reporting to the dispensary.

Inasmuch as injury incidence ratios alone have not been satisfactory indicators of safety performance, several investigators have considered other evaluative methods. (The inadequacies of incidence ratios, when used as safety performance measures, are discussed further in Appendix A.) Grimaldi and Simonds (1975), for example, propose "a means of ascertaining accident cost figures with sufficient validity and reliability . . . to warrant an executive's using the data in formulating safety policy as well as in evaluating departmental efficiency with respect to safety performance." A feature of this method is the use of dollars, rather than injury totals, for the measurement computations. (Numbers preceded by dollar signs are considered to have special significance for administrators.) The measure computes the direct and indirect expenses of work injuries plus the costs of no-injury incidents. Thus, the characteristic problem of frequency ratios, which when zero often incorrectly imply a perfect performance, would seem to be circumvented. Adoptions of this method, however, appear to be few. With respect to the limitations of Simond's method in particular, the necessary cost data often are not easily obtained. In many instances considerable time must elapse before medical bills, loss appraisals, and claim settlements can be finalized. These data usually would not be available until the events they represented had faded many months into the past. In order to keep the measurement computations timely, the method, therefore, gives a means of estimating expenses, but the uncertainties of estimations seem to dampen a prospective user's confidence in the dimensions they provide. Even when estimates can be held to close delta levels, there probably are many practitioners who are inclined to discount the measure, considering it to be only another rough depiction of performance.

A relatively unused, but nevertheless apparently useful expression of performance, is the *"three-dimensional" appraisal* (Grimaldi, 1965) of a company's experience using frequency, severity, and average severity data. It circumvents the disadvantage of employing only an incidence rate, while continuing the use of the standard measures, which in the author's view is a practical consideration.

Figure 9-3 (p. 154) plots 11 years of a company's frequency, severity, and average severity record. It will be seen that the calculated trend (that is, least squares line[3]) for the frequency rate indicates notable improvement and for the severity rate there is improvement as well, making it seem that a safety advance has occurred when in fact the third trend indicates it is more likely that an "injury control" rather than "hazard control" effect took place. (The frequency rate, it will be recalled, is improvable by controlling the temporary total cases, a variable that often is a function of management policy, availability of jobs for temporary reassignments, or the character of the medical treatment rather than strictly safety.) *Qualitative judgments of safety performance, reached exclusively in terms of the frequency, therefore, are apt to be grossly incorrect.* It is also seen that the severity rate (which may in fact climb while the frequency is reducing inasmuch as the schedule charges for permanently disabling injuries necessarily are severe) when considered simultaneously with frequency may reveal the nature of the safety performance more definitively, thus seemingly de-emphasizing the belief that "frequency is a much more valuable indicator of safety performance than severity" (Bureau of Labor Statistics, 1955). Moreover, the *average severity* trend seems to provide a third and complementing gauge of the effectiveness of the *severity* control effort, which from a safety engineering point of view would appear to be important. For the company exemplified in Figure 9-3, severity control was a prime objective after the year 1957. The effect is indicated in Figure 9-4 (p. 154). The data reflected in Figures 9-3 and 9-4 are actual and represent approximately 1000 disabling injury (Z-16.1) cases annually. The substantially large samples involved here are not replicated ordinarily. However, similarly revealing plots may be obtained, for smaller numbers of events, by using logarithmic plots or a "rolling average" to dampen the possible affect of large differences between the rates, year to year.

To return to the objective of severity control for the company illustrated, Figure 9-4 shows the effect of the goal the following ways:

1. Frequency tends to improve (that is, reduce) year to year.
2. Severity tends to improve (that is, reduce) year to year.
3. Average severity tends to remain flat (where F and S have been small and essentially are 1 to 1) but reducing where S has been generally higher.

The plot reflects the disabling injury control experience with more definition than is obtained from frequency rates alone or in combination with the sev-

Figure 9-3
Disabling Injury Trends 1947–1957.

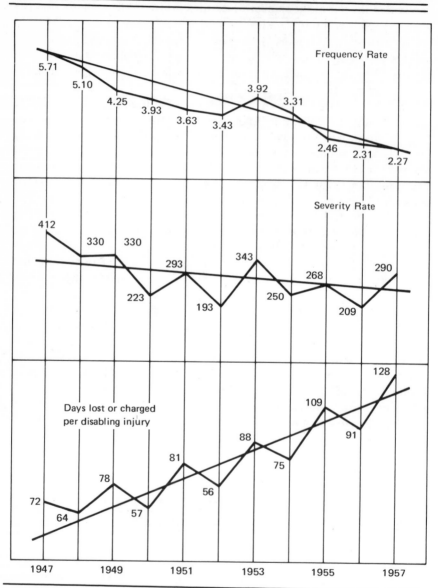

Figure 9-4
Disabling Injury Trends 1953–1963.

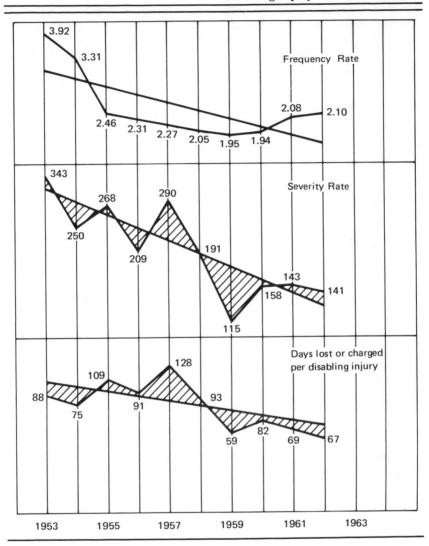

erity rate and seems particularly relative to appraising the wanted progress in controlling very severe injuries. (The Z-16.1 scale of time charges, applied to permanent disabilities and fatalities, weights the severity rate heavily. Usually in practice the severity rate, therefore, reflects predominantly the time charges for permanently disabling injuries—and fatalities—where they occur.) The appraisal method used for Figure 9-4 is confusing, however, for a number of reasons, particularly by its reliance on "accident" data to indicate safety performance.

Because safety effectiveness measurement scales whose limits are injury (or "accident") related do not seem to express competently the quality of the performance effort, measurement methods that do not employ accident data have been proposed. These generally attempt to evaluate the state of the safety climate on the assumption that where there is evident attentiveness to the safety precepts recognized as essential to prevention, progress in injury reductions will follow. Such a "pre-accident" safety performance measure is expected to persuade preventive action *before* unwanted events have occurred. One approach (Pollina, 1962; Rockwell, 1959) employs *work sampling*[4] techniques to secure data about the number and classes of unsafe acts and conditions observed in the shop. Clearly, this procedure, which provides meaningful results (inasmuch as the measurements may be made with a preassigned degree of reliability), is advantageous.

An analogous approach (Schreiber, 1957) appraised the performance of educational methods in accident prevention. It is of interest when considering the development of a safety engineering effectiveness measure since the familiar industrial safety programs are notably concerned with improving worker attitudes toward safety, through education. The investigator used a sampling procedure to observe types of unsafe behavior associated with industrial injuries at three shops of the Brooklyn Navy Yard and selected aspects of motor vehicle driver behavior at Mitchell Air Force Base (New York) and at Fort Dix (New Jersey). The observations were plotted in the manner for statistical quality control, and the study evaluated the merits of behavior sampling and statistical control techniques for measuring the influence of safety educational programs on behavior. It was concluded that the techniques were suited to the purpose.

Although not concerned specifically with the development of a safety performance measure, but responsive nevertheless to the need for establishing effectiveness in controlling accidental occurrences, Tarrants (1963) also employed work sampling methods. His interest, however, was an application of the *critical incident technique*. This technique originated during World War II as a result of studies to determine the likely causes of military aircraft accidents. In this method, as modified and applied by Tarrants, an observer interviews a stratified random sample of persons, eliciting their recall of unsafe acts and conditions occurring in their work. The stratifications are selected according to degree of hazard, type of exposure, degree of exposure, and other factors considered significant to the investigation. The purpose is to identify the critically unsafe acts and conditions in the operation and correct them before they occasion accidents. The systematic determination and quantification of critical causal factors, if accomplishable economically and reliably, could enable substantially the achievement of the accident prevention ideal and provide a valid measure of performance. Furthermore, the employment of observer interviewers, in the pursuit of the critical incident (as well as some other sampling approaches), has the advantage of reviewing periodically with employees certain accident-producing mechanisms that the refresher may assist in correcting.

There are certain disadvantages of observer sampling methods, however, which should be considered in their application:

1. Ordinarily work sampling methods are not economical for studying operators or machines located over wide areas.
2. Even if we assume uniform perceptiveness, ability, and diligence among observers, it is, nevertheless, unlikely that every operator will be a wholly reliable source of information for the observer and that uniformity of interestablishment data will obtain.
3. Sampling studies obviously present average results. There is no information as to the magnitude of the individual differences.
4. It is doubtful that an observer sampling method for performance measurement can secure the universal acceptance and widely uniform application necessary if a new standard is to be established for industrywide use.

The critical incident technique is discussed in detail in Chapter 17.

The Meaning of the "Meaning"

It will have been evident from the number and variety of considerations afforded safety performance measurement, which are noted in the foregoing pages, that the subject is unsettled but not ignored and that the approaches appear to be characterized by their concern with quantifying some aspect of "accidental" occurrences or the potential for "accidental" events.

While it may seem reasonable to measure effectiveness in terms of quanta of those elements that are the target of the effort, in the case of safety performance it seems as difficult to do this so simplistically as it is to measure productivity simply in terms of units produced. There are subtleties to be considered such as the following:

1. The meaning of the term "accident," which unless understood uniformly does not enable a constantly clear focus for an achievement effort.
2. The meaning (or objective) of safety for which one can realistically strive.
3. The implementation influences (such as management persuasiveness) that directly and indirectly must be harnessed to optimize the performance effort and to which the performance measure must be responsive since it provides the feedback that maximizes such influences.

The efforts up to now at developing an acceptable measure of safety performance have reflected their designers' views of what could or should be measured. Without general understanding and agreement on the safety needs to be fulfilled, however, the various proposals, though worthy, may be unrealistic and meaningful only to their creators. The posture of the proposers often seems reminiscent of Humpty Dumpty when he said about the use of words, "It's just a question of who is going to be master!"—the words or their meanings. Alice might have answered, "You can be the master of the meaning of your words if you like, but who will understand you? If you want to be understood, you have to use the meanings that are generally agreed to." Lewis Carroll doubtless could have written a squelching reply for Humpty Dumpty, but he might have criticized, nevertheless, tendencies to develop expressions that may be meaningful only to their originators. Numbers, for example, communicate information when what they represent has real meaning. One may do as one wishes with numbers and words, but to enable the communication that measurements should provide, one is permitted to do only what reality allows.

Evidently, safety is influenced by several interpretive factors. In fact, safety accomplishment seems to depend so strongly on personal values, which interplay and tend to affect one another according to given circumstances, that it may be impossible to establish a generally acceptable level of safety perfection. This seems to be the issue that entangles familiar safety performance measures and to which the Z-16.1 standard perhaps intuitively reacts, but in doing so it fails to measure reliably the extent to which a safety performance approaches perfection.

Z-16.1 opts a particular class of unwanted events, the disabling injury, for surveillance. It may be assumed that its originators, in choosing the disabling injury, had recognized that this was the type of event that held the greatest potential for receiving general interest in its prevention. Possibly the rationale was that work injuries of less consequence were commonplace and thus were likely to be regarded more tolerably. Attempting to develop concern about them, therefore, might have been considered unrealistic.

Some observers, notably Heinrich (1959), judged as significant small events that by the grace of good fortune were not severe but could have been. Among the safety fraternity there then occurred a great concern for tabulating and evaluating minor injury and no-injury cases with the intention of uncovering correctable causal factors and thereby shutting off the occurrence of more serious cases. Heinrich's objective often proved more hopeful than realistic. For one reason, there could be no severely harmful event if the system that produced a minor case did not have inherently a potential for a severe consequence. Unless the expanded effort involving lesser incidents was discriminating, with respect to the degree of hazard existing in the system, it was not likely to have the significant influence expected. As a consequence, a general regard for such approaches and any performance measure related to them probably would not be high, as indeed seems to be the case.

It is the influence of judgment factors that a safety performance measure apparently must be able to deal with if it is to win universal acceptance. Obviously, several considerations affect the quality of safety performance; therefore, it cannot be evaluated simply on the basis of unwanted occurrences. Such underlying considerations with respect to accepting hazardous exposures, as the degree of tolerableness that the hazards present and the consequences that may derive from them, are not regarded in the measures suggested so far. Yet decisions are made under the influence of such variables. Consequently, what may be thought to be relatively safe by one individual may not be similarly considered or attainable by another. Safety performance in the absence of common understanding of what the level of achievement should be realistically, is practically without meaning.

APPARENT CRITERIA

From the foregoing discussion it seems that a number of qualifications need to be satisfied if a safety engineering performance measure is to develop. Some of the most important considerations appear to be the following:

1. The measurement data probably should not be drawn exclusively from injury-accident situations inasmuch as safety engineering performance may not relate significantly to the occurrence of injuries—particularly over short time spans.
2. The measure probably should relate to the operation's effectiveness in fulfilling its mission generally and particularly with respect to its ability to limit severe consequences from any threatening deficiency in its systems.
3. The quantified values obtained from the measuring method should be replicable by anyone instructed in the use of its technique.
4. If possible, the measurement should enable valid comparisons of hazard control effectiveness between the operating components of an establishment as well as between classes of establishments.
5. The variables quantified should be realistic and, if possible, familiar and uncomplicated so that a minimum of instruction will be required to introduce the measure and obtain its acceptance.
6. The measure should be able to obtain national consensus approval so that it can meet the need for a *standard* measure of safety performance effectiveness.

Obviously, the construction of such a measure cannot be done routinely. The issues appear too complex for a simplistic resolution. Moreover, the suggested desirability of measuring *hazard control* effectiveness, rather than injury reductions, to determine the quality of safety performance probably requires the development of unusual measurement designs.

It may have been noticed, in the review given earlier of the standard and heretofore explored evaluative methods, that recent investigators indeed have been attacking the safety problem unconventionally, but no approach seems to have enjoyed wide acceptance as yet. This may be attributable in part to the obscureness of any relationship between effectiveness in controlling threats to an operation's mission and the dimensions of present and investigated methods of measuring safety performance.

Yet it is competence in controlling the failure possibilities, which imperil the operation and its contributor claimants, that is basic to safety effectiveness. The quality of a system's control capability would seem to be a major concern when undertaking the advancement of safety achievement, inasmuch as when the control climate is unfavorable the strongest safety techniques, devices, and personnel are not fully attended and they become relatively powerless. The meaningful measurement of safety performance, therefore, may be essentially an appraisal of the control (that is, managerial effectiveness) current for the operation generally, even though the measure's specific function is to assist the prevention of unwanted occurrences.

Feasibility of Intermodular Measurement

The writer for a number of years examined a variety of corporate performance data (drawn from annual reports to shareholders) to determine if there might be a relationship with common safety performance measures.

For the initial company studied a relatively regular distribution of year-to-year changes in sales, earnings, and the severity rate was noted. Percent differences in annual sales tended to be inversely related to percent differences in annual earnings and severity rates. No other accepted safety performance measure appeared to develop a meaningful pattern relative to the corporate business data. Figure 9-5 depicts the relationships for the first company studied.

Later, 38 companies representing a variety of industrial classifications were similarly reviewed. The corporate performance data were expressed in this review as "total costs and expenses applicable to sales as a percent of sales." These data were matched, year to year, with the corporate severity rate, and it was observed that there was a somewhat better-than-chance distribution of matching changes between the severity rates and the computed cost figures. Again, no other accepted safety performance measure appeared to have a meaningful association with the cost data. The strongest relationships were noted in the case of those injuries where severely hazardous expo-

Figure 9-5
Company-wide Percent Difference of Improvement for Sales Billed and Severity Rate, 1953–1958.* In 1956 the revised Z-16.1 (1954) standard was first applied for a full twelve months. The standard redefined reportable back injuries and hernia cases so that in some instances these relatively frequently reported injuries were counted no longer. The revision therefore may have excluded cases reported in earlier years, with a consequent improvement in 1956 rate, compared to 1955.

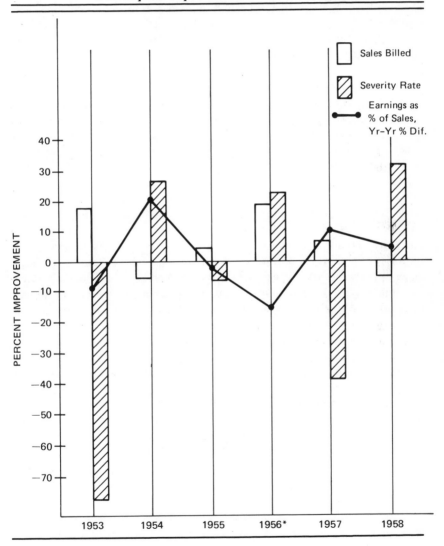

sures generally are inherent, such as petroleum and chemicals. These observations were reported in 1962 (Grimaldi, 1962).

It may be assumed that increases in costs with respect to sales ordinarily are unwanted and unexpected; therefore, they are a kind of "accident." Rising costs, of course, can be a severe threat to the corporate mission. And a management failure to control costs relative to income may be but another instance of deteriorating effectiveness in identifying and limiting the hazards that threaten the enterprise and its contributor claimants. The converse also may apply. Cost data, therefore, may be highly sensitive to changes in an establishment's control effectiveness and, when correctly analyzed, could signal the need for more effective action. To some extent this is the purpose of cost accounting departments, but their intelligence is applied parochially usually with respect to the experience for individual cost categories. Certainly, it can be expected that alert managers will be concerned with limiting threats to their systems' budgets. Failure to do so, when income is declining, will be viewed seriously. Any deficiency in maintaining such control may reflect on the control effectiveness in the establishment generally, and so a concomitant rise in the disabling injury severity rate may be explained.

If significant cost variables were identified and their relationships quantified, it seemed reasonable that an index reflecting the levels of change in effectiveness in an establishment could be computed, and it would enable the measurement of the establishment's effectiveness in controlling unwanted change. This is an important consideration, it was believed, since—for the company initially investigated—the changes in severity rates occurred while its safety organization performed relatively uniformly. Whatever the influences on the safety severity experience, they seemed to be external to the safety function. Safety achievement appeared dependent, consequently, on the establishment's general performance in controlling the many common factors affecting its welfare.

A study (Grimaldi, 1959) was designed, therefore, to investigate the possibility of a relationship between work injuries and operating variables whose control ordinarily would occupy managers as much or perhaps even more than safety. It was hypothesized that injuries are only one type of contingency that concern administrators. There may be others (such as increased scrap, shop costs, and customer complaints) that, when they occur, are unwanted, unexpected, and, therefore, accidental—and that may arise from deficiencies that affect as well the controls in safety's domain. If a statistical intercorrelation were found to exist, a measure of effectiveness in controlling unwanted events, with implications for safety performance, would seem possible.

Appendix B of this book reports the study's design, data, and the conclusions drawn from them. In order to enhance objectivity and to secure expert factorial analysis the investigation was given under contract to Max

Woodbury (then Professor of Research Mathematics at the School of Engineering and Science, New York University). The data were processed with the aid of the university's computer. The following observations were drawn from the results:

1. *Comparative differences in the safety performance profiles of the 17 diverse businesses studied appear to be explained by the variations in their* severity rates *and to some extent by the* severity frequency ratio. That is, *average severity* rather than the generally accepted measure, the *frequency rate.*

2. *The results broadly indicated that in an operating component where there is fairly good control of costs a generally good safety profile will prevail.* Figure 9-6 (p. 164) depicts the relative significance of each of the management cost variables in relation to the principal safety factor, that is, the severity rate. The graph indicates that the better businesses, from a severity control point of view, kept rework costs, direct labor costs, building and equipment maintenance costs, and complaint and shop costs on a relatively low level while scrap costs may be tolerably high.

3. *The inverse relationship of scrap costs to other costs (Figure 9-6) suggests that an effort to achieve wanted results in an emphasized performance area, such as scrap cost reduction usually is, may be at the expense of other less prominent goals.* For example, scrap costs may be reduced by reworking the unacceptable products with a consequent increase in rework costs. Or quality standards may be lowered, which then could affect complaint costs adversely. The implication seems clear that the measurement of interrelated performance areas may not be satisfactorily accomplished individually, but some method that weighs the factors and combines their effects into a single value must be used.

4. *The tendency for building and equipment maintenance costs to be low, where a good safety experience occurs, appears to be contrary to the theory of the case.* Usually it is expected that a generous posture toward maintenance exists in locations where good safety records will develop. Further consideration suggests, however, that the well-managed operation (also the safer one) receives regular preventive maintenance (under a controlled budget) and thus keeps unusual expenditures for repairs and replacements to a minimum.

5. *The findings indicated that sales tended to fall off in those businesses where a better safety (severity) record occurs.* This would be expected if the injury and sales data had been expressed only as totals. However, the data were examined on an equivalent man-hour basis. Besides, the safety factor was dominated by the severity variable (severity rate), which is influenced heavily by the penalty charges (and days lost) for each injury. The comparison, therefore, is considered significant. As a

Figure 9-6
Degree of Relationship Between Management Variables and the "General Safety Factor." (Note: In the general safety factor the dominant value is severity rate.)

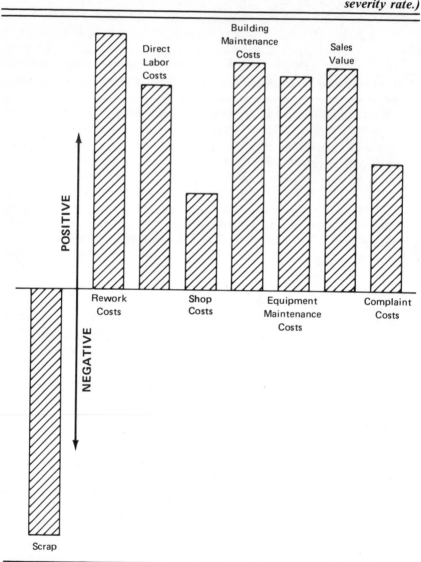

point of interest, it tends to follow the pattern noted in Figure 9-5. It may be concluded that when sales decline, managers are prompted to exert the strongest efforts to control their operations by obtaining optimum performances from subordinates. The implication seems clear that when managers control their operations most tautly, safety success, as indicated by the reduced occurrence of severe injuries, is likely.

6. *Cost accounting data, statistically analyzed, apparently reflect the operation's control effectiveness, and indications of better (or less) control seem to predict the occurrence of improved (or poorer) severity rates, with a lead time of approximately 9 months.* This may be seen in Tables 9-1 and 9-2. It may be noticed that the trend of the 1958 *safety and fire loss* experience for 15 businesses was foretold with greater accuracy than would be expected through chance. The severity experience seems to have been predicted accurately in 10 instances; fire losses in 11. Fire losses were not available at the outset of the study and so were not included with the safety variables studied. However, if the assumption is valid that the efficient control of costs is the result of effective managing (that is, maximizing the performance of people), then the statistical analysis of an operation's cost data may predict progress in controlling contingencies other than those studied, for example, fire losses in this instance. The fire loss experience noted tends to give further credence to the hypothesis.

Whether data outside of safety's realm can be employed to construct a reliable and valid performance measure, one that will advance safety effectiveness, is a question that must wait upon further studies. The possibility

Table 9-1
Operations Control Effectiveness as a Prediction
of Severity Rates and Fire Losses, Expected Improvement

Business Expected to Improve in 1958	Severity Rates		Fire Losses (cents/$100 insurance)	
	1957	1958	1957	1958
TV receiver	235	135	0.005	0
Large steam turbine	290	129	0.026	0
Locomotive and car equipment	1226	107	0.030	0.002
Home laundry	1328	8	0.004	0.027 X
Large motor and generator	458	1406 X	0.002	0.002 X
Distribution transformer	777	98	0.198	0.007
Receiving tube	40	93 X	0	0
Heavy military electronic equipment	508	46	0.067	0.003
Instrument	184	60	0.012	0

Table 9-2
Operations Control Effectiveness as a Prediction
of Severity Rates and Fire Losses, Expected Retrogression

Business Expected to Retrogress in 1958	Severity Rates		Fire Losses (cents/$100 insurance)	
	1957	1958	1957	1958
Automatic blanket and fan	363	598	0	0 X
General purpose water	248	83 X	0.007	0.020
Semiconductor	84	4 X	0	0.095
Metallurgical	116	409	0.013	0.002 X
Hermetic motor	44	68	0.008	0.011
Conduit products	104	0 X	0.002	0.124

does seem feasible, and the suggestion appears practical inasmuch as the events that safety traditionally measures apparently do not provide the necessary measurement information.

Cross-functional measuring, such as was explored by this feasibility study, may be theoretical, but it is not unfamiliar. Cross-modality matching, whereby the elements of one domain are matched to those of another domain, has been developed in sensory measurement and the practicality—even the necessity—of intermodular measurement techniques appears to have been suggested by Stevens (1968):

Back in the days when measurement meant mainly counting, and statistics meant mainly the inventory of the state, the simple descriptive procedures of enumeration and averaging occasioned minimum conflict between measurement and statistics. But as measurement pushed on into novel behavioral domains, and statistics turned to the formalizing of stochastic models, the one-time intimate relation between the two activities dissolved into occasional misunderstanding. Measurement and statistics must live in peace, however, for both must participate in the schemapiric enterprise by which the schematic model is made to map the empirical observation.

Measurement provides the numbers that enter the statistical table. But the numbers that issue from measurements have strings attached, for they carry the imprint of the operations by which they were obtained. Some transformations on the numbers will leave intact the information gained by the measurement; other transformations will destroy the desired isomorphism between the measurement scale and the property assessed.

Since the transformations allowed by a given scale type will alter the numbers that enter into a statistical procedure, the procedure ought properly to be one that can withstand that particular kind of alteration.

The view is proposed that measurement can be most liberally construed as the process of matching elements of one domain to those of another domain.

In most kinds of measurement we match numbers to objects or events, but other matchings have been found to serve a useful purpose. The cross-modality matching of one sensory continuum to another . . . supports a psychophysical law expressible as a simple invariance: equal stimulus ratios produce equal sensation ratios. (Stevens, 1968, p. 856)

A measurement method was devised according to the findings of the feasibility study, but it remains to be tested. However, by employing a modification of the study's cost accounting categories, interfunctional measurement of the control climate was continued as an experiment at the General Electric Company. Data obtained from the company's annual report to shareholders was used to compute the *ratios of costs to sales* for the period 1956–1966. (The method of calculating used resulted in a cost-sales ratio larger than unity. The company of course was making a profit during this period.) The ratios were plotted (Figure 9-7, p. 168) with the *severity rate* and *fire losses* for matching years. Although the sample is small, the relative regularity with which the severity and fire loss curves track the cost-sales ratio seems impressive and tends to confirm the hypothesis that safety achievement is a function of the general control climate. The severity and fire loss varied, it would seem, under the influence of factors outside of the safety functions.

An observation that may have significance is the apparent relevance of the severity rate to the models of general corporate control effectiveness. The knowing administrator may be expected to be vigilant for conditions in his operations that, in the event of a control failure, could threaten the mission. It may be for this reason that in business organizations operations such as marketing, production, and finance are attended closely by management. Their high correlation with mission accomplishment is easily recognized. For safety, however, the relationship to managerial achievement may be seen only when it is clear that hazards exist in the system and that they are so severe the mission may be jeopardized. Lesser injurious events in general rarely stimulate a strong administrative response, unless their number reaches an intolerable level. This is infrequent in our relatively well-informed society. Thus, the severe consequence is both the stimulus and the target for administrative control efforts with the result that where the managerial control is strong the severe event occurs less often and vice versa.

The emphasis on the severe occurrence with respect to safety performance may seem improper, in the light of currently popular views, even though some thoughtful study has occasioned it. Generally, the safety objective seems to be to prevent all accidents, that is, reduce the frequency rate to zero. But familiarity with a number of industries indicates that as frequency approaches zero, severity does not necessarily follow proportionately. The very severe occurrences may continue at an unabated rate or may, in fact, increase.

Figure 9-7
Cost-Sales Ratio versus Severity Rate and Fire Losses.

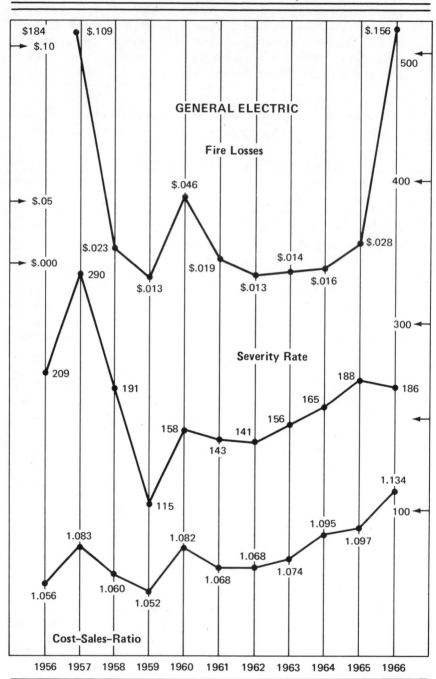

It would seem that the identification of hazards and their evaluation with respect to their relation to the mission, the consequences of an event, and the applicable methods of control must precede the preparation of means to either design the hazards out of the system or, if this is not possible, to engineer limitations of their effects. It appears also that such a practice is most needed where individual hazards have a potential for a severe consequence or for causing multiple events of a relatively less severe order. This, it is believed, is the principal function of safety engineering and, therefore, is a proper focus for the measurement of its performance.

In the writer's experience, with a variety of enterprises, only relatively few injury cases usually are responsible for the major effect of the "accident" experience—whether this be calculated in medical and indemnification costs, down-time, or degree of harm to the corporate image. Usually, the causes are familiar hazards or agents that are not too difficult to identify, with the aid of appropriate analyses. Normally, an approximate 10 percent of the disabling injury cases seem to have occasioned roughly 90 percent of the losses observed due to work accidents.

It would appear, therefore, that for practical as well as humanitarian reasons the safety engineering function should cut off the scope of its effort at some determinable level of severity and concentrate on the more severe exposures. This seems to be where the need is, and it is a level of activity that management understands and is apt to be highly responsive to, with implications that seem to be clearly advantageous. After the severe hazards are controlled with good evidence of certainty, a considerable widening of the safety engineering effort to attempt to control lesser (and often the more elusive) harmful agents would seem to be appropriate.

Whether these notions on the practice of safety engineering will have general approval is uncertain, but there seems to be some reason to believe that the measurement of safety engineering performance may be a function of the success with which severe consequences are limited. It appears also that this may be measured by employing cross-modality techniques, which may have the additional advantages of detecting a faltering in the control climate with sufficient lead time to correct it and enable limitation of the otherwise prospective severe event.

SUMMARY

Generally, it seems that the value and greater challenge of safety engineering occurs in its potential for eliminating the causes of severe events or limiting their possible consequences. Its practice in such a light is seen to be concerned with making reliable analyses of operations to identify failure points, to evaluate their significance in the system as well as their likely effects (and within the limits of practicality), and to employ creativity in the application of a number of disciplines, including engineering, to devise control methods. However, the effectiveness with which the practice is engaged appears to be a function of the general administrative control climate in which the work is performed. This is a troubling consideration when the measure's expected use as a performance stimulator is weighed. One must question whether specialist performance can be measured except as a function of the climate in which it is offered. Also it appears that injury ratios, the traditional units of safety effectiveness measures, may not describe performance reliably. Other data seem necessary. These measurement complexities may be resolved by constructing a measurement model of administrative control effectiveness that couples concretely with safety engineering performance. Such intermodular measurement has been explored, seems feasible, and appears to merit further investigation.

In the final analysis, however, the measurement method that will be used is the one that is acknowledged to be "standard." It would seem advisable, therefore, to make the most of these standards until a new measurement technique is designed and becomes generally accepted. The severity rate is the most likely indicator that most closely relates to the safety engineering objective of identifying and controlling the severe hazards in the systems served.

Notes

1.   American National Standards Institute (1967, r. 1973) *Method of Recording and Measuring Work Injury Experience, Z-16.1* has been the acknowledged standard method of evaluating industrial safety performance in the United States. However, modifications and in some cases significantly different approaches have been proposed by practitioners who have not been satisfied with Z-16.1. The revisions may be found in use from time to time according to the practitioners' interests. The OSHA-BLS incident rate is gradually replacing ANSI Z-16.1 as the standard for measuring occupational injuries and illnesses.

2.   The Z-16.1 standard qualifies temporary total disability according to whether the injured employee (if not totally or permanently disabled) is able to perform effectively throughout a full shift of a regularly established job that is open and available to him, without loss of one or more days of work due to the injury. The decision as to whether he is fit enough to do the assigned work is up to the employer's physician. This provision often encourages continuity of employment and is beneficial to the workman since it enables him to maintain his earnings. It may also have certain therapeutic values. But the employer with numerous and varied jobs is more likely to have open regularly established occupations than a small business operator. Also the physician engaged by the employer probably specializes in occupational medicine and may be more informed on desirability and opportunities for maintaining injured employees' work continuity. Consequently, the number of temporary total cases (which are the substantial component of occupational injury totals) will vary between employers and often even among their locations. Such differences, as they affect the frequency ratio, distort the rate's reflection of the safety effort. They seem more descriptive of an operation's ability to control the workday losses associated with injuries.

3.   One of the attributes of the mean of a group of measurements is that it is the value about which the sum of squares of deviations of the individual measurements is a minimum. Similarly,

with data that can be correlated by a straight line, there is one straight line from which the sum of squares of deviations of one of the variables is a minimum. This is the *least squares line* and is expressed with data as

$$y = a + bx$$

where:

y = estimated value of y for observed value of x

a = intercept, giving estimated value of y at $x = 0$

b = slope of line (identical with regression coefficient)

4. Work sampling was used first in the British textile industry and was introduced in the United States as "ratio delay" in 1940. The technique is based on the laws of probability: that is, a sample taken at random from a large group tends to have the same pattern of distribution as the large group. In its simplest form the procedure consists of making random observations at random intervals of operators or machines and noting their activity. At the outset the statistical probability ("confidence level") is decided so as to establish the likelihood that the random observations are factually representative of the normal population.

References

American National Standards Institute. *Method of Recording and Measuring Work Injury Experience: Z-16.1.* New York: The Institute, 1967, r. 1973.

Attaway, C. D. Computing a true accident frequency rate. *Journal of the American Society of Safety Engineers, 11*(10):9–10, 12–13, 1966.

Bureau of Labor Statistics, U.S. Department of Labor. *Safety Subjects.* Bulletin No. 57. Washington, D.C.: U.S. Government Printing Office, 1955.

De Reamer, R. *Modern Safety Practices.* New York: Wiley, 1958.

Drucker, P. F. *The Practice of Management.* New York: Harper & Row, 1954.

Grimaldi, J. V. A study of the feasibility of using cost accounting data to appraise safety performance. Unpublished study circulated to General Electric Co. management in an *Information safety* memorandum, December 11, 1959.

Grimaldi, J. V. Another look at stimulating safety effectiveness. *Journal of the American Society of Safety Engineers, 3*(4):20–23, 1962.

Grimaldi, J. V. Measuring safety effectiveness. In H. W. Fawcett and W. S. Wood (Eds.), *Safety and Accident Prevention in Chemical Operations.* New York: Interscience, 1965, pp. 546–549.

Grimaldi, J. V., and R. H. Simonds. *Safety Management,* 3rd ed. Homewood, Ill.: Irwin, 1975.

Heinrich, H. W. *Industrial Accident Prevention,* 4th ed. New York: McGraw-Hill, 1959.

Pollina, V. Safety sampling. *Journal of the American Society of Safety Engineers, 7*(8):19–22, 1962.

Rockwell, T. H. Safety performance measurement. *Journal of Industrial Engineering, 10*:12–16, 1959.

Schreiber, R. J. *The development of procedures for the evaluation of educational methods used in accident prevention.* Unpublished doctoral dissertation, Columbia University, 1957.

Stevens, S. S. Measurement, statistics and the schemapiric view. *Science, 161*:855–856, 1968.

Tarrants, W. E. *An evaluation of the critical incident technique as a method of identifying industrial accident causal factors.* Unpublished doctoral dissertation, New York University, 1963.

Tippett, L. H. C. Statistical methods in textile research. Uses of the binomial and Poisson distributions. A snap-reading method of making time studies of machines and operatives in factory surveys. *Shirley Institute Memoirs, 13*:35–93, November 1934.

Chapter 10

TOWARD MORE EFFECTIVE SAFETY MEASUREMENT SYSTEMS

Herbert H. Jacobs

Problems of observation, measurement, evaluation, and inference are probably more acute in that strange area of science, engineering, and management that deals with accidental human trauma than in almost any other field of investigation. The basic reasons for this are inherent in the common properties of those otherwise diverse events that are classified as *accidents*.

In the first place, accidents are inadvertencies or errors—uncontrolled and unplanned happenings. This is simply another way of describing chance-caused events. In addition, it is in the nature of things that man will control his environment and his exposure to hazard in such a way that the perceived risk of accidental trauma will always be extremely small. Taken together, these properties of chance and improbability characterize a class of incidents known to statisticians as "rare events." Since accidental happenings cannot be controlled or predicted, they cannot generally be produced or reproduced under laboratory conditions. Furthermore, because of extremely infrequent and unpredictable times of occurrence, they are not generally susceptible to direct observation.

For these reasons, most techniques for investigating the causal or preventative aspects of accidental trauma are based on postmortem inquiry and retrospective analysis. In general, conventional procedures for evaluating the degree of hazard or the level of safety performance associated with a given behavioral situation or physical environment are based on measuring and comparing the number or frequency of accidental occurrences. As most workers in this field have come to learn, such comparisons are difficult to interpret—first, because of the statistical "flukiness" that governs the occurrence of rare events; next, because each occurrence is different in its nature and its consequences; and finally, because the expected frequency of occurrence is the product of the risk or probability of occurrence and the magnitude or amount of human exposure. The last of these elements, exposure, represents a variable that generally receives little attention even though it is accounted for on a purely arbitrary basis. What all of this means is that the field of accident control—better thought of, perhaps, as safety management—has found itself constantly beleaguered by the difficulty of measuring, comparing, interpreting, evaluating, and, hence, controlling the physical and human situation of interest.

Before we consider further the special aspects of measurement associated with accidents, it might be well to give some thought to the subject of measurement alone. It is probably not an unwise generalization to suggest that the formal process of measurement, observation, sampling, and inference can be assessed only within a larger framework of purpose. What is it that is to be known? What clues are to be sought? What hypotheses are to be tested? How is knowledge to be applied? What level of precision is required? What risks of erroneous inference are acceptable? What evidence is needed for persuasion? What are the costs of not taking action when action is required or, conversely, of taking action when it is unwarranted? These and related questions are unanswerable as generalities; they can be applied only to specific problems of scientific or operational inquiry. This leads us to the central issue, which is to define an appropriate "framework of purpose" for considering measurement problems as these relate to accident prevention or safety performance measurement.

MEASUREMENT PURPOSE

Since a major goal in this area is the preservation of human life and limb, it is obvious that sooner or later we will be interested in assessing absolute or relative progress or change over time, generally, or changes during specific time periods that, in turn, may relate to other changes, events, or activities known to be associated with these time periods. Alternatively, we may wish to compare operational experience between two environments, areas, or activities as a means of assessing relative performance or progress.

A number of examples will help to illustrate these ideas. First, there is some value in making gross comparisons of the overall degree of hazard associated with various environments or activities. These comparisons can serve as a basis for establishing priorities for the application of remedial effort and can help to set target levels for improvement. For example, it is desirable to measure the relative safety experience of various modes of travel as a first step in allocating investigative and engineering resources among them. It is desirable to measure the relative safety experience of coal mines in various countries for purposes of establishing potential target levels for improvement. There is also value in gross comparisons from the standpoint of reflecting on the relative effectiveness of total systems of engineering or management.

Next, there is much to be gained from making periodic measurements over time (for example, monthly, yearly) for the purpose of assessing changes for the better or worse in the gross safety experience of a given facility, function, operation, or environment. A time stream of such gross measurement data can provide an effective basis for regulating the application of remedial effort and control measures. For example, an increasing accident rate may suggest the need for budgeting more funds for safety management. Depending on the way in which measurement data have been aggregated, it may even be possible to confirm or refute the effectiveness of particular programs and, possibly, to evaluate quantitatively the relative cost and effectiveness of such programs. Along these lines, it would be desirable to follow changes in vehicular accident experience of teenagers in conjunction with the introduction of a high school driver training program as a basis for comparing the magnitude of improvement against the costs of the training system.

A summary, then, of measurement purposes or objectives can be presented as follows:

1. Evaluating the relative "level of safety" of various activities or operations as a basis for
   - Measuring the relative effectiveness of safety management practice
   - Determining the level of remedial effort to be applied
   - Allocating effort among competing (for attention) operations
2. Evaluating over time the degree of progress or retrogress in specific hazardous situations as a basis for
   - Appraising progress in safety management effectiveness
   - Determining changes in the type or level of remedial effort being applied
   - Assessing the cost/effectiveness relationship of alternative countermeasure systems.

This summary of measurement objectives leads one to ask several specific questions: What are some of the requirements of a measurement system that might be established to serve these general purposes? What must be measured and how well must it be measured? What are the corollaries of this statement of purpose?

The Requirements of a Measurement System

First, this discussion of purpose suggests that the ultimate object of measurement is not "accidents." Rather, it is some intrinsic property that might be thought of as "safety expectation," which, depending upon the generally strong role of chance, governs the probability of occurrence of any given number and severity of human accidents during some specific future time interval or some particular amount of human exposure. This property is, of course, not directly measurable but must be inferred from the measurement of other attributes or events that are observable—including, but not necessarily limited to, current accident rates.

Second, it would appear that it is important to measure those attributes of an operating system that relate to the specific occurrences that we are attempting to control. A useful and efficient measurement scheme need not cover all occurrences, but it should at least cover those occurrences that are the subject of particular control interest. Also, we need not confine ourselves to the observation or measurement of the events of ultimate control interest; it may be preferable to focus measurement attention on precursor events or other attributes of the system that bear some appropriate relationship. The nature of this relationship should be such that effective control of the measured properties would result in comparable degree of control of the ultimate events of interest. The important thought here is that an effective measurement system aimed at the control of accidental injuries need not be concerned with the direct measurement of injury rates but should at least be responsive to controllable forms of unsafe activity. Also, since most control activities or interests tend to be focused on specific types and kinds of accidents, an efficient measurement system ought to have a similar sharpness of focus.

Third, it would appear that, for most purposes at least, relative measurement is more important than absolute measurement. We are much more interested in evaluating the magnitude of change over time or in the comparative evaluation of two similar situations than we are in establishing the absolute level of accident risk in a given situation. This implies that any measurement concepts and techniques adopted must be capable of a reasonable degree of standardization over time and across the full range of situations, unit sizes, safety management personnel quality, and so forth, likely to be of comparative interest. Standardization is important in that it provides a constancy of meaning or interpretation of measures to be compared and that it ensures that observations of performance reflect only on the observed not on the observer.

Fourth, we require measurement systems capable of sensing those changes or differences that represent a proper reflection of safety management effectiveness. This means that we want, if possible, to avoid measuring the effect of uncontrollable or exogenous variables (outside influences). We also want to avoid the confusion that would result from using measures of effectiveness that respond merely to differences in the quality or quantity of human exposure. This requirement represents the rationale for attention to the variable of exposure, which, it should be noted, possesses both qualitative and quantitative aspects.

Finally, we require measurement systems capable of achieving a known and acceptable degree of precision with regard to inferences to be drawn about relative levels of "safety expectation." This means that the random error or chance residual associated with statistics derived from those attributes observed must be capable of statistical estimation and must fall within an acceptably narrow range. This degree of precision will, in turn, be a compromise between three kinds of factors: the allowable degree of uncertainty, all other elements of purpose such as specificity and timeliness; and the cost of measurement.

Other Dimensions of Measurement Purpose

There are other dimensions to the question of purpose that deserve discussion because they tend to impose further requirements on an effective system of measurement. These relate to what Tarrants (1966) has described as the macroscopic versus microscopic approach to industrial safety performance measurement, referring to broad industrywide accident performance measures, on the one hand, and internal or within-establishment performance measures on the other.

The first of these dimensions might be thought of as *the dimension of homogeneity of exposure.* At one extreme, there is the possibility of pursuing any or all of the previously outlined objectives with little concern for situational similarity—that is, with measurement data that have been aggregated over many diverse operations and environments without regard to the composition of causal factors. For example, it is possible to establish as the subject of interest the relative accident experience of all shipyards in the nation, even though shipyards vary in types of construction activity, in facilities, in skills employed, in equipment, and in all manner of operating detail.

At the other extreme, it may be desirable to hold situational factors as constant as possible in order to make comparisons both more meaningful and more specific. For instance, rather than simply comparing accident data among shipyards, it might be considered much more valuable to make comparisons between heavy machining operations in a specific shipyard or even between lathe operations in several machine shops among multiple shipyards. Quite clearly, the more homogeneous the situations being compared, the more likely that accident differences will reflect differences in environmental or behavioral aspects of safety as compared with differences in the type or quality of exposure to risk. From this standpoint there is increasing value in narrowing the scope of measurement in the direction of favoring homogeneity of exposure.

A second dimension of purpose relates to *the homogeneity of events* to be considered for study. Accidents may be classified and compared on the basis of particular causal properties or severity consequences (for example, all eye injuries, all vehicular accidents, all accidents resulting from equipment failure). Again, it is possible to be interested in all occurrences without regard to such details as what happened, how it happened, to whom it happened, when it happened, or where it happened. To do so might provide some sort of total view of overall safety effectiveness or degree of hazard that should not be influenced by more detailed consideration. For example, it is of value to know that pedestrian fatality rates in New York City are greater than they are in Salt Lake City. At the same time, there is even greater value in measurements that contain more detailed information because of the possibility of relating these results more directly to actual or prospective countermeasure systems. Thus, comparing rates of pedestrian fatalities involving alcohol between the two cities would be of both explanatory and, potentially, remedial value.

A final dimension of measurement purpose is that of *time*. Since the choice of measurement time interval is essentially arbitrary, occurrences can be cumulated over weeks, months, or years of time. Long periods of time tend to average out those exposure variables that have seasonal components, such as the influence of seasonal weather, and variations due to purely random elements, such as work tempo, but which are more susceptible to those extraneous causes of change that develop over time and are neither random nor periodic. For example, comparison of annual accident experiences of air carriers tends to eliminate the need for detailed consideration of seasonal weather, periodic variations in air traffic density, load factor, daylight/darkness ratios, and the like, as short-term factors that must be analyzed and allowed for. On the other hand, the purpose of measurement may be to study the effects of short-term factors, and effects of this kind may be lost over longer measurement periods. For example, it may be desired to assess a new operating procedure in terms of both initial impact and subsequent decay— an objective that surely would require a measurement base of months, if not weeks.

As a crude generalization covering these three dimensions of measurement purpose, one could argue that it is a desirable objective to develop as much specificity, as much detail, and the shortest base time periods as possible since data can always be aggregated over situations or types and cumulated over time to meet any cruder needs. Toward this end, measurement practices and techniques should always be oriented toward the needs of the microscopic level because this automatically provides for meeting needs at the macroscopic level. The major, and probably inseparable, obstacle to fulfilling this objective for most safety management purposes—with any measure dependent upon actual accident results—is the problem of statistical unreliability. As a result, most measurement conventions in this field have evolved in a direction that tends to aggregate dissimilar events over rather long periods of time (a practice which has resulted in the forsaking of much important detail) in the hope of establishing an index that will be statistically reliable for purposes of gross comparisons.

To develop this point, we shall consider some of the specific measurement problems that present themselves in moving from gross analysis to fine detail, along with some of the measurement conventions developed as a part of the evolution of safety management practices.

CONVENTIONAL MEASUREMENT PRACTICES AND PROBLEMS

Most of the measurement questions that have arisen in this field have been concentrated on the appropriate statistical treatment of accident reports. By and large, these issues are the result of a conflict between wanting to pack as much information as possible into a few simple statistics while desiring that these statistics possess similar descriptive qualities and sufficient reliability so that they can be meaningfully compared.

The following questions reflect the kind of perplexing issues that the American National Standards Institute (ANSI) and the BLS–OSHA measurement systems were partially designed to resolve:

How do we combine inquiries resulting from a wide range of nonhomogeneous accident situations?

How do we count and combine accidents that range in severity from paper cuts to fatal injuries?

How severe must an accident be to count at all?

How do we aggregate accidents that involve only property damage with those that involve human damage?

How do we count and combine accidents that involve the simultaneous injury of more than one person?

How do we allow for the fact that during different periods of time and in different situations the quantity and quality of human exposure to risk varies?

What statistical treatment is required for inferring real differences and interpreting and acting upon observed changes in accident counts?

Although it would be difficult to describe the full range of conventional practice or thought regarding these problems, it is possible to take up each of them in turn for brief review and appraisal.

HOW CAN ACCIDENTS THAT DIFFER IN NATURE AND CIRCUMSTANCE BE COMBINED? Typically, accidents are combined for measurement purposes if they involve human injury or death, regardless of similarity. For example, simple lifting accidents would, ordinarily, be combined with electrocutions. While this procedure has value for gross measurement purposes, it forecloses the possibilities of more detailed examination of accident situations, circumstances, or causal factors. The principal rationale for this form of aggregation is to obtain the largest possible "sample size" for statistical purposes and thus to facilitate the detection of real changes or differences in the gross environment. In effect, in the absence of standard reporting categories, established procedure tends to discard important specificity and detail for the sake of statistical power and gross description.

HOW CAN ACCIDENTS THAT DIFFER IN SEVERITY OF HUMAN INJURY BE COMBINED? Three conventions apply in the customary treatment of accident data. One is to ignore severity aspects entirely and to aggregate all events regardless of severity into an overall frequency count. Another is to introduce some measure of severity, such as the number of man days lost, and then to cumulate these as a measure of accident experience. The third is to separate out and report fatal accidents only. The first of these procedures has the disadvantage of being oversensitive to minor incidents, since these are always in the preponderance. However, it has the advantage of greatly amplifying sample size and statistical power. The second measure, which includes severity information, is obviously more sensitive to severe and fatal injury accidents, but at the cost of effective sample size and statistical power. The third, cumulating fatal accidents only, is a still further step in this direction, focusing entirely on the most important class of events but with very severe statistical limitations because of the small number of events included in most samples.

To the extent that it can be argued that all injury-producing events are the consequence of the same causal structure and that the degree of injury is merely a matter of chance, there are clear advantages in a measure that includes all events in the sample. On the other hand, and more typically, perhaps, the actual situation is often quite the opposite. Not only is a paper cut and a fatal electrocution not of equal importance, but it is hard to imagine that they are equivalent events insofar as reflecting safety attitude, supervisory effectiveness, education, environment, or most other aspects of safety management are concerned.[1]

HOW SEVERE MUST AN INJURY ACCIDENT BE TO COUNT AT ALL? The customary answer to this question is to count an accident as an injury-producing accident if it is disabling (costs a human workday). This is, it seems, an arbitrary standard of inclusion and can only be defended on grounds of reporting uniformity and convenience, which, of course, is the argument for standardization. Clearly, the lower the criterion of severity, the larger will be the sample size—but the more sensitive will any measure be to the more trivial class of accidents and the more difficult will it be to establish a standard reporting basis. Recent modifications to the established reporting standards have tended in the direction of lowering the severity threshold in order to augment the number of reportable events or incidents.

HOW CAN NONINJURY-PRODUCING ACCIDENTS BE COMBINED WITH THOSE RESULTING IN HUMAN INJURY? In most safety management practice these two classes of accidents are not combined since safety management by its nature is concerned only with human damage. There are, however, some who argue that noninjury accidents do indeed reflect the same type of failure experience as injury accidents, and since they are closely related events, they should be included in any measurement system for the purpose of boosting sample size. There are circumstances where this argument might be shown to be valid (such as vehicular roll-overs), but for most applications it is likely to produce serious distortions of interpretation. For example, it would be equivalent to combining parking lot fender dents with high-speed turnpike collisions. The implications of this issue are similar to those discussed above.

HOW SHOULD ACCIDENTS THAT PRODUCE MULTIPLE-PERSON INJURIES BE COUNTED? The usual treatment of multiple-person injuries is to allow the measure of severity to reflect the aggregate loss. The effect on severity rates is to decrease effective sample size by weighting most heavily those events that produce multiple damage. For example, a single automobile accident resulting in fatal injury to two passengers and a driver would be weighted three times as heavily as it would have been in the absence of passengers. This issue would tend to have greatest interest for those activities where multiple-person accidents are not uncommon (mine cave-ins, maritime collisions, bus accidents, airplane crashes, and the like).

HOW CAN VARIATION IN THE QUANTITY AND QUALITY OF HUMAN EXPOSURE TO ACCIDENT RISK BE ALLOWED FOR? The problem of exposure measurement is one of the most difficult aspects of accident measurement to deal with and, consequently, has received very little attention. The standard and most common technique is to adopt an arbitrary but common-sense measure of exposure, for example, man-days worked, vehicle miles driven, and so forth. To the extent that this measure does not reflect uniform quality or degree of exposure over time or between situations, it will not respond proportionately to real changes in the hazard situation. Beyond this, it is commonly assumed that there is a linear relationship between exposure and safety expectation. There is at least one safety area, that of traffic safety, where it is more plausible to assume nonlinearity (accident expectation is proportional to the square of traffic density as measured by vehicle miles).

Consideration of some of the questions posed by the exposure issue will help to illuminate the problem. Is it reasonable, for example, to compare accidents per mile driven for city drivers and rural drivers? Is it reasonable to use man-hours as an exposure base for comparing the accident experience of establishments or industries without regard to work-force composition, tasks, equipment, working environment, and so on? Is there meaning in following changes in accident frequency rates from year to year even though job structure and content are constantly changing? There are many questions of this kind that are much easier to ask than to answer. Fortunately, this component of the measurement problem is well subordinated in most situations by the problems associated with the imprecision or variance of the accident statistic.

WHAT STATISTICAL TREATMENT IS REQUIRED FOR INFERRING REAL DIFFERENCE? Standard safety measurement procedure ignores entirely the question of statistical inference. Performance statistics are almost always reported without estimates of an appropriate confidence interval. Furthermore, it is generally impossible to impute such a confidence interval retrospectively to a ratio statistic. It is equally impossible, of course, to make confidence statements regarding the difference between two or more of these ratio statistics.

In summary, then, and reviewing together all of the foregoing questions, one issue appears repeatedly. This is the need to consider aggregating or combining as many events as possible—regardless of similarity of cause or importance—in an effort to enlarge the sample size as much as possible by extending the range of incidents included in the measurement system.

The compulsion to compromise other measurement interests in order to build sample size is readily explainable for any "rare-event" process. Statisticians liken the time stream of accidental occurrences to a stochastic process and, for purposes of convenience, tend to regard it as a Poisson process— that is, one in which the likelihood of occurrence at any instant is infinitesimally small and is independent of the previous history of occurrences.[2] Table 10-1 illustrates the fundamental importance of chance-caused variation in such processes and the difficulty in seeing through such "noise" for purposes of inferring change or difference.

Table 10-1

Illustration of the Importance of Chance-Caused Variation to Accident Rate Interpretation*

	Period or Situation 1		Period or Situation 2		Period or Situation 3	
	Expected or Mean Number of Occurrences	*95% Probability Interval*	*Expected or Mean Number of Occurrences*	*95% Probability Interval*	*Expected or Mean Number of Occurrences*	*95% Probability Interval*
Fatal Accidents	20	(13 to 29)	15	(9 to 23)	10	(5 to 17)
Differences in number of fatal accidents	*Expected Difference 1 and 2* 5 (25%)	*Approx. 95% Probability Interval* (−6 to 16)	*Expected Difference 1 and 3* 10 (50%)	*Approx. 95% Probability Interval* (0 to 20)		

*The illustrative numbers in this table have been developed on the assumption that the accident processes on which they are based can be characterized by a constant Poisson process. Statistical control chart studies suggest that this assumption is in error and understates the real uncertainty of inference in dealing with accident data.

A review of the numbers in this table indicates the limitation of statistical power in attempting to measure real change or difference in accident situations. Imagine that these illustrative numbers represent fatalities over full-year time periods and that it was desired to infer from actual fatality experience whether or not improvements had been effected during the three year period. Over the full-time interval of year 1 to year 3, the fatal accident expectation had in fact been reduced by half, from 20 per year to 10 per year. Despite so great a real change, the 95 percent probability interval covers the range from no apparent change to a 100 percent apparent improvement.

It is immediately apparent that the heart of the uncertainty problem is the small number of occurrences being compared. Yet, as small as these illustrative numbers appear to be, they are still very large for any real-life accident situation. For example, at typical industrial accident fatality rates (0.10 deaths per million man hours), 20 fatalities per year would correspond to 200 million man-hours of exposure or to the exposure of 200,000 men over one year.

It is reasonably clear that except for the crudest kinds of comparison— such as comparing the safety experience of entire industries or cities over relatively long periods of time—fatality measurement as a yardstick of safety management is an exercise in futility. If measurement is extended to include all disabling injury-producing or recordable injury-producing accidents, a somewhat more reliable statistic is generated, even though it may be a less valid measure of safety performance.

Consider first the improvement in reliability that would result from this extension. During one year, the Bureau of Labor Statistics estimated the national work injury rate at 14.7 injuries per million man-hours of exposure, which is equivalent to 29 injury accidents per 1000 men per year. Although the expected number of injury accidents then would be much higher than that for fatalities (147 times higher by estimate), it is still too low to serve as a reliable statistical index for most purposes—that is, for all but the largest operating units. Pure chance would cause the number of injury accidents for a work force of 1000 men with a (nationally) average safety level to fluctuate between 18 and 40, 95 percent of the time. It has been reported that two-thirds of all occupational injuries occur in operations with fewer than 100 workers. The consequence of this fact is that for most units or operations the expected number of disabling injury or recordable injury accidents during one year of exposure is so low that even minimally precise statistical inference regarding the level of safety performance is out of the question.

The issue of statistical reliability, arising as it does out of the limited sample size associated with a rare-event process, has proven deeply frustrating. It is clearly evident that it must continue to serve as the major impediment to the achievement of suitable levels of precision for any measure based on injury-producing accidents. Concern over this problem has resulted in a number of proposals for new measurement techniques, ranging from minor modifications in existing standards to radical innovations in concept.

PROPOSED MODIFICATIONS OF STANDARD MEASUREMENT PRACTICE

Experience with the current standards for reporting and measuring accident experience has led to widespread dissatisfaction on the part of many officials responsible for safety management activities and performance because of the statistical limitations just described. One of the consequences of this has been the development and application of various reporting modifications, generally proposed for application concurrently with standard reporting. Some of the modifications which have been put forth in recent years are described below.

The Serious Injury Index

One popular index employed currently is the serious injury index (Batchie, 1962; Gilmore, 1965; Haier, 1964; Voland, 1962). It is an attempt to enlarge the sample of occurrences beyond those which are defined as disabling injury accidents. In effect, it includes certain types of nondisabling injuries along with those disabling injuries that meet the lost-time criterion. These include specified kinds of nondisabling eye injuries, fractures, lacerations, and other injuries for which work restrictions are prescribed.

The proponents of this modification (which has apparently found substantial useful application) argue that it results in a more meaningful as well as a more sensitive measure of safety performance. Other advantages claimed for the serious injury index include the greater degree of specificity and detail that can be developed regarding type of accident, place of occurrence, and so forth, because of the substantially greater numbers involved. It should be noted, however, that while the higher frequency of occurrence of these less severe injuries would surely tend to produce a more sensitive measure, it would also be a measure that would probably tend to reflect a different class of events and that would be extremely sensitive to such situational factors as propensity to report minor injuries and consistency of medical treatment standards. These issues ought to be susceptible to some kind of objective evaluation and clearly deserve further study.

The High Potential Accident Index

Enlarging the universe of events beyond disabling injury accidents alone while at the same time generating a safety performance measure that remains sensitive to serious rather than minor hazards is the goal of the high potential accident index (Allison, 1967). In practice, a reporting system was established to encourage the reporting of all injury-producing accidents, whether disabling or not. These events are then studied for the purpose of identifying those few that, in the judgment of investigators, had the potential for serious human or property damage (estimated at 4/10 of 1 percent of all accidents versus 7/1000 of 1 percent for lost-time accidents). These high potential accidents represent the measurement base of interest. In support of this proposal it has been argued that indexes based merely on raw minor injury data tend to reflect reporting rather than incidence, result in pressure on reporting and minor injury reduction, and do not reflect the overall level of accident cost performance.

On the other side of this case, however, is the issue of whether the index tends to reflect changes in the level of safety or changes in the standards by which minor accidents are judged to have been potentially serious. Although there is no logical reason for excluding the role of human judgment in any measurement process, it is important to develop rules for applying judgment that can result in verifiably consistent performance. There is much intuitive appeal in this proposal and it clearly warrants intensive development, refinement, and evaluation.

Property Damage Accident Index

The proponents of a measurement system that includes property damage accidents (Bird, 1966) point out that the ratio of property to disabling injury accidents is at least 500 to 1. The measurement index most commonly mentioned is that of total realized dollar cost for replacement, repair, lost production time, and so on. The plan as it has been applied in practice requires a workable scheme for enforced reporting of all property damage accidents. Although this proposal has the least surface appeal (it is not clear that property damage costs should bear any relationship to human safety), it too should be susceptible to some kind of objective evaluation.

BLS–OSHA Recordkeeping System

Chapter 2 explained in detail the BLS–OSHA Recordkeeping System. The system's definitions broaden the scope of recordable work injuries and illnesses over the ANSI Z-16.1 standard. Most of the same statistical disadvantages remain.

PROPOSED MODIFICATIONS IN STATISTICAL TREATMENT

In addition to reporting modifications, numerous proposals have been made from time to time with regard to techniques of statistical interpretation. These include:

1. Statistical confidence intervals
2. Statistical control chart techniques

The impossibility of comparing standard accident frequency ratios for purposes of statistical inference was suggested earlier. There is a strong argument for at least trying explicitly to recognize the uncertainty of interpretive comparisons as an adjunct to standard reporting. One way to do this would be to provide estimates of confidence intervals along with standard frequency and severity reports. Appropriate confidence interval formulas and charts have been developed to facilitate application of this idea (How to Test Performance, 1963).

For the purpose of identifying real rather than random changes in the level of safety performance, attempts have been made to adopt statistical control chart techniques, analogous to those which have proven successful in manufacturing quality control. Although none of these efforts have been tested over long enough time periods to establish the merit of this application, there is no reason to believe that statistical control theory cannot be of value (Littauer and Irby, 1957).

While these techniques can contribute importantly to the correct interpretation of statistical indexes that have much imprecision associated with them, they cannot contribute to the improvement of the fundamental reason for imprecision, which is the use of any measurement index based on small numbers of actual injury accidents.

OTHER POTENTIAL MEASUREMENT APPROACHES

In an effort to circumvent the troublesome problem of statistical reliability in accident measurement, attempts have been made to introduce other measurement techniques for characterizing the relative safety or relative danger inherent in a given situation. (For a more complete review see Tarrants, 1966.) Whereas most of the more traditional measures deal with retrospective accident experience as a basis for estimating actual accident expectation—or for gauging prospective accident experience—it is appropriate to consider possible means for estimating prospective accident experience through some direct measure of *accident propensity*. To be useful for this purpose, it is desirable, but not necessary, to be able to transform measures of propensity directly into estimates of future accident experience, or at least insofar as the objective is to make relative rather than absolute assessments.

Near-Misses as a Measure of Accident Propensity

One possibility for such application is the *critical incident technique*. Basically, this technique attempts to obtain reports of "near-misses" or accidents that, in the opinion of those reporting, almost happened. It is intuitively apparent that much of the accident-avoiding behavior of individuals is based on near-miss experiences, and it is surely sound in principle to expect that operating groups and organizations can similarly profit from the study of near-miss occurrences.

The critical incident, or near-miss, technique has many values. It offers much promise in particular for investigating causal aspects of accidents, but until now it has found only limited application as a measure of accident propensity. The major reason for this is that near-miss data are totally dependent on reports from those who are part of the situation and, in fact, close to the scene. This is a requirement that is likely to result in intrinsic deficiencies from the standpoint of reliability. As a measurement technique, critical incident recording may be less a reflection of accident propensity than of propensity to report. Beyond this, near-miss measurement is not independent of the quality or quantity of human exposure and contains the same exposure measurement problems as actual accident data.

Nonetheless, although it is far from certain that the various problems of definition (what events are to qualify as near-misses), of reliable reporting, and of exposure measurement could be solved well enough to develop a continuing and widely applicable measurement system, there appears to be enough promise to justify serious developmental effort. Of all of the proposed measurement techniques that do not depend upon actual injury or fatality data, the critical incident measurement possesses the strongest face validity as an index of accident propensity. The issue of statistical reliability will surely be less acute since near-misses probably range anywhere from tens to hundreds of times more frequent than actual accidents. This greater frequency should make it possible to categorize more finely types of occurrence and to avoid intermixing potentially trivial with potentially serious incidents.

A major contribution to the potential applicability of the critical incident technique has been the development of a data collection procedure that employs a stratified random sample of participant-observers who are asked to recall and describe unsafe errors or conditions that have come to their attention (Tarrants, 1963). Employing periodic samplings of this kind should lead to the development of a useful index of safety performance. Clearly, much work remains to be done to establish that this measure is not only one that is statistically reliable, but also one that constitutes a valid reflection of safety expectation.

Without attempting to spell out the details of the work that might be undertaken along these lines, one should be able to establish a number of experiments in various installations to study various forms of near-miss data collection procedures. The key procedural stumbling block will be the development of a stable operating method for collecting data so that any near-accident that falls within a specified description and a specified "closeness" will be equally likely to be reported. In essence, what is proposed is the initiation of experiments to establish whether or not the critical incident technique has potential application, not simply as a means for studying accident causation but as a measure of accident propensity oriented toward the larger purpose of safety management. The question is not whether this technique has potential but rather for what situations its potential is greatest. There are some applications, near-collisions of commercial airliners, for example, where the combination of professional reporting and self-interest surely favors its successful application.

Unsafe Behavior Frequency as a Measure of Accident Propensity

There exists another, little known, proposal for measuring accident propensity that deserves further testing and development. As originated and tested by Schreiber (1957), it includes three basic properties. First, it is based on the direct sampling of specified types of deficiency or discrepancy with regard to human action or behavior. Second, the sample size is directly controllable through the number of observations or opportunities selected. Third, it allows for the identification of change in the level of accident propensity through a form of statistical control chart technique analogous to that employed for statistical quality control.

In practice, this technique consists of using trained observers to examine or inspect operations under safety surveillance on a carefully scheduled basis (with respect to time, locale, and so forth). Each observer is required to note the behavior of a given sample of operatives and to record each instance in which a specific breach of safe practice is seen to occur. From these observations a discrepancy fraction statistic can be developed on a daily or weekly basis as the root measure of accident propensity. Schreiber has shown how appropriate statistical confidence intervals can be calculated and statistical control charts developed. He has demonstrated the actual application of this basic procedure to a variety of hazardous operating situations over time periods lasting as long as several months and has indicated the utility of this type of measurement system for identifying changes in the level of accident propensity and for correlating these with known or suspected changes in the social and physical environment.

There would appear to be promise in extending this basic measurement idea to a wider range of applications and in exploring a variety of observational and statistical techniques. For example, it should be possible to develop sample measures of accident propensity for purposes of comparing two or more situations of similar nature. To illustrate, the frequency with which established safety rules (such as not wearing safety glasses, standing under operating hoists, smoking in unauthorized places), are violated could readily and meaningfully be compared between divisions of an organization or between shipyards. Sample measurements of safety knowledge or skill could be compared between drivers or operators from various installations. Proper condition and functioning of equipment can be sampled in exactly the same way and used as a basis of accident propensity measurement. In fact, it should even be possible to develop composite indexes that combine sample observations of many aspects of operator behavior, equipment condition, safety knowledge, and so forth. These indexes would have known statistical properties, would not be of limited reliability because of sample size restrictions, and, most importantly, could be constructed to reflect the controllable aspects of safety management.

The potential advantages of this procedure include the following:

1. It could be applied equally as well to situations in which the accident expectation is high because of large human involvement and exposure, as to those situations where accident risk, human exposure, and, hence, accident expectation is very small.
2. It could be applied with considerable confidence for comparing situations that are vastly different from the standpoint of external variables. It could be focused on specific types of safety deficiencies or discrepancies or generalized to particular composites or combinations of these.
3. Although it would produce statistics that, at best, constitute only approximate measures of accident propensity, these could still represent direct measures of those behavioral or environmental factors that safety management aims at controlling. In this sense, this technique is capable of yielding direct measures of the effectiveness of safety management.
4. Furthermore, these measures have known statistical properties and, to the extent that time and effort are available for sample observation, can be developed to any desired level of statistical precision. The problem of exposure measurement is not present so that the usual problems of interpretation disappear. Statistical control chart procedures can be applied for assessing changes in accident propensity over time and for associating these changes with known or suspected environmental disturbances or programmed control measures.

There are also a number of shortcomings and problems to be explored. The most consequential is, of course, the problem of determining what kinds of behavioral deficiencies and environmental discrepancies to observe and how to combine these, if they are to be combined. This relates to the basic question of validity; in the final analysis any statistic that purports to measure accident propensity must have some power for predicting future accident experience. The handling of this issue is highly dependent on the existence of some body of knowledge concerning known or suspected causal factors. To this extent, any operational application of this procedure would have to be combined with a continuing program of direct accident postmortem studies.

Another problem, very much of unknown magnitude, is the problem of distorted measurement because of the behavioral response of most people who believe themselves to be under observation. This could be especially acute in a punitive working environment. Schreiber circumvented this problem through disguised surveillance, but this would only work in the short term and in the absence of any related disciplinary system.

There would also be problems deriving from different standards of application, both for the same observer from day to day as well as between observers, especially when they are operating remotely. It is possible that standard procedures and specialized training could help to reduce the magnitude of this difficulty. Considering all of these possible problems, there is still probably enough net potential for a workable scheme of measurement to justify serious efforts toward developing detailed techniques and procedures for trial application and evaluation. An adaptation of this measurement technique, in combination with the industrial engineering work measurement method of work sampling, is described in detail in Chapter 16 under the title "Behavior Sampling."

SUMMARY OF PROGRESS TO DATE

In summary, then, it is seen that the conventional standard measures of safety effectiveness have serious, perhaps insurmountable problems, insofar as the principal purposes of measurement are concerned. Foremost among these is the problem of statistical unreliability due to the variability of a rare-event process in combination with small to very small sample sizes. This problem is basic and cannot be bypassed by refinement of interpretive procedure.

Attempts to circumvent this obstacle by such means as aggregating nonsimilar events, extending base measurement periods, lowering the severity threshold to include nondisabling injuries and even noninjury accidents have been proposed. While modifications of this kind tend to relieve some of the problems posed by statistical unreliability, they necessarily tend to cast doubt on the validity and significance of the results reported and to reduce the meaningfulness of the entire measurement process. Despite these problems there are reports of satisfactory experiences with some of these modifications.

In addition to questions on accident classification and reporting, there are complicated problems of exposure measurement that differ from situation to situation and that importantly detract from useful interpretation. There has been relatively little discussion on this issue, yet anyone with experience in this field appreciates its contribution to the confusion produced by the "accident numbers game."

The major alternative route to a concept of measurement based on the recording of actual injuries is one that deals instead with the concept of accident propensity. Accident propensity is, of course, only a concept and as such cannot be measured directly. But it can be considered to have many components or ingredients, which are measurable. Some, perhaps many of these ingredients, are measurable with considerable precision, and it may well be that these can be combined to form some kind of an acceptable index of accident propensity.

One type of accident propensity measure is the frequency of occurrence of various kinds of near-misses. These can be sampled within specific categories or classes of occurrence or can be combined into aggregate indexes. This measure, of course, still requires consideration of the exposure factor and is still not entirely free of reliability and sample size limitations. Some work has been done in attempting to apply this idea, although it is not yet possible to establish fully its merit.

Another type of accident propensity measure is that of unsafe behavior frequency. This can be applied to any specified type or form of unsafe behavior, unsafe condition, or unsafe equipment. It is independent of exposure variation, possesses controllable precision, and is directly related to the effectiveness of both general and specific safety management programs. This, too, has had some intermittent application—enough at least to establish feasibility if not value.

It should be pointed out that neither of these latter measurement concepts has been developed much beyond the pilot test stage. Both require considerable conceptual and practical refinement as well as extensive practical testing and evaluation. In view of the increasing dissatisfaction with conventional methods, further exploration of these possibilities seems warranted. And, of course, there will be other schemes put forward from time to time that may offer superior promise and which will require careful test and evaluation.

FURTHER DEVELOPMENT OF SAFETY MEASUREMENT CONCEPTS AND TECHNIQUES

Any reasonably objective outsider reviewing the developments in this field cannot help but feel a certain sympathy with the dissatisfaction and frustration that conscientious practitioners of safety management must be experiencing. In most small and medium-size operating units (which account for the preponderance of all accidents) it is virtually impossible to apply standard safety measurement criteria and come up with reliable estimates of safety effectiveness. In larger units it may be possible to develop gross comparisons with some precision, but it is difficult to do this with enough detail and specificity to provide strong guidelines for action.

At the same time there are a number of proposals for modifications of standard practice and even for entirely different measurement concepts (still oriented toward the basic measurement purpose of safety management) that hold some promise for fruitful application. Unfortunately, there exists no reasonable scientific evidence that would make it possible to assess the validity of claims and counterclaims for these. In addition, what practical support there is tends to consist of unsupported statements by the authors of these proposals asserting that successful application has been made. Beyond this, of course, some of these proposed measurement schemes are still extremely crude and primitive with much refinement and further development required before any useful application is likely to result.

Where do we go from here? How are these refinements to be accomplished and tested? How are these and possibly other conceptual innovations for safety measurement to be tested and evaluated in such a way as to replace innovators' claims with the much more convincing weight of objective and technically sound evaluation? Consideration of these questions suggests that *the fundamental research challenge in this field is not that of developing more sophisticated statistical procedures or more imaginative measurement system proposals. It is the need for some plan to develop, refine, and definitively to assess the abundance of existing procedures, proposals, and possibilities.*

For example, consider just one of the more promising of the recent proposals, the high potential accident index (HIPO). How are standard procedures for judging the potential seriousness of recorded accidents to be established? How can the reliability of these investigative judgments be evaluated? How can the relationships between the HIPO index and other known measures of safety expectation or accident propensity be determined? How can the areas for possible useful application be established? How can needed validation data for management purposes be generated?

Ordinarily, it might be assumed that this problem, like most of those of an applied nature, would be solved in time. Decades of trial-and-error experience, gradual accumulation of successful case studies, occasional student dissertations, evaluation of successively more powerful procedures and refinements probably would, within a half century or so, begin to weed out ineffective proposals and bring the stronger measurement system possibilities to the surface. Can this tedious, drawn-out, costly route of improvement be accelerated? Perhaps, although at this time there does not exist any effective means for doing so. Yet, if this is an area that has room for scientific investigation, then what better place to apply disciplined experimental research and development than to the fundamental measurement process.

One possible scheme for doing this would be to establish a national industrial safety research "laboratory"—under either federal or private auspices. The laboratory could readily be assembled since it need not consist of $200 million linear accelerators or $90 million radio telescopes. The basic laboratory requirement is simply met—real people functioning normally in existing work situations. It could, for example, be organized to consist of cooperating plants, factories, facilities, and installations belonging to the federal government or to other cooperating agencies or private organizations. The laboratory staff might be organized to include experienced safety engineers as well as representatives of appropriate disciplines such as psychology, statistics, medicine, and so on, all combined into an integrated central research group. Possible facilities could include shipyards, ammunition plants, defense plants, overhaul shops, motor fleets, naval ships, manufacturing plants, and so forth.

The mission of this "laboratory" would be to conceive, develop, refine, test, apply, and evaluate alternative measurement schemes and proposals to form the basis for future safety management application. Since many of these proposals could, and in fact should, be tested concurrently it would be possible to study them in side-by-side application. These studies could be continued over many years if necessary so that longer term deficiencies and problems would have an opportunity to reveal themselves. In time, perhaps, the scope of interest might be broadened to include the investigation of countermeasure effectiveness as well. The National Institute of Occupational Safety and Health (NIOSH) laboratory at Morgantown, West Virginia, might be a candidate as a coordinator of the proposed measurement research program.

Some of the specific areas of measurement study and research might include the following:

1. Refinement of techniques for applying various measurement concepts
   - Exposure measurement concepts and techniques
   - Sampling and interview schemes for near-miss recording
   - Surveillance techniques for unsafe behavior sampling
   - Standard judgmental methods for high potential accident analysis
2. Research to validate measurement concepts
   - Statistical relationship between disabling and nondisabling injury rates
   - Statistical relationship of near-miss frequency rates, unsafe behavior frequencies, and disabling injuries
   - Correlation of ratings between investigators judging high potential accidents
3. Research on interpretation techniques
   - Evaluation of statistical control chart techniques for identifying assignable causes

- Evaluation of statistical problems as a function of unit size
- Appraisal of techniques of classification of occurrences to provide homogeneous categories

This list of possible areas of inquiry could, of course, be extended in order to facilitate both development and evaluation of all of the aspects of both new and established measurement ideas. Open publication of all results would help to communicate data of interest both to the safety professional and to the research community.

Finally, if competent scientists and engineers sponsor and conduct these studies, it should be possible to establish a technically authoritative publication. This would have convincing impact on the professional safety management community at large with the result that useful findings might win much faster acceptance and might find their way into successively more valuable standard measurement and reporting systems.

Notes

1. *Author's note:* The examples cited represent the extreme ends of the potential severity continuum. There are many situations where the severity of a given accident may appear over a wide range of magnitude, depending somewhat on chance events. For example, dropping a heavy object from a scaffolding onto a floor where other workers are present has a potential for resulting in a range of severity consequences from no injury to fatality, depending on the chance position of the workers below (and perhaps other factors such as warning of the impending drop, worker density, the wearing of protective equipment, such as hard hats). Similarly, the frequently occurring "fall from the same level" has a potential for producing no injury, minor injury, major injury, or fatality, depending on a number of variables such as the part of the body striking the surface or object, the hardness of the object struck, the total area involved in the body-object impact interface, the energy exchange occurring at the point of contact, and so forth.

2. The Poisson distribution has a frequency function of the form:

$$f(x) = \frac{m^x \, e^{-m}}{x!}$$

$$\text{for } x = 0, 1, 2, \ldots$$

where:

    m = a parameter that is both the mean and variance, the mean and variance of the Poisson distribution being equal.

This distribution often appears in observed events that are very improbable compared to all possible events but occur occasionally since so many trials occur, for example, traffic deaths, industrial accidents, and radioactive emissions.

References

Allison, W. W. How to foresee tragic accidents by use of the high potential accident-prone situation hazard control method. *National Safety Congress Transactions,* 2:4, 1967.

Batche, M. R. Understanding and purpose of the serious injury index. *National Safety Congress Transactions, 11*:30, 1962.

Bird, F. E., Jr. Property damage, safety's missing link. *National Safety News,* September 1966, p. 24.

Gilmore, C. L. Statistics—A crystal ball approach to accident prevention. *National Safety Congress Transactions, 25*:38, 1965.

Haier, O. C. How accidents can help your safety program. *National Safety Congress Transactions, 12*:24, 1964.

How to test performance. *National Safety News,* April 1963, p. 38.

Littauer, S. B., and T. Irby. Analytical and mathematical studies of the analysis of accident data: II. Application of statistical control methods for the reduction and control of industrial accidents. *Traffic Safety Research Review, 1*:6–15, 1957.

Schreiber, R. J. *The development of procedures for the evaluation of educational methods used in accident prevention.* Unpublished doctoral dissertation, Columbia University, 1957.

Tarrants, W. E. *An evaluation of the critical incident technique as a method for identifying industrial accident causal factors.* Unpublished doctoral dissertation, New York University, 1963.

Tarrants, W. E. Measurement of industrial safety performance—Research. *Transactions, National Safety Congress, 12*:124–127, 1966.

Voland, L. J. Beyond frequency and severity. *National Safety News,* December 1962, p. 28.

Chapter 11

TWO APPROACHES TO A NON-ACCIDENT MEASURE FOR CONTINUOUS ASSESSMENT OF SAFETY PERFORMANCE

Thomas H. Rockwell and Vivek D. Bhise

For those close to the accident prevention problem, the difficulties in measuring safety performance are well known (Rockwell, 1966). It is sufficient to point out that the present measures are neither stable nor sensitive to changes in system inputs and have dubious reliability. Moreover, and most importantly, safety measures depend on loss-type accidents to occur before they can be applied. Such a postmortem approach means a long lead time between action and subsequent effects, both when action is viewed as negative forces (the introduction of unsafe practices) and as positive forces (such as the introduction of accident prevention programs). It is analogous to a man taking a shower with 50 feet of pipe from controls to the shower head. It is little wonder that safety directors are unable to predict performance and are usually unable to explain either upturns or downturns in accident experience.

The purpose of this chapter is to take an abstract view of the safety measurement problem, indicating several useful research tools of the past to arrive at a nonaccident measure for continuous assessment of safety performance. The authors have attempted to move from an academic posture to a quasi-practical approach for individual application. Two approaches based on the same premise are outlined—a passive one in which workers need not cooperate in the elements of the measure and an active one that depends upon employee cooperation and participation. Both approaches are based upon past research experiments and thus represent measurement proposals rather than de facto or historically proven safety performance measurement techniques.

The theory of accident causation has remained elusive even with the advances of physical and behavioral sciences. We are forced to live with a causal model of accidents that, in essence, insists that the causes of accidents can be identified and their subsequent removal will diminish the likelihood of similar occurrence. The causal model rests minimally on the following basic assumptions.

1.   Causes can be dichtomized into unsafe acts and unsafe conditions.
2.   Every accident has one or more causes that can be identified.
3.   Elimination of past accident causes will eliminate similar and related future accidents.
4.   Contributory subcauses for the proximate causes can be determined for most accidents. Common causes that have no objective measure or easy countermeasure lie behind the occurrence of many types of accidents, while unique causes with efficient countermeasures account for only a small percentage of total accidents.

The drawback of the causal model is that it is deterministic, and frequently it is found that accidents do not occur even in the presence of unsafe acts and unsafe environments. Moreover, accidents can occur when either known unsafe acts or unsafe conditions are absent.

At present there are two approaches to safety performance measurement being used in industry in order to determine the level of plant safety. The first approach involves the use of available historical data, that is, accident cost, accident frequency rates, severity rates, incidence rates, and so forth. The other approach involves indirect indexes based on the occurrence of events that are believed to be related to accident causation and are statistically much more frequent in their occurrence. Measuring unsafe acts and unsafe conditions is important in evaluating accident prevention methods, training programs, safety procedures, and inspections without waiting for accidents involving loss to occur.

It is this latter approach that will be treated in this chapter, with emphasis on methodology for determining existing hazards, weighing the relative criticality of such hazards and developing operational procedures for their measurement and analysis.

THE GENERAL PLAN

Both the active and the passive procedures developed in this chapter are basically similar in structure. Each procedure involves eight major steps and the corresponding steps in each of the procedures are functionally similar and have identical precedence relationships. The flow diagram in Figure 11-1 shows in detail these eight important steps and their necessary precedence relations that constitute the basic structure for the procedures.

The first step involves identification of major unsafe acts and unsafe conditions that can occur in the work area under consideration for the assessment of safety. The person or persons involved in this task should be very familiar with the operations performed in this department, the type of workers involved, their habits, the unsafe conditions that can take place, and

Figure 11-1
Procedure for Measuring Safety Performance.

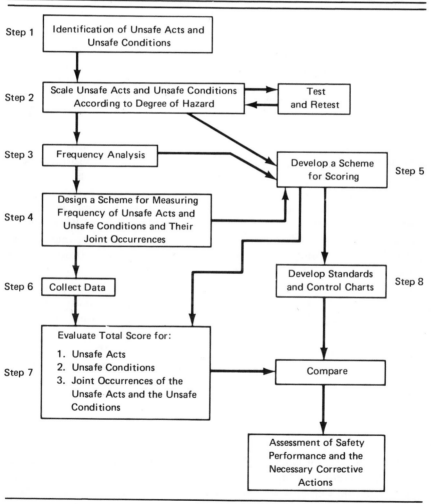

accident reports from the work area. Since high accuracy in identifying the unsafe acts and unsafe conditions is necessary for successful applications of both the active and the passive procedures, it is proposed that the critical incident technique be used for the former and detail analysis of previous accident statistics for the latter. The approaches are discussed in detail later in this chapter.

The second step in the procedure is to scale all the identified unsafe acts, unsafe conditions, and their combinations (sometimes called interactions) with due consideration given to the process, equipment, working conditions, and the potential injury severity and loss that can occur in case of accidents. Each item (unsafe act or unsafe condition or their possible combination) will be given a hazard index and the variance associated with such an index will be computed.

After the scaling, all the items are screened on the basis of expected frequency (Step 3). The methods used for obtaining the estimates of frequencies in both procedures are different. For the active procedure the method is based on subjective judgments of persons familiar with the work area; in the passive procedure the method used is analytical in nature.

The fourth step is to design a scheme for measuring (sampling) the frequency of unsafe acts, unsafe conditions, and their combinations. For the active procedure this scheme consists of designing checklists, which will be distributed to plant employees, and for the passive procedure the scheme involves designing a behavior and environment sampling study. The next step (Step 5) is to develop a scheme (method) for obtaining scores for safety performance based on measured frequencies of occurrences of unsafe acts and unsafe conditions and the hazard indexes associated with each of the unsafe acts, unsafe conditions, and their combinations. The scheme to be developed yields three scores for each of the procedures. The scores are denoted by the letters A, B, and C in the active procedure and the symbols $\overline{A}$, $\overline{B}$, and $\overline{C}$ in the passive procedure. The scores A and $\overline{A}$ are associated with the occurrences of unsafe acts only, and the scores B and $\overline{B}$ are associated with the occurrences of unsafe conditions only. The scores due to combinations of unsafe acts and unsafe conditions are denoted by C and $\overline{C}$. A score in the active procedure represents the sum of weighted hazard indexes (expected loss) for all employees in the department or work unit per period; a score in the passive procedure corresponds to the sum of the weighted hazard indexes for all employees in the department or work unit, per observation.

The sixth step consists of collecting the data according to the scheme developed in Step 4, and the next step (Step 7) involves evaluation of the three scores from the collected information about the occurrences of unsafe acts, unsafe conditions, and the joint occurrences of both. Control charts are constructed in Step 8, similar to statistical quality control procedures used by industrial engineers, to assess the safety of the system. Inferences about levels of safety performance within an organization can then be drawn from these control charts, and necessary corrective actions can be generated to keep safety performance under control.

In the following sections of this chapter the two procedures—(1) active and (2) passive—are developed and discussed as methods for assessing safety performance in any work situation. Both procedures are based on the same fundamental concepts previously discussed. The active system evaluates safety performance based on information obtained through the cooperation of people in actual work situations, whereas the passive procedure evaluates safety performance on the basis of the behavior and environment sampling studies developed from an analysis of historical data.

ACTIVE PROCEDURE

As discussed in the general plan, the active procedure is divided into eight main steps. The first involves the use of the critical incident technique to identify unsafe acts and unsafe conditions (described in detail in Chapter 17). This technique has been successfully employed by Rockwell (1968) in pilot error studies and by Tarrants (1963) in industrial applications.

The interviewer in this case interviews workers, leadmen, and foremen from the area selected for evaluation of safety performance to obtain details about all unsafe acts observed during working hours and the different possible unsafe conditions that have or could have existed in that work area.

In obtaining relevant information from the critical incident technique, it is necessary to make the following assumptions:

1. The full cooperation of all individuals is obtained, when they are interviewed.
2. The interviewer is familiar with the detailed operations of the department under study and also the various associated unsafe acts, unsafe conditions, and the department's accident-injury history.
3. The workers are knowledgeable enough to point out human errors and can recall them when asked by an interviewer.
4. The workers are fully aware of the objective of the safety assessment program and thus are not inhibited in disclosing unsafe acts and the unsafe conditions.

The second step is to scale the relative danger of each unsafe act, unsafe condition, and their combinations. Several psychometric approaches are possible. Thurstone's Method of Successive Intervals is a scaling technique that appears to be well suited to use in this case. Selected persons (perhaps all workers), who are familiar with the work area and can thus serve as expert judges, are asked to rate each of the unsafe acts and unsafe conditions and their combinations. This method also can be easily adopted to handle large numbers of experts. The procedure for scaling is described as follows:

Scaling Unsafe Acts, Unsafe Conditions, and Their Combinations

From application of the critical incident technique a list of unsafe acts and unsafe conditions is obtained in the first step of the active procedure. Each of these and their possible combinations are rated separately for the degree of hazard present. This is done by a number of persons familiar with the work and workers.

The details of the suggested method are as follows. Since seven levels is considered to be an approximate number of levels that can be accurately discriminated, seven equal appearing levels (intervals) are best suited for this purpose. Further, each level should be given a successive odd number value from 1 to 13 to provide a wide psychological separation between levels.

After obtaining scaling judgments from each of the individuals for each of the unsafe acts (UA), unsafe conditions (UC), and their combinations (UAUC), a distribution curve for ratings can be obtained. Then Thurstone's Method of Successive Intervals can be used to determine the hazard index. The value of the abscissa, for which the ordinate of the distribution curve has the value 0.50, is defined as a hazard index—also called a preference index (PI). The value of $(b - a)$ is defined as the discrimination index (DI) (see Figure 11-2). The PIUA (i) is used to denote the hazard index associated with the ith unsafe act and PIUC (j) is used as the hazard index for the jth unsafe condition. DIUA (i) and DIUC (j) indicate the discrimination index for the ith unsafe act and the jth unsafe condition, respectively.

The distribution curves similar to the one shown in Figure 11-2 are obtained for the following:

1. Each unsafe act UA (i)
2. Each unsafe condition UC (j)
3. Each possible combination of an unsafe act and unsafe condition UAUC (ij)

From these curves the hazard indexes and discrimination indexes for each of the listed unsafe acts, unsafe conditions, and their possible combinations can be obtained.

The assumptions made in scaling are as follows:

1. The total number of hazard levels used for scaling are small enough so that a worker has no difficulty in discriminating between various levels. (Anchor effects suggest seven levels.)

Figure 11-2
Distribution Curve for "Scaling."

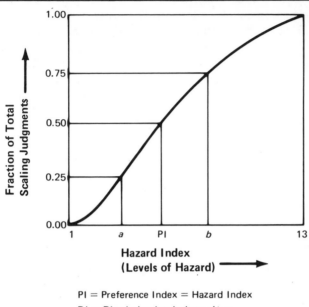

PI = Preference Index = Hazard Index
DI = Discrimination Index = $b - a$

2. Workers are consistent in their scaling ability. [This has been verified in past studies (Rockwell, 1968).]
3. A worker has enough knowledge about his work so that he understands the unsafe acts, the unsafe conditions, their consequences, and the nature and severity of the hazards associated with each of these and their combinations.
4. The Thurstone method assumes that the individual scaling judgments are independent and normally distributed.

   One of the outputs of this scaling technique is a measure of agreement called the discrimination index (DI). Large values of DI suggest that the description is either too vague and must be restated or that the item simply received widely different judgments as to its seriousness. In the latter case, this is a reason to discard the item for future use in a checklist. In order to develop reasonably stable hazard indexes, this scaling procedure should be tested and retested until the hazard indexes and their associated DI values are found satisfactory.

   The third step of the active procedure is rating each of the unsafe acts, unsafe conditions, and their combinations as to their frequency. Individuals experienced in the work area are requested to scale each according to their relative frequencies of occurrence. This frequency analysis is necessary in designing a useful checklist.

The fourth step involves designing a checklist. The checklist should include unsafe acts and conditions that have high probability of occurrence, a wide range of hazard indexes, and low discrimination indexes. A sample checklist is illustrated in Figure 11-3. These lists are distributed periodically in the department, and the workers are asked to participate in completing them. The following assumptions are made concerning their use:

1. The worker is interested in participating in the safety activities and is willing to cooperate in filling out the checklists.
2. The worker understands what the checklist means and its purpose in connection with the study.
3. The worker can read and interpret the checklist with no difficulty and has no hesitation in giving correct information.
4. The worker does not forget the unsafe acts that he and others in the department performed and the various different unsafe conditions that existed during the period for which he is asked to fill out the checklist.
5. The checklists are filled out according to instructions, and they contain correct information.

The following instructions are given to the workers for completing the checklists (similar to one presented in Figure 11-3).

> Before completing this checklist familiarize yourself with the unsafe acts listed in the rows of the checklist and the unsafe conditions that are listed in the columns.
>
> 1. Place check marks in the first row under the appropriate unsafe condition that you have observed during this period in your department if no unsafe acts were performed by either you or anyone else in conjunction with these unsafe conditions.
> 2. In each of the other rows, indicate the number (1, 2, 3, etc.) of incidents that you recall where the specified unsafe acts (rows) were performed by either yourself or someone else in the department under the specified unsafe conditions (columns) during this period.

The fifth step is to develop a scheme (method) for obtaining safety performance scores. A method similar to that suggested in Figure 11-4 (pp. 206–207) on the left hand side was used by the author in one of the studies conducted by the Department of Industrial and Systems Engineering of the Ohio State University. This method gives three separate scores A, B, C for (1) unsafe acts, (2) unsafe conditions, and (3) joint occurrence of unsafe acts and unsafe

Figure 11-3
Checklist for the Active Procedure.

Active Procedure	Safety Checklist for Dept. No.					Date:
	Unsafe Conditions					
Unsafe Act Description	0. No unsafe conditions	1. Hot parts left with no sign			j. Oil on shop floor	
0 No unsafe act						
1 Did not use goggles						
2 Did not use gloves						
i. Operated forklift with fork up		▨				
		Not meaningful				

conditions, respectively. The scheme for scoring involves obtaining mathematical expressions for calculating each of the scores. The mathematical expressions are derived by using the hazard indexes obtained in Step 2 and the information obtained from the checklists.

Let us assume that there are N number of employees working in the department, and at the end of a period, the total number of checklists col-

Figure 11-4
Method for Obtaining Safety Performance Scores.

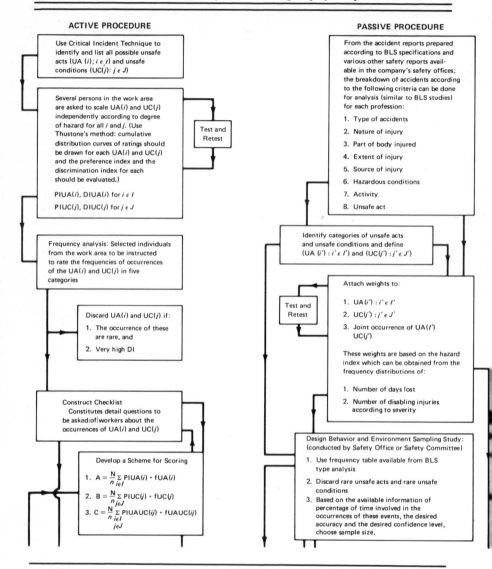

ACTIVE PROCEDURE

Use Critical Incident Technique to identify and list all possible unsafe acts (UA (*i*); *i* ∈ *I*) and unsafe conditions (UC(*j*): *j* ∈ *J*)

Several persons in the work area are asked to scale UA(*i*) and UC(*j*) independently according to degree of hazard for all *i* and *j*. (Use Thustone's method: cumulative distribution curves of ratings should be drawn for each UA(*i*) and UC(*j*) and the preference index and the discrimination index for each should be evaluated.)

PIUA(*i*), DIUA(*i*) for *i* ∈ *I*
PIUC(*j*), DIUC(*j*) for *j* ∈ *J*

Test and Retest

Frequency analysis: Selected individuals from the work area to be instructed to rate the frequencies of occurrences of the UA(*i*) and UC(*j*) in five categories

Discard UA(*i*) and UC(*j*) if:
1. The occurrence of these are rare, and
2. Very high DI

Construct Checklist
Constitutes detail questions to be asked of workers about the occurrences of UA(*i*) and UC(*j*)

Develop a Scheme for Scoring

1. $A = \frac{N}{n} \sum_{i \in I} PIUA(i) \cdot fUA(i)$

2. $B = \frac{N}{n} \sum_{j \in J} PIUC(j) \cdot fUC(j)$

3. $C = \frac{N}{n} \sum_{\substack{i \in I \\ j \in J}} PIUAUC(ij) \cdot fUAUC(ij)$

PASSIVE PROCEDURE

From the accident reports prepared according to BLS specifications and various other safety reports available in the company's safety offices; the breakdown of accidents according to the following criteria can be done for analysis (similar to BLS studies) for each profession:

1. Type of accidents
2. Nature of injury
3. Part of body injured
4. Extent of injury
5. Source of injury
6. Hazardous conditions
7. Activity
8. Unsafe act

Identify categories of unsafe acts and unsafe conditions and define (UA (*i'*) : *i'* ∈ *I'*) and (UC(*j'*) : *j'* ∈ *J'*)

Attach weights to:

1. UA(*i'*) : *i'* ∈ *I'*
2. UC(*j'*) : *j'* ∈ *J'*
3. Joint occurrence of UA(*I'*) UC(*j'*)

These weights are based on the hazard index which can be obtained from the frequency distributions of:

1. Number of days lost
2. Number of disabling injuries according to severity

Test and Retest

Design Behavior and Environment Sampling Study: (conducted by Safety Office or Safety Committee)

1. Use frequency table available from BLS type analysis
2. Discard rare unsafe acts and rare unsafe conditions
3. Based on the available information of percentage of time involved in the occurrences of these events, the desired accuracy and the desired confidence level, choose sample size.

Figure 11-4
(Continued)

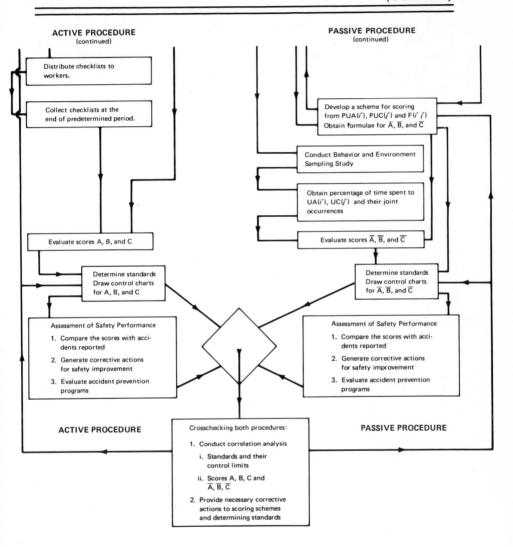

lected is equal to n. The values in the corresponding cells (except in the first row) of each of the checklists are then added to obtain the total frequency for each of the unsafe acts and the joint occurrences of unsafe acts and unsafe conditions. Since the workers are instructed to put check marks in the first row, which corresponds to no unsafe act, to obtain frequency of only the unsafe conditions, we add up the check marks for each unsafe condition.

The data are then collected (step six) and the scores are evaluated (the seventh step of the procedure) by the following formulas:

$$A = \frac{N}{n} \sum_{i \in I} PIUA(i) \cdot fUA(i)$$

$$B = \frac{N}{n} \sum_{j \in J} PIUC(j) \cdot fUC(j)$$

$$C = \frac{N}{n} \sum_{\substack{i \in I \\ j \in J}} PIUAUC(ij) \cdot fUAUC(ij)$$

where:

$A =$ safety performance score for the occurrence of unsafe acts

$B =$ safety performance score for the occurrence of unsafe conditions

$C =$ safety performance score for the occurrence of combination of unsafe acts and unsafe conditions

$I =$ set of unsafe acts

$J =$ set of unsafe conditions

$PIUA(i) =$ hazard index for i^{th} unsafe act

$PIUC(j) =$ hazard index for j^{th} unsafe condition

$PIUAUC(ij) =$ hazard index for the joint occurrence of ith unsafe act and j^{th} unsafe condition

$fUA(i) =$ total number of incidences involving i^{th} unsafe act obtained from the n number of check lists collected

$fUC(j) =$ total number of checkmarks counted for the jth unsafe condition from the n number of checklists collected

$fUAUC(ij) =$ total number of incidences involving the joint occurrence of ith unsafe act and j^{th} unsafe condition obtained from the n number of checklists collected.

Note that A/ N, B/ N, and C/ N correspond to the average weighted hazard index (loss) per worker per period of collection of checklists due to unsafe acts, unsafe conditions, and the joint occurrences of unsafe acts and unsafe conditions, respectively. Thus, by using the scoring scheme and the information obtained from the checklists the scores A, B, and C can be evaluated.

After the scores A, B, and C are evaluated (Step 7), they should be tested for significant correlations with (1) first-aid (minor injury) frequency, (2) lost-time accidents, (3) recordable incidents, and (4) production efficiency. These should be found satisfactory before proceeding further with the procedure.

Since the estimates of the variances of A, B, and C can be statistically determined from the collected data, separate control charts for the three can be drawn (Step 8). The new estimates of A, B, C should be plotted periodically to assess the safety performance of the department. Figure 11-5 (p. 210) contains a control chart for score A, and it also illustrates briefly the effects of various control actions on safety performance. The procedure can be called "continuous" in the sense that the workers will be asked to complete the checklists and the information about unsafe acts and conditions is essentially fed to the safety controller regularly and compared over time. The success of the procedure lies entirely in the cooperation of the workers with the organization's safety department. Therefore, it is *essential* that the workers understand the importance of this procedure. It is not likely to succeed, to repeat, if proper cooperation from the workers is not obtained.

THE PASSIVE PROCEDURE

The passive procedure is given in detail in the right-hand schematic diagram appearing in Figure 11-4. Information required for this procedure can be obtained by using the various accident reporting standards along with the information available from safety inspections. The Bureau of Labor Statistics (BLS) and the National Safety Council (NSC) have published several reports presenting an analysis of accident causal information that can be used for developing the passive procedure. This procedure, therefore, assumes the availability of such information in detail, prepared according to different accident reporting criteria. It further assumes that categorization of the unsafe acts and the unsafe conditions according to different hazard and accident criteria is possible.

As for the active procedure, the passive procedure involves eight steps. First, the categories of unsafe acts and unsafe conditions are identified. The data in each of the categories will be presented in a format similar to the ANSI specifications. The second step is to assign weights or to scale the different categories of unsafe acts and unsafe conditions and their possible

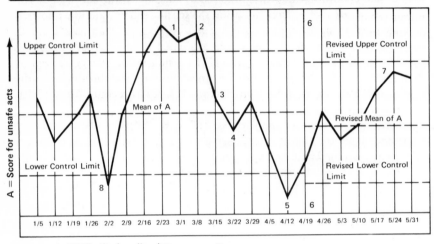

TIME: Week ending date ⟶

1 = Score A is out of control (above upper control limit).
2 = On March 8, the management acts to control safety performance by introducing
    safety programs and new schemes for installation of accessory safety device.
3-4 = Score A drops considerably due to new safety plans.
2-5 = Downward trend indicates steady improvement in safety performance.
  5 = Plant management decides to lower the control limits and mean of A.
6-7 = Upward trend indicates that the workers are steadily neglecting the new safety
    program.
  8 = "Safety inspection" week.

combinations based on criteria such as number of days lost due to accidents
identified with particular unsafe acts or unsafe conditions or the joint occurrence of both of these, assuming that each accident is reported giving the following details:

1. Unsafe acts responsible for the accident
2. Unsafe conditions present at the time of the accident
3. Number of days lost due to the accident
4. Severity of injury according to ANSI specifications

    In assigning weights, hazard indexes are obtained for each category of
unsafe acts, unsafe conditions, and their combinations. To determine the
hazard index for unsafe acts of ith category, we obtain a frequency distribu-

tion for the number of days lost by considering only those accidents reported to have been associated with the presence of unsafe acts of ith category. The hazard index then can be derived from the parameters of the distribution. Either the median of the distribution or the second moment about the axis passing through zero number of days lost may be used to define hazard index. After the distribution curves of accidents with the criterion of days lost due to accidents are obtained for each of the ith unsafe acts, j^{th} unsafe conditions, and for each of their possible combinations, then a suitable method can be arrived at to obtain the corresponding hazard indexes.

In the third step a frequency analysis for an unsafe act and unsafe condition is made. In the BLS and NSC accident causal reports such analysis has been conducted for certain industries and operations.

The fourth step involves designing an activity sampling study to determine the percentage of time a worker is engaged in unsafe acts and unsafe conditions. This technique has been used successfully in the past (Rockwell, 1959; Tarrants, 1965). A study very similar to behavior sampling can be conducted, but since we are recording unsafe conditions (hazardous environment) and the joint occurrences of unsafe acts and unsafe conditions, as well as unsafe acts (worker behavior), it seems more appropriate to identify this technique as "behavior and environment sampling." This method consists of making a number of observations of behavior and environment at random points in time. Unsafe acts and unsafe conditions are recorded according to a previously established classification scheme (for example, according to ANSI specifications).

It is important that every point in time has an equal chance of being selected as an observation time, for this allows a valid inference to be made concerning the state of the entire population from which the sample was chosen. Thus, it is possible to compute either the percent of workers involved in unsafe acts or the percent of time the observed workers are behaving unsafely.

In this application the above method is suggested to obtain estimates of the percent of time a worker spends in each of the categories of unsafe acts and unsafe conditions. In Table 11-1 (p. 211) an example of categories of unsafe acts and unsafe conditions is presented. The categories described are applicable for assessing the safety performance of hospital employees (Bureau of Labor Statistics, 1967).

The percentage of time spent in the joint occurrence of unsafe acts of i'^{th} category and unsafe conditions of j'^{th} category is defined as PUAUC $(i'j')$, for $i' \neq 0$ and $j' \neq 0$. The technique of behavioral sampling is explained in detail in Chapter 16.

Four basic assumptions are made in designing and conducting a behavior and environment sampling study:

Table 11-1

i'	Description of i'^{th} Category of Unsafe Acts	Percentage of Time Spent
0	No unsafe act	—
1	Improper use of body parts	PUA(1)
2	Inattention to footing	PUA(2)
3	Failure to secure or warn	PUA(3)
4	Failure to use protective equipment	PUA(4)
5	Taking unsafe position or posture	PUA(5)
6	Unclassified	PUA(6)

j'	Description of j'^{th} Category of Unsafe Conditions	
0	No unsafe condition	—
1	Defects of agencies	PUC(1)
2	Inadequate help for lifting	PUC(2)
3	Restive patients	PUC(3)
4	Placement hazards	PUC(4)
5	Unclassified	PUC(5)

1. The observations are independent.
2. The observer has the ability to dichotomize between "safe" and "unsafe," is familiar with the operations of the study area, and can recognize and categorize the observed unsafe acts and unsafe conditions.
3. The person designing the behavior and environment sampling study has knowledge of statistical sampling theory and can determine the following:
   a. Type of sampling plan
      (1) Simple random
      (2) Stratified random
   b. Precision and confidence level desired
   c. Number of observations required
   d. Economic considerations
   e. Scheme for randomization of observations over working period according to type of sampling plan
   f. Scheme for taking observations
4. The workers do not change their behavior when an observation is made. This can be assured by either of the following two methods:
   a. The workers are unaware that they are being observed.
   b. The workers are informed that they are being observed and are convinced about the objective of the study and are interested in safety to the extent that they are willing to participate in a cooperative manner.

The most important aspect of designing the behavior sampling and environment study is the design of the observation sheet. A specimen of an observation sheet is illustrated in Figure 11-6 (p. 214). Since the sampling study consists of actually observing the workers and the work area at random points in time, it is often a difficult task to observe all workers. In such cases only a few workers can be selected for observation. The workers can be stratified according to their job description, and a proportional sampling plan can be used for defining the stratas so that the sample consisting of selected workers will represent the population of workers in the entire department under study.

In Step 6 the observer observes selected workers and completes an observation sheet for each observation. The number of observations, the time of observation, and the total period during which the observations are taken are decided before the study. After the study is conducted, the observations can be summarized and recorded on a Summary Record Sheet similar to that presented in Figure 11-7 (p. 215).

The fifth Step, conducted concurrently with step 6, deals with the development of a scheme (method) for scoring. The scores obtained from this scheme are denoted by $\overline{A}$, $\overline{B}$, and $\overline{C}$, which correspond to the unsafe acts, the unsafe conditions and the joint occurrences of unsafe acts and unsafe conditions, respectively. The scoring method is similar to that used in the active procedure except that the percentage of time in which employees are engaged in unsafe acts and the percentage of time in which unsafe conditions exist is the base instead of actual frequencies. The method for scoring is as follows:

N = total number of workers in the department

n = total number of workers observed

$HUA(i')$ = hazard index for i'^{th} category of unsafe acts

$HUC(j')$ = hazard index for j'^{th} category of unsafe conditions

$PUA(i')$ = percentage of time spent in i'^{th} category of unsafe acts

$PUC(j')$ = percentage of time spent in j'^{th} category of unsafe conditions

$HUAUC(i'j')$ = hazard index for joint occurrence of i'^{th} category of unsafe acts and j'^{th} category of unsafe conditions

$PUAUC(i'j')$ = percentage of time spent in joint occurrence of i'^{th} category of unsafe acts and j'^{th} category of unsafe conditions

I' = set of categories of unsafe acts

J' = set of categories of unsafe conditions

Figure 11-6
Observation Sheet for the Passive Procedure.

Sheet No.

Observer:

Date:

Time:

Obs. No.

Department: Section:

Unsafe Condition	Worker Code Number	Unsafe Acts										
		0. No unsafe act	1. Improper use of body parts	2. Inattention to footing	3. Failure to secure or warn	4. Failure to use protective equipment	5. Taking unsafe position or posture	6. Working under suspended loads	7. Improvising unsafe ladders and platforms	8. Misuse of air hose	9. Wearing improper or loose clothes	10. Other (unclassified)
0. No unsafe condition	XYZ											
	ZNK											
	OT											
	ABC											
1. Defects of agencies	XYZ											
	ZNK											
	OT											
	ABC											
2. Inadequate help	XYZ											
	ZNK											
	OT											
	ABC											
3. Placement hazards	XYZ											
	ZNK											
	OT											
	ABC											

Figure 11-7
Summary Record Sheet for the Passive Procedure.

SUMMARY RECORD SHEET								
Period Ending								
Unsafe Act	Unsafe Condition	Periodic Total	Total to Date	% to Date	Periodic Total	Total to Date	% to Date	
UA (0)	UC (0)							
1	0							
2	0							
3	0							
4	0							
5	0							
6	0							
7	0							
6	2							
7	2							
8	2							
9	2							
10	2							
0	3							
1	3							
2	3							
3	3							
4	3							
5	3							
6	3							
7	3							
8	3							
9	3							
10	3							
Total								

The three scores would be:

$$\overline{A} = \frac{N}{n \times 100} \cdot \sum_{i \epsilon I'} HUA(i') \cdot PUA(i')$$

= score for occurrence of unsafe acts

$$\overline{B} = \frac{N}{n \times 100} \cdot \sum_{j \epsilon J'} HUC(j') \cdot PUC(j')$$

= score for occurrence of unsafe conditions

$$\overline{C} = \frac{N}{n \times 100} \cdot \sum_{\substack{i \epsilon I' \\ j \epsilon J'}} HUAUC(i'j') \cdot PUAUC(i'j')$$

= score for the joint occurrence of i'^{th} unsafe act and j'^{th} unsafe condition

Note that the values of $\overline{A}/N$, $\overline{B}/N$, and $\overline{C}/N$ obtained by using this scoring scheme correspond to the average weighted hazard index (loss) per worker per observation made in the behavior and environment sampling study due to unsafe acts, unsafe conditions, and their joint occurrences respectively.

Step 7 is evaluation of the safety performance score. This is done by using the Scheme for Scoring with the data recorded on the Summary Record Sheet. The three safety performance scores $\overline{A}$, $\overline{B}$, and $\overline{C}$ are then checked for significant correlations with existing safety performance measures—for example, (1) minor injury frequency, (2) lost-time accidents, (3) recordable injuries, and (4) production efficiency—and if found satisfactory, they are plotted on the three different control charts (Step 8). From these control charts inferences are drawn, and the necessary corrective actions are taken to improve accident prevention programs.

LINKING ACTIVE AND PASSIVE PROCEDURES

The important aspect of developing the type of dual procedure approach discussed in this chapter is the advantage achieved in obtaining more reliable information about safety performance. The scores obtained from applying both procedures serve as a check for each procedure. Since the active procedure relies on current rather than historical data for developing the scoring scheme, it should be favored for decision-making, and the passive procedure should be used as a check. The frequency of using such checks depends upon the support and accuracy of the active procedure. Thus, to insure that the active procedure yields scores that reflect true safety performance, a cross-correlation of the three sets of scores (1) A and $\overline{A}$, (2) B and $\overline{B}$, and (3) C and $\overline{C}$ should be computed. Another important factor is that since a behavior and environment sampling study will require a very large number of observations to obtain sufficiently high precision for the small percentages of time in which unsafe acts and unsafe conditions exist, it should be used sparingly compared to the active system which is comparatively inexpensive.

INTERPRETATION OF THE SCORES

As discussed in the previous section, the active procedure will be more practicable in assessing safety performance, and therefore, it is important to make a few comments on "deriving inferences" from the scores A, B, C. The main purpose in having three separate scores is to provide a better indication of safety performance for use by the evaluator. The score A, being evaluated by measuring only the unsafe acts, mainly reflects the unsafe working habits of the workers in the individual jobs and the effectiveness of safety programs. If it is found that the score A remains out of control limits even after introducing new safety programs, one should investigate the work methods, the equipment, and the design of the work place, since the causes of unsafe acts may be related to those factors where safety was possibly overlooked. Problems of this type can be detected by observing the score C, which would be sensitive to the effects of safety programs; workers; and engineering, maintenance, and housekeeping. Changes in score B would be more sensitive to changes in plantwide general safety activity.

The safety engineer should use caution in drawing inferences and initiating control actions on the basis of the scores A, B, and C on the control charts. Any interpretation should be based on detailed investigation. Perhaps the simplest way to investigate would be to review the checklists individually and then track down the occurrences of unsafe acts and the unsafe conditions that contribute the most to pushing the safety score outside the control limits.

The safety engineer should also be alert to any changes in worker behavior toward safety that occur after introducing an active procedure or countermeasure. Since the workers are asked to fill out the checklists, they are made aware of what is unsafe every time they review the list. It may be discovered that simply providing this checklist will improve the safety of the department. Therefore, it may not be surprising to find that after repeated use, the checklists and the whole active procedure will become ineffective. It is, therefore, advisable to be continually on the lookout for this effect and to make changes in the checklists from time to time to assure that they provide useful information and are current with existing unsafe practices in the work area.

CONCLUSIONS

The procedure of continuous assessment of safety performance without waiting for accidents to occur has several advantages. The active procedure provides for worker participation and awareness. The method of evaluating three separate scores will help in identifying specific weaknesses in the safety program. The effectiveness of the active procedure, as mentioned earlier, was demonstrated in a study conducted by Rockwell (1968) with the cooperation of the 166th Tactical Fighter Squadron of the Ohio National Guard. The results were encouraging. It appears that the active procedure coupled with occasional cross-checking with the passive procedure provides an effective tool for assessing and controlling safety performance.

References

Bureau of Labor Statistics, U.S. Department of Labor. *Work Injuries and Accident Causes in Hospitals.* BLS Report No. 341. Washington, D.C.: U.S. Government Printing Office, 1967.

Rockwell, T. H. Safety performance measurement. *Journal of Industrial Engineering, 10*:12–16, 1959.

Rockwell, T. H. *Problems in the measurement of safety performance.* Unpublished paper presented at the Symposium on Measurement of Industrial Safety Performance, May 1966.

Rockwell, T. H., and T. R. Clevinger. *Development and application of a nonaccident measure of flying safety performance,* Unpublished paper, The Ohio State University, 1968.

Tarrants, W. E. *An evaluation of the critical incident technique as a method for identifying industrial accident causal factors.* Unpublished doctoral dissertation, New York University, 1963.

Tarrants, W. E. Applying measurement concepts to the appraisal of safety performance. *Journal of American Society of Safety Engineers, 10*(5):15–22, 1965.

Tarrants, W. E. *Research.* Report on the Proceedings of the Research Workshop of the Symposium on Measurement of Industrial Safety Performance, Sponsored by the Industrial Conference of the National Safety Council, Chicago, May 1966.

Chapter 12

SAFETY PERFORMANCE MEASUREMENT: MEASURING SAFETY ATTITUDES

Harold M. Schroder

Safety performance, like all other performance, is a function of the interaction between P (person) and E (environmental) factors. In this chapter the focus is upon P variables. This emphasis is not meant to reduce the importance of E factors; however, the very *achievement* of a "safe" environment rests on certain characteristics of persons at all levels in the industrial enterprise. It will be argued that the achievement of safety may be understood developmentally as the evolvement of more mature safety attitudes that are expressed in the search for safer environments and the reduction of unsafe behavioral acts. Like other attitudes, safety attitudes are viewed as information-processing structures. Such attitudes act much like a computer program in determining the way an individual selects, stores, processes, and transmits information in the safety domain.

What do we mean by the development of more mature safety attitudes? Following the studies of the development of moral concepts by Piaget (1932) and Kohlberg (1958, 1964), it is not simply a question of the value of safety. Most people express dedication to the value of increasing safety in industry, on the highways, and so forth. It is rather a question of the *nature* of this belief about the value of safety. If the belief is absolutistic (expressed by such slogans as "act safely"), if safety is viewed as residing either in the person *or* in environmental conditions (instead of some interaction between the two), if it does not imply internal causation (the conception that one can have an effect on enhancing safety), and if it does not imply individual responsibility—then it is more immature. "Immaturity" refers to the conceptual properties that evolve for processing information, in this case, in the safety domain. It has been described as the level of complexity of conceptual structure for processing information in a given domain (Schroder, Driver, and Streufert, 1967).

In any domain (for example, safety) the level of conceptual development of attitudes is defined in terms of two parameters—(1) differentiation and (2) organization of information. These will be dealt with in turn.

DIFFERENTIATION

Differentiation refers to the number, nature, and weighting of attributes, categories, dimensions, or kinds of information a person uses in making judgments and in thinking in the safety domain. It refers to the number of kinds of information the person takes into account. For example, one person may consider worker carelessness, worker dullness, and negativity of attitudes toward foremen as the three basic attributes. Persons scoring high on each dimension would be viewed as decreasing the safety level. Another person may use a different set of attributes of information in making safety judgments, for example, organizational distractions, unsafe machines, or atmospheric characteristics.

Thus, the initial question in the measurement of safety attitudes is to determine the number of independent classes (or scales) of information the person selects as being relevant to the safety issue in any situation. If possible, estimates of the weighting of such dimensions should also be assessed. In this way we can discover how different persons and groups in industrial situations perceive the safety problem. We can identify differences and compare the dimensionality of perception of any person or groups of persons with the results of scientific studies which attempt to specify objectively the factors relevant to safety performance in a given area.

The Measurement of Dimensionality

In measuring the number of dimensions involved in performance, it is erroneous to proceed by the traditional method of presenting the subject with a large number of scales and then requesting him to indicate the degree to which each is relevant. In such procedures the presentation of the attributes may influence the selection and consequent ratings. This problem can be overcome, to some degree, by using methods that force the subject to select or generate his own dimensions. A number of projective or sorting procedures should be designed to estimate the kind of attributes persons use in a particular safety situation. These procedures are fairly well known and will not be described in detail here. They suffer from a lack of objectivity, they are difficult to score, and when different scores are involved, comparative analyses are usually not legitimate. Two additional but more objective methods for measuring dimensionality are *multidimensional scaling* and *free-adjective description.*

MULTIDIMENSIONAL SCALING. Individuals make global judgments about the degree of similarity (or dissimilarity) between pairs of stimuli when they use multidimensional scaling (MDS). In the safety domain certain relevant stimuli would be selected—machines, people, production processes—and individuals would be asked to judge the degree of dissimilarity between pairs of such stimuli on the basis of safety. The resulting matrix of judgments are analyzed in order to discover the number of dimensions necessary to generate the similarity judgments (Gulliksen, 1964; Kruskal, 1964; Shepard, 1962; Torgerson, 1958; Tucker and Messick, 1963).

In this geometric model a set of stimuli or objects are viewed as points lying in some multidimensional (usually Euclidean) space. The basic notion is that judged psychological similarity between pairs of objects may be likened to a "distance" between the points representing these objects in space. With a set of objects that may vary along an unknown number of perceptual dimensions, multidimensional scaling attempts to determine the location of points in a space of minimum dimensionality such that the distances between points in this space correspond closely, in some sense, to the judgments of similarity or dissimilarity among the objects. The goal of this procedure is to provide a description of the set of objects in terms of empirically discovered, rational dimensions that in some sense represent the "true" underlying aspects of the stimuli.

An example taken from the international domain might make this clear. In this example (Schroder and Blackman, 1965) we were interested in discovering the dimensions used by individual A in making judgments about national governments. He was presented with all possible pairs of 10 governments—$\frac{N(N-1)}{2}$—and requested to judge how dissimilar each pair was using a nine-point scale extending from very similar to very dissimilar. For $N=10, \frac{10(9)}{2} = 45$ pairs. These 45 judgments were then analyzed using the Torgerson (1958) program. The latent root data demonstrated that a two-dimensional space accounted for 92 percent of the variance in the judgmental space and that the dimensions were approximately equally weighted. The plot is reproduced in Figure 12-1 (p. 222). The projections of the nations onto each dimension gives the ordering of these stimuli along each dimension. The dimensions are interpreted as democracy-autocracy and capitalism-socialism. Dimensions can be interpreted in a number of ways including (1) by interviewing the respondents afterward and (2) by having the subject list descriptive adjectives for each stimulus during the testing period and using

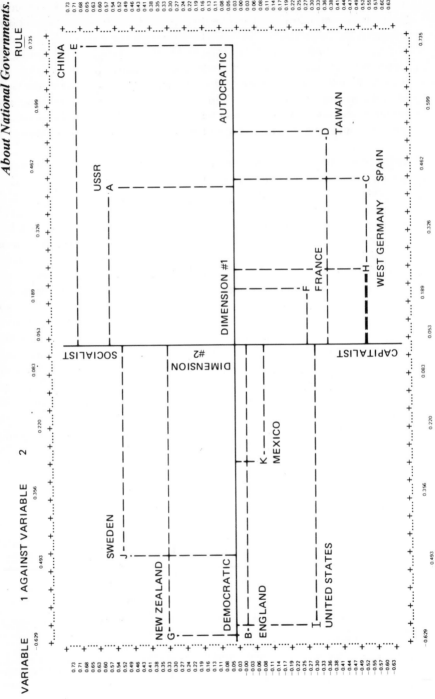

Figure 12-1
Multi-Dimensional Scaling: Dimensions Used in Making Judgments About National Governments.

222

these adjectives as a basis for interpretation. Interpretations made on this basis (or on any other basis) can be tested by asking the subject to order the stimuli according to the hypothesized attribute and then correlating this with the ordering obtained by MDS. The higher the correlation, the more confidence we can place in the label given to a dimension.

This method can be used to discover the dimensions that people use in making judgments and choices in safety situations. For example, the stimuli could be industrial tasks or roles, types of industry, or kinds of machines or persons. Respondents would be asked to judge degree of dissimilarity between pairs of persons or pairs of roles in terms of the level of safety present. The purpose of the study would determine the class of stimuli to be used.

In the past, MDS has been employed mainly in studies of the way physical *stimuli* are judged. More recent developments, however, have shown the technique to be potentially very valuable in other fields. Validation work is reviewed by Jackson and Messick (1963), Kennedy et al. (1966), and Schroder, Driver, and Streufert (1967). A study by Faletti (1968) also demonstrated that when all stimuli are judged along all dimensions, the method provides a reasonably accurate representation of the number, nature, and weighting of dimensions used in making judgments in person perceptions.

FREE-ADJECTIVE DESCRIPTION.   A second approach to the measurement of structural properties is a modified method of categorization (Triandis, 1964). In the Adjective Description Test (Warr, Blackman, and Schroder, 1968) individuals are asked to respond to each member of a stimulus set (persons, components) with as many relevant descriptive adjectives as possible. Each person is also asked to rate each stimulus in terms of the dimension expressed by the adjective. For example, if he lists reliability as a characteristic of stimulus A, he is asked to rate stimulus A on its degree of reliability.

A dictionary is then constructed that defines and labels single categories into which all adjectives judged to have similar meanings are placed. The dictionary can then be used to translate a set of free-adjective descriptions into a set of dimensions, each of which is judged to represent a different meaning.

In a study of the dimensionality of political judgments (Warr, Blackman, and Schroder, 1968) respondents were asked to note and record five aspects or adjectives describing each of the 10 governments in the MDS example given above. After all responses had been made, the subjects were asked to place a number (on a scale from 1 to 11) against each description to indicate the importance of that aspect to their judgment.

It will be noted that this too is an open-ended instrument in that we do not specify the dimensions that judges are expected to use. From it we obtain 50 descriptions (10 stimuli, 5 responses to each), which are then classified. Once classification has been completed, we can examine a subject's responses from a number of standpoints. For example, we can ask how many dimensions the individual is employing. This index could simply be the number of classes of responses in the person's descriptions. It may be that his 50 descriptions are really somewhat redundant, so that we conclude he is only utilizing 6 separate dimensions. Or if he makes responses which differ more widely, he may be using 20 different dimensions.

As with MDS, we can also obtain from this measure estimates of the relative importance of a judge's dimensions and, of course, of their nature. Free-descriptive responding (FDR) differs as a technique from MDS in that MDS will only generate general dimensions in the way described above, but FDR also identifies stimulus-specific dimensions.

For example, if we conclude from the MDS analysis that strong-weak is a dimension that a particular subject uses to conceptualize governments, we have to make clear that this is a dimension that he applies to *all* the stimuli in the set. FDR, on the other hand, can tell us about dimensions that a judge applies only to one or a few governments. Thus, he might respond that the British government is "keen to retain NATO" or that the Spanish government is "not dominated by a king." These dimensions are applied specifically to individual stimuli and would not emerge at all from the MDS analysis. Thus, estimates of general dimensions (those applied to all stimuli) but also of specific ones (those applied to less than all stimuli) can be derived.

However, a system is needed for classifying the free responses. As part of this research, a manual was developed for coding data from the FDR task. This code embraces nine major categories of response, each of which is subdivided into several dimensions of judgment. The nine broad categories are:

1. Governmental structure in general
2. Specific aspects of structure
3. Internal policies in general
4. Internal policies in specific
5. Internal characteristics
6. Position in the world
7. Foreign policies
8. Alliances and ties to other governments
9. External characteristics

The total number of specific dimensions covering these nine groups is 53. For example, there are seven dimensions in the first group (governmental structure in general): (1) democratic-autocratic, (2) constitutional status, (3) capitalist-communist, (4) nature of leadership, (5) nature of legislative body, (6) internal organizational characteristics, and (7) presence of ideology.

In analyzing the FDR material, each of the 50 descriptions is coded into one of these 53 dimensions. The dimension democratic–autocratic, for example, includes all responses having to do with the extent of governmental control. In this way it embraces most comments about elitism, oligarchy, despotism, repression, dictatorship, and so on. We have found that the manual employed is satisfactory for British and American subjects. Pilot studies with French and German samples have also been run, and in all cases every response was readily coded according to this system.

This method could be suitably modified for application in the safety field in order to discover the kinds of information individuals use in making safety (risk-taking) judgment. Having selected a set of stimuli relevant to the purpose of the study, respondents would be requested to note and record words, adjectives, or short phrases describing safety factors associated with each stimulus. These data could then be processed in order to discover the number and nature of unique dimensions used by individuals or groups of individuals and their relative importance.

Implications of Measures of Dimensionality

Measures of dimensionality that do not impose dimensions on the respondent provide methods for discovering the kinds of information (dimensions) that persons actually use as a foundation for thinking and judging in the safety domain. The successive applications of these methods over time can provide accurate information about the nature and kind of change occurring in safety attitudes over time—or as a function of the introduction of safety programs, new production methods, organizational changes, and so on.

Change may be objectively plotted in the following terms.

1. *The number of dimensions used in judgment.* Following a given program, we might expect (and use the above methods to check) the emergence of a new dimension—a new kind of information—in making safety judgments. In other cases irrelevant dimensions might be expected to drop out. Only when the population involved uses relevant information in judgment can we expect improved safety performance.
2. *The weighting of dimensions used in judgment.* Measures of dimensional properties can be used to study the effects of policies or programs on the way persons weight certain kinds of information in making safety judgments.
3. *The discrimination of stimuli along relevant dimensions.* Dimensions are ordering systems or scales along which people rate stimuli. Under some conditions attitude change may be expressed simply as the change in scale value of certain stimuli. For example, a certain machine or process may move from high to low on one or more dimensions due to the introduction of some safety feature or procedure.

Information about the dimensional properties of safety attitudes is critical for planning and policy making in the safety field. Only when we are appraised of the number and nature of dimensions used in safety performance by a particular population are we in a position to plan and design specific goals. When goals can be stated specifically (such as adding a new dimension and reducing the weighting of others), the effectiveness of programs can be evaluated and their effects observed on specific aspects of safety performance.

Ideally, we would like a continuous assessment of dimensional properties of safety attitudes and continuous measures of safety performance. If we assume such measures of performance are sufficiently sensitive, this would provide the setting for longitudinal validity studies in any industry. Fortunately, techniques are available for reliable and sensitive measures of safety performance. Tarrants (1964) has described the critical incident technique for studying environmental safety hazards (E factors) and the behavior sampling method for identifying unsafe behavioral acts (P factors). These two methods yield data that can be categorized into a set of attributes or dimensions relevant to safety performance in a particular situation.

Successive measures of both the actual person and environmental factors involved in safety performance and of the dimensional properties of individuals' safety attitudes provide relevant information for policy and planning. Such information does three things:

1. Pinpoints the match or mismatch between the information relevant to individuals in their perceptions of safety and to the actual factors involved in safety performance
2. Specifies the precise nature of the mismatch and defines a specific program goal
3. Provides the basis for carrying out validity studies showing the effects of attitude structures on safety performance

ORGANIZATION OF INFORMATION

The second major parameter of attitude structure refers to the way dimensional scale values of information are organized. Obviously, the number and weighting of relevant kinds of information are important as a foundation for performance, but no amount of relevant information can be effective if it cannot be combined for effective and adaptive use. A high level of safety performance requires the development of increasingly complex conceptual properties for processing safety information.

Conceptual complexity may be defined as extending from single combinatory rules at one end of the continuum to multiple connected combinatory rules at the other. (Schroder, 1968).

At the simplest level of organization only a single principle of organization exists: the individual generates but a single perspective of the safety domain—an egocentric perspective anchored in one's own needs and goals. Single perspective thinking might be characterized by viewing safety simply as a function of persons or as some inevitable consequence of the environment. It is characterized by closeness to conflicting information, avoidance of search that could produce conflicting information, and the resolution of conflict or dissonance by a reaffirmation of the single perspective. This level of information processing in a domain is more basically expressed as

1. Minimal feelings of internal causation (that is, one can influence the achievement of safety by one's own efforts or by exploratory means)
2. Maximal tendency to externalize blame (accidents are inevitable or due to causes beyond one's control)
3. Warding off conflict (for example, misinterpreting data or feedback in order to bolster one's belief)

A person with such simple and closed conceptual processes may profess a high degree of dedication to safety, but the direction and aims of safety policies that are likely to evolve are self-defeating in that they (1) are narrow and constrict the range of search and inquiry and (2) fail to develop machinery for scientific study and follow-up.

At more complex levels of organization the person is capable of combining attributes of information in different ways. He is capable of generating multiple perspectives. Such multiple perspectives are expressed in the ability to organize and process information from another person's point of view. Higher levels of information processing involve the consideration of the related implications of conflicting perspectives. Conflicting perspectives may evolve from different scientific points of view from the production line, management, the community, and so on.

This development from a more absolutistic to more conceptually complex stages has been studied and documented by Piaget (1932, 1954), Kohlberg (1958, 1964), and Brunner et al. (1956). The progression has been studied particularly in reference to moral concepts. More advanced stages of conceptual development in any domain are reflected in

1. The internalization of the concept of "cause" in the safety domain. The ability to generate alternative and conflicting perspectives opens up a problem-solving orientation.
2. The assimilation of a greater range of safety information into various perspectives that are available for use in thinking.
3. Exploration, search, and the formulation of new perspectives.

Measurement of Organizational Properties

Many measures of "cognitive complexity" (Bieri, Atkins, Bieriar, et al. 1966; Kelly, 1955; Scott, 1963) are more precisely measures of differentiation—namely, the number of dimensions used in judging or thinking. Research by Vannoy (1965), Faletti (1968), and Anderson and Gardiner (1968) demonstrates the independence of measures of differentiation and the complexity of concepts for organizing scale values of information.

Objective tests of organizational properties have not proved fruitful. It appears that we must engage persons in information processing and utilize process measures in order to assess their properties. At this time the most effective method of assessing structural properties in the interpersonal domain is the paragraph completion test. In this test persons are asked to complete sentences that express interpersonal conflict or uncertainty ("When I am criticized, . . . :" "When I am unsure, ..."). Respondents write at least three or more sentences, and the responses are scored by raters instructed to ignore the context of the response and to use a manual to rate the degree to which the response was generated by (1) multiple perspectives and (2) the degree to which the perspectives are connected (as opposed to being compartmentalized). A number of studies in the area of complex social interaction have demonstrated the validity of this scoring procedure (Schroder, 1968).

The paragraph completion test could be modified to assess the general level of conceptual development in the safety domain. The initial problem would be to produce stems that would generate responses representing samples of information processing. An example of such a stem might be, "When safety procedures are changed, ..." Responses would be rated in terms of (1) the number of different perspectives a respondent produces, (2) the presence of conflict between perspectives, and (3) the degree to which these discrepancies are connected in thinking about the safety domain.

Implications of Organization of Information

The importance of considering the organizational properties of safety attitudes can be illustrated by the following example. Assume that in an industrial setting the dimensional properties of safety attitudes held by the population involved accurately match the actual safety hazard and unsafe behavior categories discovered in performance. In this situation programs involving safety attitudes can have only minimal effects. The remaining strategy is to preach the value of safety and attempt to induce people to act more safely. Such programs soon become boring and may even desensitize the recipients. The point here is that while the particular industrial setting is more or less accepted as inevitable—as fixed, that is, while it is viewed from one vantage point—interest in and responsibility for safety are reduced. When the organizational properties of attitudes are more complex, production and organizational practices are viewed from many different perspectives; alternative methods, procedures, and goals are more central; search and exploration are broad; and diverse options are generated.

At this level of development the broad manifestations of safety become integrated with overall political-industrial policy. Production methods, organizational and personnel practices, organization planning, and so on, are considered in the framework of man's interaction with his environment—both in and out of the work situation. Measurement of the level of conceptual development of safety attitude of decision makers is particularly relevant for predicting the extent to which the individual's well-being will be enhanced in future work environments.

CRITERIA FOR MEASURING SAFETY PERFORMANCE

Just how higher levels of conceptual development are to be accomplished in the safety domain is not of concern here. However, conditions that enhance such development have been studied by Piaget, Kohlberg, Schroder, and others. One of the major prerequisites for such development is provision of the opportunity for persons to experience accurate and reliable feedback as a consequence of their actions. In the safety area this would require the development of sensitive criteria against which safety performance can be continuously judged.

Criteria such as accidents—or other rare events—do not provide a responsive environment conducive to the development of a "mature safety conscience." Such development would require the continuous availability of sensitive, scientifically based feedback against which performance could be evaluated. The criteria suggested by Tarrants would appear to begin to meet the requirements of such development. However, broader criteria measuring the effects of the industrial or work environment on all aspects of human well-being should also be taken into account.

References

Anderson, C. C., and G. S. Gardiner. *Some Correlates of Cognitive Complexity.* 1970.

Bieri, J., A. L. Atkins, S. Bieriar, et al. *Clinical and Social Judgment: The Discrimination of Behavioral Information.* New York: Wiley, 1966.

Bruner, J. S., J. J. Goodnow, and G. A. Austin. *A Study of Thinking.* New York: Wiley, 1956.

Faletti, M. V. *An experimental validation of some measures of cognitive complexity.* Unpublished senior thesis, Princeton University, 1968.

Gulliksen, H. The structure of individual differences in optimality judgments. In M. W. Shelly and G. L. Bryan (Eds.), *Human Judgments and Optimality.* New York: Wiley, 1964.

Jackson, D. N., and S. Messick. Individual differences in social perception. *British Journal of Social and Clinical Psychology,* 2:1–9, 1963.

Kelly, G. A. *The Psychology of Personal Constructs.* Vol. 1, *A Theory of Personality.* New York: Norton, 1955.

Kennedy, J. L., B. L. Koslin, H. M. Schroder, et al. Cognitive patterning of complex stimuli: A symposium. *Journal of General Psychology, 74:*25–49, 1966.

Kohlberg, L. *The devlopment of modes of moral thinking in the years ten to sixteen.* Unpublished doctoral dissertation, University of Chicago, 1958.

Kohlberg, L. Development of moral character and moral ideology. In J. Hoffman and L. Hoffman (Eds.), *Review of Child Development Research,* Vol. 1. New York: Russell-Sage, 1964, pp. 383–431.

Kruskal, J. B. Nonmetric multidimensional scaling: A numerical method. *Psychometrika, 29:*115–129, 1964.

Piaget, J. *The Moral Judgment of the Child.* New York: Free Press, 1932.

Piaget, J. *The Construction of Reality in the Child.* New York: Basic Books, 1954.

Schroder, H. M. Conceptual Complexity and Personality Organization. In H. M. Schroder and P. Suedfeld (Eds.), *Information Processing: A New Perspective in Personality Theory.* 1968.

Schroder, H. M., and S. Blackman. *The Measurement of Conceptual Dimensions.* Technical Report No. 16. Princeton, N.J.: Office of Naval Research, Princeton University, 1965.

Schroder, Harold M., M. J. Driver, and S. Streufert. *Human Information Processing.* New York: Holt, Rinehart and Winston, 1967.

Scott, W. A. Conceptualizing and measuring structural properties of cognition. In O. J. Harvey (Ed.), *Motivation and Social Interaction.* New York: Ronald, 1963, pp. 266–288.

Shepard, R. N. The analysis of proximities: Multidimensional scaling with an unknown distance function. *Psychometrika, 27:*125–140, 1962.

Tarrants, W. *Applying measurement concepts to the appraisal of safety performance.* Paper presented at the 33rd Eastern Regional Safety Convention, New York City, 1964.

Torgerson, W. A. *Theory and Methods of Scaling.* New York: Wiley, 1958.

Triandis, H. C. Cultural influences upon cognitive processes. In L. Berkowitz (Ed.), *Advances in Experimental Social Psychology.* New York: Academic Press, 1964.

Tucker, L. R., and S. Messick. An individual difference model for multidimensional scaling. *Psychometrika, 28:*333–367, 1963.

Vannoy, J. S. Generality of cognitive complexity-simplicity as a personality construct. *Journal of Personality and Social Psychology, 2,* 385–396, 1965.

Warr, P. W., S. Blackman, and H. M. Schroder. The structure of political judgment. *British Journal of Social and Clinical Psychology,* 1968.

Part III

CONCEPTS AND METHODOLOGIES OF SAFETY PERFORMANCE MEASUREMENT

Chapter 13

MAJOR ISSUES AND IDENTIFIED PROBLEM AREAS

Numerous major issues relative to safety performance measurement are presented in list form and discussed briefly in this chapter. While approaches to resolving the issues are not always identified or specifically defined, insights are often introduced that provide "food for thought" and perhaps suggest directions for future research on this subject. Some issues are conceptual or methodological statements. Collectively, these statements serve to highlight the safety measurement problems and issues and provide guidance in the search for solutions.

THE MAJOR ISSUES

1. Measures should be developed that will permit the identification of accident-prone work situations and that will lead to the redesign of such situations.
2. One measurement approach is to define various operations in terms of the number of opportunities for error in human behavior that could lead to some type of injury or damage.
3. A good denominator needs to be developed. It is easy to obtain the numerator in an index fraction, but it is much more difficult to define a good denominator. It is one thing to identify the number of accidents, but it is even more important to identify the number of opportunities for accidents. In the safety field very little effort seems to be devoted to measuring the number of opportunities to have an accident. Accurate predictions cannot be made without knowing the denominator.
4. One approach to accident prevention is to measure the accident potential of conditions and task requirements and to reduce the accident potential through improved design of man-machine-environment systems.
5. Safety professionals are overly preoccupied with a narrow kind of mea-

surement precision for interplant comparison purposes. More important is the need for measures to aid managers in making in-plant safety decisions.

6. There is a need to start examining the individual's entire life-style to determine where in other areas of his life he is likely to act unsafely. There is some indication that the safer the job is made in the industrial setting, the more likely it will be that the person with an accident propensity will have accidents outside the plant. Out-of-plant accidents are related to what goes on inside the plant and should be measured also.

7. The topic of measurement should be expanded to include not only the measurement processes or models that use accidents as dependent variables, but also the more orderly processes that, it is believed are related to accidents, such as job satisfaction.

8. Any measurement technique must fit into the existing system of power and power relationships. Consideration must be given to who is going to make decisions and take the action that comes out of a measurement result. Within the power structure of how things get done, safety measurements will either be adopted and used or be quietly put into a file and forgotten. When measurement techniques are considered, one should also consider what is to be done with the results.

9. Measurement should help one define a class of technical problems that can be translated into the models, procedures, or both, that will ultimately result in management decisions for safety improvements.

10. The major issue in safety measurement is the criterion problem. What do we want to measure?

11. While it is necessary to measure loss-producing accident events after they happen, it is more important to measure problems before an accident occurs so that corrective action can be taken in advance of loss.

12. Measurement is conducted for the purpose of prediction to reduce risk.

13. The kinds of predictions currently being made are straight actuarial ones, and the lowest level variety of actuarial at that. Frequencies are determined, and one projects that the frequency that will occur tomorrow or during the next time period is going to be similar to the frequency existing today. A better form of prediction is based on a model of the particular process with which we are dealing.

14. There are many ways of predicting that involve generating an index of the vulnerability of an individual to an accident. These predictions take the form of probabilities for an individual to have an accident.

15. Insurance companies continually attempt to reduce their liability on the basis of actuarial analyses of the accident liabilities of certain classes of people. Measuring the idiosyncratic phenomena of individual behavior should be stressed.

16. A measure is needed to detect the occurrence of transitory accident proneness that makes the person incapable of avoiding potential acci-

dents built into the work situation. The measure should identify the combined incapacity of the individual plus the accident proneness of the work situation.

17. Since accidents occur with far greater frequency when a person is under personal stress or the influence of alcohol, methods of measurement are needed that enable one to determine when these states exist and their severity as well.

18. Strategy for reducing the likelihood that people are going to have accidents should involve the measurement of probable events at the man-machine interface. The kinds of things that should be measured at that interface are intrinsically different from the kinds of things that should be measured if we begin by conceptualizing some other kind of accident prevention strategy. Operational objectives must be specified before we can begin to apply improved measurement techniques.

19. Psychologists have developed measurement methods that enable the screening out of certain high accident risks. Based on the work of Jenkins and others, it has been determined that if the number of applicants for a job is large enough and the selection ratio is low enough, it is possible to identify a meaningful percentage of those in the work force who, if they were hired, would be likely to have accidents. This application of preemployment measurement techniques should substantially benefit the accident problem. The issue here is not to eliminate people from the worker population but, instead, to match a worker to a work environment so as to minimize accident risk.

20. Multiple sets of measurement techniques are needed that address both the man-machine interface problem and the unstable employee problem at both the pre- and post-employment stages.

21. One approach to measuring the accident problem is to rate the circumstances of an accident in terms of the likelihood of its recurrence. Prevention resources could then be applied to those problems with a relatively high likelihood of recurrence.

22. The modern quantitative techniques that enable managers to predict the next quarter's incidents of very unique events could be applied to the accident prediction problem.

23. Measures are needed that predict, not simply record, accident occurrences. Historical records of accidents mean nothing unless they can be used for prediction and control purposes.

24. As new social and technological problems appear, new measurement techniques need to be developed and applied that will assist in identifying and solving safety-related subproblems.

25. Emphasis should be placed on determining the cost of accidents. Costs of both personnel and material losses should be measured.

26. Measures of safety performance should lead to the establishment of a scientific basis for safety programming decisions. There is nothing in

the literature that allows one method of prevention to be compared with another or to determine whether there is any possibility that any of the existing accident prevention methods can, in fact, be made to work. The merits of proposed new measurement methods should be scientifically assessed.

27.   Multiple measures should be established that are useful to decision makers in lowering accident risk.

28.   It should be kept in mind that neither the measurement technique nor its products, in themselves, produce safety improvements. Someone must prescribe countermeasures and then take the necessary action to implement them, using the measurement results as a decision-making, evaluative tool.

29.   Since there is a need for a plurality of measures, this implies that there is a plurality of dimensions to the problem. It is not known today how to measure these dimensions simultaneously on different scales. One should begin by listing the dimensions of the problems and scaling them if possible.

30.   It is possible to measure the incidence of accidents themselves, precursor events such as unsafe acts (errors) and unsafe conditions, and the risk of accident inherent in a process or task.

31.   The risk of accidents should be assessed when a new process is contemplated.

32.   Measurement of accident potential (predicting the probability of a catastrophe, for example) can be accomplished by working backward from the speculated accident, examining probable human behaviors and system events and their interactions, and assessing the probability of each of these by means of a "tree" diagram. One can work backward and estimate the probabilities of occurrence of these systems' events and human behaviors, then identify particular pathways with excessively high accident probabilities. This technique should lead to changes in the man-machine-environment system that reduce accident probabilities.

33.   Our culture encourages a certain level of risk taking. People assume a finite probability of accidents that is acceptable in a given situation. Measurement techniques are needed to assist people in assessing the true (objective) probability of accidents on a conscious level, thus helping them to avoid the harmful consequences of inaccurate probability (risk) assessment. People are going to take risks. Valid and reliable measurement data need to be provided to assist them in achieving an optimum level of risk behavior.

34.   Achieving a zero accident frequency rate is a meaningless goal. Managers accept risks of accidents when they accept cost-benefit trade-offs in making decisions at the design and operational stages of production processes. Improved measurement techniques will assist managers in

assessing the merits of various alternatives (trade-offs) and will ultimately lead to better management decisions.

35. Measures of safety performance should assist in *problem solving,* not in fixing blame.

36. A measurement problem exists in that low injury frequency rates tell nothing about the potential for catastrophe. For example, one plant experienced a $45 million accident loss, with its walls covered with safety awards received based on a low frequency rate. Better measurement techniques are needed to identify loss potential.

37. Human errors (near-misses) that have an accident-producing potential need to be measured. The family of probable errors most associated with a particular task or process needs to be identified and measurement techniques applied that will tell when these errors occur.

38. There are management errors inherent in every error-prone work situation. Measures should be introduced to identify both planning and operational errors in the work situation. A measurement system based on an assumed sequence of management events needs to be established that will lower the risk of accidents.

39. The behavior variance associated with human errors that lead to injury and damage or near-injury and near-damage events needs to be measured.

40. Severity rates are heavily weighted by deaths, which are much more rare than other accidents; therefore, these rates are much less stable as measures of safety performance.

41. The measurement process itself may change the outcome of events by measuring potential accident problems before they produce loss; the loss may be prevented by influencing behavior changes or process changes. *Behavior sampling* and the *critical incident technique* might be useful in this regard. Applying the critical incident technique, for example, may produce a positive Hawthorne Effect on the general behavior of all employees.

42. Measurement implies the existence of some sort of scale based on a predetermined criterion. One criterion that might be used in scaling accidents is relative human suffering. Another might be dollar value of losses.

43. Cost benefit or cost effectiveness information should be involved in all safety decisions made by management.

44. In a more sophisticated concept of exposure, some individuals are a lot more "exposed" than others. What about the idea of a scaling of severity in relation to the types of exposure? An exposure index could be computed based on the relative danger of the work environment, task or process. "Exposure" as a concept does not relate exclusively to elapsed time; it includes all the conditions involved.

45. When rates are used as measures, the same technology should be ap-

plied in the denominator of the ratio as is applied in the numerator.

46. Certain kinds of measures can be used by researchers (scientific or cause-and-effect evaluation, for example); other types should be used by management (administrative evaluation, for example). A distinction should be made between the two kinds of measures.

47. The measurement unit should be based on opportunities for errors instead of straight time exposure in the denominator. A base rate might be defined in terms of "per opportunity" for accidents. The point is that different individuals or groups of individuals have different inherent exposures, and these differences should be considered in the measurement approach.

48. The concept of "change" should be considered as a basis for measurement. When a process or operation goes out of control, there is obviously a change. Both planned and unplanned changes occur. Change is pervasive and continuous. Most accident causal spectrums involve change. Changes provide opportunities for accidents as well as for a safer work situation. Change could be used as a management device for diagnosing cause and effect. "Change" should be studied as a means of assessing accident potential. In revising processes or building new plants, known hazards can be designed out of the system. Planned changes present many opportunities for reducing accident potential.

49. Mistakes, errors, and unsafe maneuvers should be measured on a sampling basis. One might observe 10 trials of an operation or task per day or week as an index of safety performance. The times of these observations should be selected at random from all of the possible times when the operation or task is performed. With this measurement technique inferences can be made concerning the "safety state" of the entire universe from which the random sample was selected, within a predetermined, acceptable level of confidence.

50. Criteria contained in state workers compensation laws represent one form of measure. Unfortunately, there are 52 variations in these laws among the states. Nevertheless, data produced by this measure may have value, particularly as applied to operations within a given state.

51. There are many contaminating factors in existing measurement systems. For example, many large companies operating with the ANSI Z-16.1 standard have physicians on the premises who provide immediate treatment. In many instances the injured person can return to work after such treatment, thus avoiding a reportable accident. In small shops an injured person is sent off the premises to an outside physician. The victim may or may not return to work on the same day, depending on the judgment of the physician performing the treatment. Thus the physician can control the disabling injury frequency rate by determining who can return to work. The ANSI Z-16.1 system has this defect.

52. Measures should assess what effort has been made to create safe condi-

tions within the working environment. Present measures do not reflect the efforts or equipment improvements incorporated in the work environment to make it as free from danger as is reasonable.

53. The, method of measurement should provide a basis for taking specific action to correct identified problems.

54. Rates are too crude. They fluctuate too much, they are too unstable, and changes in the rate do not necessarily reflect changes in the level of safety within the organization measured. The rate can fluctuate drastically on a fortuitous basis. Likewise, measures of severity can fluctuate widely by chance. For example, a person can fall and bruise an elbow. One can also fall from the same height and fracture the skull and die. The causes can be the same, but the effects and ultimate results are substantially different.

55. Injury rates vary widely among industries. For example, the frequency rate for the aircraft or aerospace industry is about one disabling injury per million man-hours. On the other hand, a furniture plant may have a good record if it has 15 disabling injuries per million man-hours. Perhaps safety professionals should evaluate types of exposure and establish systems of measurement that consider exposure differential with regard to accident potential. Perhaps different industries should have different systems of measurement. Too much may be expected of a system of measurement when it is applied equally to an industry with a rate 10 to 15 times higher than that of another industry.

56. Why have a national index of any kind? What do managers expect to do with it? What management decisions rely on the need for a national index? A company with a frequency rate above that of its industry would have an incentive for improvement. Safety programs undertaken by associations within an industry have a basis for effectiveness comparison. At the national level there is a need for an industry rate for use as a guide for legislative and regulatory action. Where should the OSHA place its emphasis? Industry frequency rates can help decide this.

57. Managers are the real users of measurement results since they must make the decisions. What kind of information does management need to make safety decisions? Measurement should be based on this need. We need to know how to trade-off X units of production versus Y units of safety. How can we best make the necessary trade-offs between safety and production?

DIRECTIONS FOR THE FUTURE

What conclusions can be reached from an analysis of these major issues?

New measurement systems should meet the needs of an industrial establishment with respect to evaluating its ongoing record. Is it doing any better?

Are improvements being made? Is the establishment maintaining good control over its hazards? Measures of both the behavior of the individual and the nature of the work environment are needed. The environment measure needs to identify the hazards and to bring about their elimination or control. The need is to measure the degree or extent of our safety control and to be able to identify the point in time when the system or process is headed out of control.

Measures must be related to the tasks performed. The potential severity of a hazard should be measured. A boiler explosion, for example, could destroy the establishment and severely injure or fatally injure workers in the immediate vicinity. At the opposite end of the severity spectrum is the hazard imposed every time a worker drives a nail (although this action could have serious consequences if a metal chip puts out an eye). The total system of measurement for hazards must be sensitive enough to include these extremes. Our measurement system is really expected to provide some kind of indicator that will permit managers to identify problems and take corrective action.

Different levels of measurement should be considered. At one level measures provide indicators of something being wrong, but they do not specifically point out what it is. These indicators could then trigger further investigation and problem analysis to identify causal factors. Another level of measurement might reveal what specific corrective action to take. Supplementary measures may vary from one level to another. Ideally, the multiple measures should provide enough information to permit managers to act. They should contain within them the suggestion for action. For example, at one level of measurement the existence of cancer can be detected. The measure does not reveal a cure for cancer, but it performs an important function when it permits accurate location of when and where it exists in the body. Similarly, a measure may reveal a hazardous condition or hazardous behavior. It is the safety professional's responsibility to prescribe corrective action. But he cannot do this if he is not aware when and where a problem exists. At this level it is quite appropriate that the measurement system be conceived and applied independently of any actions that might be taken to correct the problem. The measure can tell the manager what to worry about. National statistics do not do much at any level of measurement for problem definition or problem solution within a particular establishment.

A measurement system is needed that provides information to the line manager who has responsibility for a particular geographic region, plant location, or operation within an organization. A statistical measure should take into consideration the likelihood of chance runs of accidents. Managers frequently jump on these chance runs, get very "uptight," and take some strong actions that may actually worsen the situation. It is questionable whether safety actually is improved by a manager who jumps in and makes drastic changes based on a chance run of events. If anything, the situation may be worsened when a technically sound safety program is altered.

Many accidents are geographically related, yet comparisons are made across geographic regions. Another problem is the time lag between the identification of a problem and the implementation of a decision to correct it. A manager at a high level may make a decision to take corrective action; it may take from six months to perhaps a couple of years for the full impact of his action to be felt at the operating level. By then many managers have rotated in and out so that safety performance is actually dependent on the decisions of the previous manager. Most of the effects of these actions fall on the shoulders of the next manager who comes along. The time between the application of the measurement system and the actual taking of corrective action based on the problem identified by the measure must be shortened. A criterion for measurement needs to be established that tells what the problems are and prevents the decision maker from making statistical, geographical, demographic, or time-lag mistakes.

Let us consider the issue of trade-off. There is a problem of trade-off when too much emphasis on productivity produces a negative influence on safety. In establishments where outstanding management exists, there is both high productivity and an outstanding safety record. Over the long term companies with the best productivity also have the best safety records. The reason, of course, is that in highly productive, safe organizations safety is valued and productivity is valued, and the most effective managers are those who are able to incorporate and reflect the values of the organization. Good management practices exist at every level of supervision.

Unfortunately, the kind of motivation that controls individual operator behavior is often derived from short-term, immediate pressures. A worker reasons as follows: "Today I have X number of units to produce. If I sacrifice safety—if I don't take the time to do it safely or use the safety equipment—I can exceed today's production quota more easily, so I will take the risk." Over the long run, everyone agrees that the safe way of doing the job will produce the best results. But the pressures of the immediate situation too often control today's worker behavior. The worker is often willing to take risks because his or her managers permit or encourage it, and these managers are willing to tolerate unsafe behavior as long as no apparent losses ensue and production quotas are met. Perhaps these kinds of managers do not really want an improved measurement system! The decisions made by higher levels of management based on the results of these measures may inhibit their freedom to tolerate unsafe behavior or conditions for immediate production gains.

Multilevel measures are needed. One level will serve as an early warning indicator. It may focus on the necessity for applying the second level of measurement to describe the problem more completely. A third level of measurement may be applied to demonstrate what to do about it. At this level one asks, "Which countermeasure should I select?" Finally, a fourth level of measurement can be used to tell how effectively the problem is solved by the application of the chosen countermeasure.

The indicator measure may be a very simple one. Identification of these basic indicators may be a lot simpler than we think. A single measure may be overspecific for one purpose and underspecific for another and thus not serve either purpose effectively.

Most managers prefer to use costs as measures of safety, or the lack of it. Most line managers do not understand rates. They want measures expressed in terms of dollars. They prefer measures of damage control effectiveness to be presented in terms of dollars saved, for example. This is where loss prevention becomes an attractive concept. Loss means dollars. Loss prevention means dollars saved. Realistically, one should not expect to find one method of measurement that will meet all measurement needs. Costs of accidents are one after-the-fact measure of losses. They measure failures. We need this type of measure, but it cannot be the only one. It is like waiting until the horse is stolen before we are able to determine that the barn door lock is defective. As before-the-fact loss potential measures become perfected and more effectively applied, fewer and fewer losses need to be measured. Eventually, loss or dollar costs, per se, will become small or perhaps nonexistent. As dollar losses are reduced, costs of accidents will become less and less meaningful as measures of safety performance. Measures of loss *potential* will then take on greater significance as safety performance indicators. Indexes based on dollars lost will be limited in application to the rare event situation where control failure exists.

One approach to measurement is to set up an experimental design with proper controls, which will permit a true significance test to be conducted. Briefly stated, an experimental design involves setting up an experimental group and a control group, preferably by randomly assigning the individuals or units to be measured to each of the two groups. The experimental group receives a particular countermeasure or safety program and the control group does not (or perhaps it may receive a "placebo" or other substitute program not directed toward the safety objectives of the experiment). The safety performance of each of the two groups is measured during the same time period to determine if the group receiving the countermeasure performs significantly better than the control group that was not exposed to the countermeasure. This measurement approach is particularly useful when a pilot program is initiated on a small scale to determine its usefulness for possible application on a larger scale. The controlled experiment is strongly recommended as a measurement procedure.

Managers and safety professionals alike often expect too much from a single safety performance measurement technique. They want to know how often accidents occur, how serious they are, what did they cost, what were the causes, why did these causes exist, how can we prevent them, where and when will the next accidents occur, and on, and on. It is too much to expect that all of these questions will be answered by applying a single measure of safety performance. Multiple measures are needed, each focusing on a portion of the problem or a particular level of information.

A very crucial question in measurement is who does the reporting of accidents and what kind of "ax does he have to grind" in the reporting process? If the investigator is interested in winning a safety contest, the reporting results will probably be biased, either consciously or unconsciously. As a result of an attitude biased toward winning safety awards, one may have a low opinion of accident statistics. An ability to "doctor" the records may have been perfected. If one person does not do this, others are observed who do it every day. In many plants the process of safety or accident measurement is a psychological game aimed at winning safety contests. It is impossible to devise a safety performance measurement system that cannot be perverted in this way. The theoretical problem of devising a valid, reliable measurement system cannot be attacked without solving the psychological problem of outwitting the individual who knows how to circumvent or defeat the valid, reliable system developed to produce a meaningful measure of safety performance. Often it is a matter of judgment whether or not an accident is reported. Under pressure to look good on paper and perhaps win an award, the manager will often choose to ignore an accident and not report it if at all possible. If the pressure is there, an injured person may not report an accident and, instead, continue working. It is like playing when hurt in a football game. If a player is needed in order to win, there is pressure to continue playing even though he is injured. This situation creates a problem in reliability.

It is strongly recommended that systems of safety performance measurement be maintained separately from contest and award programs to avoid the phenomenon of "contest contamination." The primary function of all safety performance measurement should be to provide information for management decision-making. It is important to maintain the integrity of safety performance measurement in order to achieve this objective.

Chapter 14

PUTTING CONCEPTS TO WORK — PROPOSED METHODOLOGIES

Present methods of measuring safety performance are negative in the sense that they measure losses. An individual must have been injured or killed or a piece of equipment or material must have been damaged in order for the measurement techniques to be applied. Mistakes, for the most part, are measured at a point in time when these mistakes have produced a loss of some kind. These loss events are relatively small in number compared to the total events contained in a given exposure within a specific time period, which have some potential for producing future loss. Thus, two major problems in safety performance measurement are that a negative event is measured and that our criterion or standard for measurement permits events to be counted that are relatively small in number. Out of a large number of accidents or health hazards that might appear in a working environment, only a small number of the total do, in fact, produce losses of sufficient magnitude to be counted within a given time period.

The current measures are not adequate. They do not permit the evaluation of efforts being made to create safe conditions within the working environment. They do not reflect the efforts made to improve the equipment, machinery, or processes to reduce the probability or risk of injury.

The safety professional responsible for assessing the progress of his company in maintaining a safe man-machine-environment system, free from danger as the work will reasonably permit, is seeking answers to such questions as "How are we doing?" "What are my efforts and those of our supervisors and managers accomplishing?" "Are our workers behaving safely?" "Where are our next losses likely to occur?" A measurement system is needed that will provide information for use as a basis for making decisions and taking action to improve the system. Unfortunately, our present measures are unstable and thus produce unreliable information. They fluctuate widely based on fortuitous circumstances.

The same causal factors can produce different effects, and the ultimate results are substantially different. The more severe the loss or injury, the more likely an accident or occupational disease will be included in the measurement system and the more likely corrective action will be taken. Over-reactions to a fatality are common, but little or nothing is done as the event moves closer to the no-loss position on the severity continuum. Yet the no-loss event may identify a problem with the potential for a serious loss in the future that is much greater than the future loss potential of a single fatal event. A fatality or other accident is only useful in the measurement system when it permits identification of problems that *can be* corrected. Whether a single accident produces a severe result or not is of less consequence than whether it is predictive of future accidents or losses and thus leads to corrective action for preventive purposes.

LEVELS OF MEASUREMENT

There are several levels of measurement available, each associated with a particular measurement need. A national index, particularly one categorized by Standard Industrial Classification (SIC) is useful as the basis for comparing the rates in one company very broadly with the overall rate for industries within the same SIC code number. Companies with records worse than the national rate for that industry are encouraged to improve their records. Another level of measurement is based on evaluating the same company or establishment at various points in time to determine whether the rate is improving or getting worse. If the measurement procedures are applied consistently and with reliable techniques, there is some value in identifying whether that rate or unit of measure is going up, down, or remaining the same, at least within the limits of the items or events being measured, whatever criteria are asked in the systems.

A third level of measurement is based on the identification and definition of problems within a given organization or establishment. At this level losses are measured for the purpose of identifying casual factors and taking corrective actions necessary to prevent recurrence of a problem. These are loss-based measures. Within a larger organization such as the General Electric Company or the Department of the Army, this level of measurement can be very effective. A large number of accidents or occupational diseases are available on a recurring basis and a relatively sensitive measurement system can be developed. As the size of the organization—and thus the magnitude of exposure—decreases, there is a proportionate reduction in the sensitivity of this measure as a meaningful indicator of safety performance. Unless *some* losses occur within a given time period, there is nothing to measure.

A fourth level of measurement is the identification of human, equipment, and environmental problems within a given organization or establishment *before* loss-producing failures occur. In an individual establishment there are very few situations which are totally static. Things are never constant, they are always changing. New hazards appear. Old prevention techniques cease to be effective. New unsafe behaviors are introduced. A sensitive, reliable measurement technique at this level should reveal when these changes occur and show enough about the problem to permit corrective action to be taken. Emphasis at this level is on *problem identification*. The measure may not demonstrate what specific corrective action to take, but it should alert the decision maker to the fact that a problem exists and permit him to consider alternative countermeasures which are available for problem solution.

A reapplication of the measure should also permit the manager to determine whether or not his chosen countermeasure is effective, how effective it is, and whether that effectiveness is sustained over time. A sensitive, valid and reliable in-plant measure of safety performance will permit the internal assessment of multiple problems requiring management decision and action. These problems can be rank-ordered according to priority for countermeasure attention based on such criteria as magnitude of loss potential, probability of recurrence, cumulative loss potential from repeated exposure, and so forth. With this information at hand on a continuous basis, better, more cost-effective decisions can be made concerning how available countermeasure resources can best be allocated for maximum payoff.

The option that an excellent measurement system can be developed that tells when a problem exists but does not, in itself, suggest a plan for solution should be left open. To use an analogy mentioned previously, it is perfectly proper to develop a method for measuring or detecting the presence of cancer that is vital and useful but does not reveal the slightest idea what to do about it. A measure that will diagnose cancer when it exists is needed, quite apart from a measure to prescribe the best available treatment. Similarly, a measure of safety performance that only diagnoses a problem would be perfectly acceptable. A methodology for measuring an accident or environmental health problem can be very accurate, but the technology for *correcting* the problem may not yet exist or may require further research to permit it to be used effectively. A safety performance measure could be conceived quite independent of any countermeasure action that might be taken to correct a problem identified by the measure. To put it another way, the measure can reveal what to be concerned about, but the corrective action taken can be quite independent of the diagnosis made in the first place.

As another approach to in-plant measurement, one might consider applying internal measurements on at least two levels. One level would provide an early warning indicator. The second level would focus on an in-depth analysis of the problem and perhaps an assessment of the relative value of alternative countermeasures in an experimental setting. The early warning measures may indicate that other measures will aid more precisely in decision making. Many of the present measurement systems are not refined enough to provide an in-depth analytical capability and are not properly applied as an early warning indicator because key persons are not alert to their limited usefulness in this context. Measurement systems are often rejected because they are not a "cure-all" for all of our measurement problems. As they are used, it is discovered that they combine the worst of both worlds—too specific for one need and insufficiently specific for another.

MEASUREMENT PROBLEMS AND BIASES

One problem with present measures of safety performance is that managers do not understand them. Managers respond most effectively to cost criteria. Several industries have used damage control as a tool of measurement. Others use the concept of loss prevention with its measures of cost control. Managers understand the cost and loss concepts. While measures of loss in terms of dollar costs must, of necessity, be applied, it is important that the cost criterion not be used exclusively or even primarily as a basis for safety performance measurement. Costs, of necessity, involve losses. Here again is dependence on losses as a criterion for applying our measurement technique. As was mentioned previously, other measures are needed that reveal something about *potential* loss problems *before* they actually produce loss. The problems encountered and the solutions that do not work will still show up in our loss criterion. *Multiple measures* need to be applied. We should not rely on those that must involve loss before they can identify a problem.

Another problem that must be considered in measuring safety performance is the nature of the exposure. There is a tendency to combine similar operations where critical variations exist in the nature of the exposure. There may be geographical differences, differences in risk factors, differences in proportions of exposure to specific hazards, time-lag differences where the dynamics of an operation result in continuously changing situations, and innumerable other differences. There is a tendency to combine exposures with varying degrees and types of hazards in computing rates for a department, plant, or company. A literal interpretation of the data produced by these measures may be misleading since the data may not be representative of the total organization.

Who actually conducts the investigation, reporting, and analysis of accidents? This is a crucial issue. What are his qualifications and biases? Two major problems affecting the validity and reliability of accident investigations, reporting, and analysis are (1) the technical incompetence of the typical individual who performs these tasks and (2) a built-in bias on the part of a supervisor, safety engineer, or technician who wishes to avoid criticism from supervisors or perhaps produce a more favorable statistic for purposes of contest competition or other recognition or award. Certain biases of the reporter are reflected in the causal factors and recommended corrective actions associated with an accident. For example, the Supervisor's Report of Work Accident, or similar form prepared by a supervisor, tends to emphasize the unsafe behavior of the injured worker. It states that the operator failed to take some preventive action or performed some act considered unsafe. If the operator is asked what caused the accident he will emphasize the unsafe equipment, poor maintenance, poor environment, or the laxness of management in providing a safe work place. Ideally, the safety engineer should conduct a thorough, professional investigation. This would include the reporting and analysis of all significant causal factors, the identification of conditions most in need of correction, and the prescription of countermeasures or remedial action necessary to avoid a recurrence. Often it appears that even the safety engineer has a predisposition to alter the records or to introduce a strong bias in his concern for presenting himself, his boss, or his organization in the most favorable light.

In many instances safety performance measurement becomes a psychological game of wits aimed at winning safety contests or avoiding a poor accident record. The measurement becomes an end in itself instead of providing a valid and reliable index of the safety state of the organization. There is a need to attack the theoretical problem of devising a valid and reliable measurement system and build in a check-and-balance procedure to outwit the evaluator who knows how to bias the accident reporting system.

One example of bias is that the less severe accidents receive less attention in the investigation. Even though the *potential* for severe loss may be great, if by chance the resulting property damage or injury was small, the attention given to the investigation, the thoroughness of the review and analysis, the search for major causal factors, and the like, are all proportionately less. The result is that, unless there is a severe loss (a fatality or large cost consequence), the accident receives little attention from an evaluative and preventive countermeasure standpoint. Often a supervisor will say, "If there is little or no loss, we don't worry about it." On the other hand, if there is a significant loss, the supervisor may seek to "fix blame" on an individual instead of investigating to determine and correct the multiple causal factors associated with the accident. It is important to understand the motives of the individuals preparing the accident reports. With considerable internal or external pressures to "look good" or "not rock the boat," there are likely to be some biases introduced. It is important to maintain the integrity of the safety measurement system since it is useful only if it is a true indicator of "real-world" performance.

WHAT DATA ARE WE SEEKING?

An important consideration in measuring safety performance is to determine what kinds of hypotheses exist in this field. Safety is one activity that has been most flamboyantly empirical. A typical attitude is, "Let's go out and gather a lot of data and then determine what they are good for." There is a much better approach to problem identification. Before one rushes out and spends a lot of effort on data gathering, why not initially question what kinds of information are expected to be derived from this exercise? Perhaps then one can become more efficient in gathering data which are more precisely targeted on identifying the problems and possible solutions. If one wants to know whether or not our countermeasures or prevention programs are effective, then formulas need to be formulated, design experiments designed with proper controls, and the right kinds of data collected in the right way to enable us to accept or reject the hypotheses. Ideally, the measurement technique should be applied by disinterested persons whose jobs are not "on the line" based on the results of the measurement application. It is perhaps more realistic to assume that a measurement system will be internally generated and self administered. But at some level a check-and-balance system must be introduced, perhaps on a sampling basis, to assure that the system of measurement is reliably maintained.

BILEVEL ACCIDENT REPORTING

Consideration should be given to a bilevel accident reporting system. At one level is a general measurement system that can be very simple. It provides a warning or "flag" that something needs attention. At that point the investigator can choose whether or not to enter the second level of reporting. This level probes in depth the multiple causal factors associated with the accident. Not every accident requires an investigation of such depth. The basic statistical system provides numbers. The second level of investigation adds quality to the quantity. The first level of measurement does not need to do all of these things. The key to successful bilevel reporting is to build in a trigger mechanism that will provide a valid alert indicator for use in selecting those events appropriate for the second-level investigations. One does not need to wait for injuries or deaths to occur before the bilevel investigation system can be applied. Measurement tools such as the critical incident technique, near-miss measures, and behavior sampling can be used at the first level and often at the second level of the bilevel reporting sytem. With an adequately chosen sample, errors in performance, equipment defects, and environment problems that have an accident loss potential can be detected and second-level probing can be conducted to provide an in-depth analysis of the critical, loss-potential accident problems.

The Need For Multiple Predictors

The particular measurement approach used often influences the outcome of the measurement application. Two examples from classical psychology come to mind as illustrations of this point. One relates to the use of error analysis measures. From the experience with learning experiments, different information is produced when we use the criterion of "kinds of errors made" as opposed to the criterion of "how much was learned." It is assumed that learning will be highly correlated with error reduction; however, in practice, the selection of one of these criteria as the dependent variable will influence the outcome of the experiment in a biased fashion toward that criterion. Another problem is the futility of trying to use a unity criterion to measure *anything*. Very rarely, if ever, is one effect produced by a single cause. For safety measurement in particular it is important to use multiple predictors. The causal factors associated with accidents are multidimensional. According to learning theory literature, overall performance should not be measured by using only error analysis or any other single criterion.

Multiple performance criteria are necessary to reduce the biases of measurement application. Management is interested in knowing when changes occur in the system. These changes must be measured in several ways. Multiple independent variables are needed to provide a comprehensive search for the indicators that are most sensitive to system changes. A surveillance procedure is needed that is sensitive to multiple indicators of change so that the true impact of a safety program may be measured. A less detailed system of reporting a large variety of events provides a rough indicator of the existence of a problem. This should be followed by a more intensive reporting system within defined problem areas that are significant in order to answer specific questions about multiple causes and to test the effectiveness of selected countermeasures.

When the broad measurement system reports a cluster of events or an important problem worthy of attention, as a part of the same reporting system, a method of supplementary reporting should be applied to tell us why this problem occurred. The surveillance system does not reveal what action to take, it tells us that a problem exists. A second level of reporting is then applied to provide the analysis necessary to prescribe the most useful set of countermeasures. A third level of measurement should be applied to determine scientifically, with proper controls, the degree of effectiveness of a chosen set of countermeasures and perhaps when that effectiveness, if it exists, ceases to operate in a particular problem area.

The problem with most existing measurement techniques is that they are ex post facto; that is, someone must have been injured or some dollar loss must have been incurred before the measurement system can be applied. It is possible to identify unsafe conditions and behaviors which have, *by definition,* the potential for injury, death, or property damage, and then apply a bilevel system of measurement and analysis to provide information needed to take corrective action. Small plants, in particular, cannot afford to wait for a loss or series of injuries before they are able to identify a problem. A determination needs to be made prior to an accident that a problem exists. If one can predict with some degree of reliability what might occur and then take countermeasure action, it is a lot more beneficial than measuring what happened after the fact. Our current measures are not accurate enough to make them meaningful as predictors of performance.

HUMAN FACTOR TESTING

One measurement approach with new systems (or systems still in the planning and development stages) is to apply the techniques of human factor testing to identify potential problems within these systems before they become operational. Here a priori analyses are conducted that will provide a better focal point for taking corrective action at the prototype stage before the system can operate and produce losses. This technique provides an opportunity for empirically testing ahead of time when and where we might expect principal problems to occur. The systems engineer could be trained to accomplish this a priori task. He can measure human performance and identify where errors will occur, what the consequences will be, and what actions are necessary to prevent serious consequences for the whole system. He then proceeds to engineer the errors out of the system. He looks at the tasks and assesses the possible undesirable events that could happen. For every action requirement there is some probability of the operator not performing the right task or omitting the task when it should be done. What are the forces that may drive the operator to fail? Can the operation be made fail-safe? Can a man-machine task be produced that will avoid injury, death, or property damage no matter what the operator does?

COMPREHENSIVE SYSTEMS ANALYSIS

What is needed in the safety field is a comprehensive systems analysis with emphasis on the safety consequences of system performance. Then there is at least an intellectual baseline against which we can play the ex post facto statistics. Without proper attention given to this approach, everything becomes ad hoc, and we drift along waiting for the next loss to occur before we have an indicator that something needs changing. We do not really know where to put our safety emphasis. We do not really know how the events that do occur deviate from expectations because the only expectation we have is the normative one. A normative frame of reference is not needed. What is needed is an analytic frame of reference. One that says, with the total sum of our knowledge, "here is what we expect to happen," not just because we have collected statistics effectively, but because a deductive posture has been taken in our systems analysis application. We have three subsystems working here: a preventive subsystem, a measurement subsystem, and a control subsystem. This type of model is used in the field of public health almost universally. Reasoning by analogy, an active public health department does not wait for an epidemic to occur before trying to determine what people are doing or might do that is harmful. Then a decision is made concerning what preventive steps to take, for example, innoculation against a disease. Innoculation is the result of a priori analysis, and it becomes an element in the preventive subsystem. The system is monitored by measurement application and various control systems are activated when the measurement subsystem indicates a need.

The notion of a priori analysis in the introduction of preventive subsystems is a technique that safety professionals can learn from the public health professionals. This is not just making one discrete part of a piece of equipment safe by putting a guard on it! Rather, it is conducting a methodological subsystem analysis of that piece of equipment—documenting it and determining quantitative probabilities of failure or loss due to injury, death or property damage. This is more than the traditional job safety analysis approach. What is suggested is that before a job is established, before the operation is organized, before the tasks are ultimately defined, every element of that operation is analyzed from a system safety standpoint. The techniques of maintainability analysis, reliability analysis, failure-mode-and-effect analysis, fault-tree analysis, and other similar analytical methods are applied to improve the system from a safety standpoint *before* losses occur.

SERVING THE USERS OF A
SAFETY MEASUREMENT SYSTEM

Different individuals within an organization have different needs for a safety measurement system. The safety professional wants to be able to do a better job of identifying accident problems and correcting them. Management needs to be shown that the safety program will result in dollar savings, increased production, or improved efficiency. Management also wants to avoid penalties for violations of safety standards and laws. A third group of persons interested in measurement of safety performance are the insurers who are searching for some means of evaluating and rating an organization in terms of future loss potential. Insurance companies are fully aware that they are losing millions of dollars because they don't have effective measurement techniques. A systems approach is needed in safety measurement so that these various "customers" of measurement will be adequately served.

Another class of client is the court of law. Legal theories are changing. We have come a long way from negligence law, a thing of the past in safety. There now is strict liability, absolute liability, new variations of expressed and implied warranty, and product liability. Up to 99 percent of the laws regarding safety have nothing to do with statute or legislative law. They are judge or jury-made laws. They are unique to each jurisdiction or state. The final arbitrator, the one who decides whether or not a company has met a standard, exercised due care, or committed some form of negligence is usually the member of a trial jury. The members of the jury decide what is reasonable and their attitudes are rapidly changing. They are also, in a sense, the customer who must be served by a measurement technique.

MULTIPLE CRITERIA AS A MEASUREMENT BASIS

One concept that needs addressing is that of multiple criteria. One can never identify a single criterion or source that will enable us to measure effectively the safety state or the lack of it within an organization or for the nation. Multiple criteria that address various facets of the safety problem are absolutely essential.

As a start, 12 criteria for use in safety performance measurement might be identified. The first category involves disabilities. Under the "disability" classification are included injuries and diseases. "Disability" is a generic term that includes death as a unique class, long-term disability, short-term disability, and monetary disability of less than one day.

The second criterion category is undesired behavior. Altman has identified two types of undesired or unsafe behavior—the unplanned, undesired behavior and the purposive, undesired behavior. For some people, planned or purposive undesired behavior is adopted as the normal mode. For example, the punch press operator who habitually hand feeds a part into the dies from behind the sweep guard. Or the bus driver who habitually exceeds the legal, posted, safe speed limit. This is planned, undesired, unsafe behavior that is functional in terms of accomplishing an immediate objective and, most important, has become positively reinforced as a result of repeated exposure without a harmful consequence. The person may "get away with it" most of the time. This behavior becomes a potential measurement criterion. We can measure the incidence or frequency of undesired behavior.

A third category of measurement criterion is unsafe conditions. These are also quite observable and, for the most part, not transitory but relatively stable in their existence. Environmental health problems as well as physical conditions are included in this category.

A fourth category is property damage. This criterion is, of course, generally related to the previous ones in that it is an outcome of unsafe behaviors and conditions.

Dollar costs comprise another criterion category. Included here are both direct (insured) and indirect (uninsured) costs of accidents that can be identified by the accounting system within the organization.

A sixth category is workers compensation claims. Since workers compensation laws vary among the states, the criterion cannot be aggregated among many states, but within an organization located in a given state it will provide a meaningful measurement index.

Accidents produce delays in production that ultimately influence the company's profits. Thus production disruption can function as a criterion.

The eighth, ninth, and tenth criteria are scrap costs, rework costs, and customer complaints. Production schedules missed often mean sales losses. A junked or reworked unit of production can also be costly. These criteria are measurable and they can be related to accidents.

An eleventh category is the minor injury or first-aid case that does not produce disability but, nevertheless, may identify an accident problem with a much greater loss potential.

Finally, a twelfth category of accident criterion is absenteeism. An absence from the job may result from either on-the-job or off-the-job accidents. Absences are both costly and a detriment to operating efficiency. They should be carefully measured and controlled where possible.

These are some of the multiple criteria that are available for use in the measurement of safety performance. Records can be developed and maintained that will provide data for each one.

A fundamental problem in the safety field arises from the tendency to reject multiple criteria while groping for some unifying concept that will provide a single measure of total safety or the lack of it. Often a criterion is immediately rejected when it is discovered that it does not do everything an ideal measurement technique should do. Safety problems are multidimensional. Approaches to measuring these problems must also be multidimensional if effective accident prediction and control levels are to be achieved. Use of multiple criteria is vital.

USE OF INDUSTRIAL ENGINEERING TECHNIQUES

Industrial engineers have a number of techniques that are applicable to safety performance measurement. One measure is applied at the planning and design stage of systems development. Production methods are applied that will maximize production quantity and quality and minimize unsafe behavior and conditions. The techniques of human factors engineering are applied. The equipment, machinery, process, and so forth, are all designed to be compatible with the psychological and physiological capabilities and limitations of the human operator. Potential error performance is anticipated in its various forms, and, insofar as it is feasible and practical, the system is designed so that the introduction of these errors by the operator will not produce injury, death, property damage, or production loss. Once the system becomes operational, other techniques are used to measure, predict and control man-machine performance. Work sampling, also referred to as behavior sampling or ratio-delay, is covered in detail elsewhere in this text. The industrial engineering techniques of time and motion study and methods engineering are also useful as analytical tools for measuring work performance. Safety is one of the primary design criteria used in the optimization of operating effectiveness.

In industrial engineering quality control two types of errors have been defined: (1) either something is perceived as bad when it is not or (2) something is judged to be "OK" when it is actually bad. An analogy can be found in the safety field. Action may be taken to bring about change based on a run of accidents that are not significant, that is, would be expected to occur by chance. The second type of error occurs in safety when the measures are so poor that everything is judged to be "OK" when things are actually bad and in desperate need of correction. As in quality control, the safety measurement system must provide a means of identifying safety-critical variables or attributes. A number of plants combine quality control, safety, and reliability in one measurement system.

GROSS MEASURES FOR ASSESSING SAFETY AND MANAGEMENT EFFECTIVENESS

Ideally, the various components or multiple dimensions of a safety measurement system should be capable of being combined in some fashion to identify the relative contribution of each to the overall criterion of loss reduction. This concept of combining components should also help avoid duplicate measures of the same item, and thus improve the efficiency of the measurement system. This is analogous to the technique of factor analysis. A weighting system is developed for multiple components in proportion to the contribution each makes to the ultimate criterion.

Various gross measures can be applied to assess the general level of safety and management effectiveness in a given organization. For example, a walk through an industrial plant can reveal a number of items that correlate with management effectiveness or lack of it, such as housekeeping conditions, whether tools not in use are in their proper location, whether aisles are well marked and free of clutter, the condition of the lighting, how the equipment is maintained, whether there are oil spills on the floor, how material is moved and handled, and what the supervisors are doing. These are visible, gross measures that can identify very quickly the general level of management effectiveness and safety within a plant.

THE SAFETY AUDIT

Another approach is the plant or department safety audit—an inspection or inventory of behavior and conditions on a scheduled basis to determine their safety relative to a standard criterion. A safety audit conducted by a qualified safety professional can have considerable impact on conditions and performance and it can provide a quantitative measure of safety program effectiveness. Some managers have questioned the value of an audit as a measure of plant safety since individual workers are most likely to change their behavior when they are aware that they are being audited. In general, it has been found that an industrial operation remains pretty much the same whether an audit is being conducted or not. The product still has to be produced with the available equipment and people.

In addition, part of the audit involves talking with individual employees. Most employees are very candid in their responses to specific questions about the plant operations. The auditor also examines existing records. These records have been accumulated over time, with copies forwarded to higher levels of management. Changing these records to produce a more favorable audit would be very difficult. The audit involves examining foremens' reports on a sampling basis to determine the number of safety contacts made during a month, quarter, or year. The plan for safety orientation and training of employees is examined, along with a sample interview program to determine if the plan is being fully implemented. Physical conditions are observed within the plant. In preparing for a safety audit, lists of questions are prepared relating to the most important elements of a safety program. These questions are then answered by the visiting audit team. As an example, the audit team in one large corporation consists of the local safety director, the corporate safety director or his representative, an industrial hygienist if called for, and one or two operating people from another plant. These team members know the operation and what to look for since they have considerable exposure to similar operations within their own plants.

Another type of safety audit involves a team of multidisciplinary specialists consisting of the various engineering disciplines (mechanical, electrical, chemical), an industrial hygienist, a safety engineer, and perhaps a physician and an industrial psychologist. This team moves into a plant when a safety problem is identified, perhaps as the result of a serious accident. The team may also conduct a periodic audit on a sampling basis to determine if safety problems existed before serious accidents occurred. A report is prepared for management containing the team's findings and recommendations for corrective action. A followup is made by the team leader to determine what corrective action was actually taken. The audit is most effective when the report of the auditing team is presented to the top company official.

COST-BENEFIT ANALYSIS

Another approach to safety performance measurement is cost-benefit analysis. This technique provides an objective means of allocating various countermeasures to the solution of particular safety problems. When various loss prevention approaches are being considered, one basis for decision is to examine the cost of each versus the benefits of its application. The alternative countermeasure that produces the greatest benefits for the least cost is normally selected. This technique assumes the availability of valid cost data for both the countermeasure and its benefits in terms of reducing future accident losses. One problem with this approach is the difficulty in assessing dollar costs of such intangible accident correlates as pain, suffering, and loss of life. What is a life worth?

Also, under workers' compensation laws, the plant manager does not really have to pay anything extra for the personal injury. In one sense workers' compensation laws serve as a barrier to progress in occupational safety because the company's losses are restricted. The injured employee has recourse in third-party suits to sue the equipment manufacturer if faulty equipment contributed to the accident. In this case the company where the accident occurred does not pay the costs of the injury. It is the manufacturer of the turret lathe, the punch press, the grinder, or other equipment involved who incurs the loss.

ENERGY TRANSFER CONCEPT OF ACCIDENT CAUSATION

Any measurement system should provide information about accident cause and effect, preferably with proper control of extraneous variables. Most accident data bases that identify characteristics associated with accidents leave out those same characteristics or conditions when they are associated with an activity or condition that does not lead to disabling injury, death, or property damage. This is one of the real serious deficiencies in present measurement systems. It is the familiar "population-at-risk" concept. What about the many exposures to so-called "accident causes" that do not produce injury or other losses?

A different approach to accident prevention should be considered in which an accident is defined as an event involving energy exchange that exceeds the injury or property damage threshold—the threshold of human or materials integrity. This approach subdivides or classifies accidents according to the energy involved. (This concept was suggested by William Haddon, Jr., President, Insurance Institute for Highway Safety, and former Director, National Highway Safety Bureau, U.S. Department of Transportation.) Management decisions concerning which countermeasures to apply are based on an assessment of potentially injurious or damaging energies. Accident potential can be described in terms of the injurious energy involved, the part of the body affected by the energy, the relative severity of the energy impact on the human or material, and other energy-related influences on the system. Management decisions concerning accident control are based on preventing the marshalling of hazardous energies; preventing, modifying, or controlling the release of energies; and separating the release of energies from susceptible structures (human or material) in time or space.

One safety aspect of the energy transfer concept is that a *point* will be more injurious, given the same amount of energy, than when the energy is spread out over a wide area. We can control unwanted consequences of energy exchanges by releasing energy over a greater time period, by producing a larger area for its dissipation, or by separating the location of its release from the people or property needing protection. Ideally, the energy control system must be introduced automatically without deliberate action being required of an operator or other person. While energy control, per se, is not a measurement technique, it does involve measurement and could eventually lead to a design manual or handbook that a safety engineer could use to design out deficiencies and provide energy controls which will keep the system within tolerable limits. A study of the feasibility of the energy transfer concept of accident causation is needed.

QUANTIFYING SAFETY JUDGMENTS

In this discussion of major issues in safety performance measurement, three distinct subsystems are addressed: (1) a preventive subsystem that may or may not involve measurement, (2) a measurement subsystem that attempts to quantify the problem and its consequences, and (3) a control subsystem that introduces countermeasures designed to maintain the system within acceptable tolerance limits. It is important to quantify the judgments necessary to move from measurement to control. One way to improve the reliability of these judgments is to obtain independent judgments from several individuals. Higher priority for action would be placed on those similar judgments that have been made by multiple individuals. In order to avoid the problem of trying to achieve intelligence by pooling ignorance, it is important that each person making countermeasure judgments be highly qualified in the subject area involved. One study was conducted in a munitions command in the U.S. Army, involving 20 plants (chemical, metal fabricating, and other operations) engaged in munitions production. About 5000 workers were asked about unsafe acts and unsafe conditions that they had observed in their own work area during the past month. Both operators and first-line supervisors were included in the interviews. A total of 48 different types of unsafe acts and conditions were described during the interviews. This measurement technique provides descriptions of behaviors and conditions that identify the hazards in the total production environment. It provides data for use in prescribing prevention programs before accident losses actually occur. This is an indirect measure of safety performance similar to the critical incident technique.

The basic model of experimental design, with its emphasis on hypothesis testing, should be considered as a means of measuring the effectiveness of alternative countermeasures, once the causal factors have been identified by the critical incident technique, behavior sampling or other methods. The methodologies of this approach can be found in Chapter 15. Demonstration or pilot projects can be designed to test the feasibility or effectiveness of proposed countermeasures in solving a particular problem, in advance of a costly, broad application of a new countermeasure throughout the organization. If the company has an operations research group, a quality control group, or a systems analysis group, these experts in experimental design should become involved in designing a well-controlled experiment and properly analyzing the data. This approach is particularly useful when the countermeasures under consideration are expected to impact on human behavior. An experiment will enable the safety engineer to determine if the countermeasure changed behavior, how much it was changed, and how long the behavior change lasted. This can be done on a sampling basis or with a pilot program. Thus with reasonable cost, the effectiveness of the program can be determined before it is implemented on a large scale. For example, a new education program can be tested to determine its effectiveness in achieving specific performance objectives, using experimental and control groups, and measuring the impact of the training on behavioral change. If proven successful, the training program can then be expanded to a larger target population. The procedures for conducting experimental studies are presented in the following chapter.

MEASURING ENVIRONMENTAL HAZARDS

Another aspect of the measurement problem is the environmental hazards affecting the health and safety of the worker. Examples include ventilation, illumination, sanitation, atmospheric contamination, noise, and the like.

Detection of health hazards in the environment depends upon measurement techniques that involve special equipment. Various threshold limit values have been established. These serve as criteria for acute and chronic exposures to environmental health hazards. Instruments are available for measuring the degree of health hazards present in the environment. Industrial hygiene surveys are conducted by specialists in this field who are qualified to use these instruments. This type of measurement should be applied to detect environmental problems as a vital component of an organization's safety performance measurement system.

One approach is to consider the environment as a potential source of stress on the human. Measurement techniques are introduced to detect the presence and determine the intensity of these stresses. Examples of environmental health stresses include carbon monoxide, heat, cold, vibration, noise, ionizing radiation, dust, and so on. Measures designed to detect and evaluate these stresses are not simply performance measures; instead they involve their physiological impact on workers. Measures of the physiological effects of these stressors, including their presence in the environment, should be a part of every measurement system. Research is needed to provide a more scientific determination of the effects of environmental health hazards on humans.

A scientific base for threshold limit (TLV) values needs to be established. For many contaminants the TLV was established based on animal studies. It is essential that research be conducted to determine the impact of the environmental contaminants on humans. Studies of this nature are very rare. In some cases a longitudinal study covering several years is required. Also, exposure to various combinations of toxicants and their combined effects on humans should be measured where it is present. There is often a synergistic effect when more than one contaminant exists at once.

In terms of the total work environment in the United States there are many more traumatic accidents than there are long-term health hazards. The effects of health hazards on workers (for example, Kepone, asbestos, ionizing radiation) are usually widely publicized so that more people in the general population become aware of health problems. The "garden-variety" traumatic injury does not receive the same publicity, and, therefore, it is often assumed not to be as serious or important. This is one reason why more sensitive measurement of potential injury-producing environmental and behavioral problems is needed. Small organizations with relatively limited exposure often assume that because no serious injuries are occurring within a given time period, no accident problems exist that deserve attention. This does not mean that health problems should be ignored. Both longitudinal and cross-sectional studies of environmental contamination as well as potential traumatic injury exposures that will aid in problem identification and problem solution are needed.

ASSESSMENT OF MAN-MACHINE ACCIDENT POTENTIAL

Another type of measurement is the assessment of the relative safety of an item of equipment, machine, or tool. How much safety was actually designed into the products used? For example, it would be fairly simple to take a table saw and conduct a functional event analysis, a systems analysis, a fault-tree analysis, a failure-mode-and-effect analysis, a job hazard study, or some other type of analysis to detect unsafe conditions, operator errors, or risk probability. Based on the results of these measures, a 10-point checklist could be prepared to provide an objective basis for rating a particular table saw and its use in a man-machine-environment situation on a 10-point scale. This scale could be used to evaluate new equipment before purchase, in terms of its relative safety, as well as the safety of the equipment as it is being operated in the shop.

A measure of how well a piece of equipment or environment was designed in the first place is also needed. The measure would provide an index of relative safety which could be applied to discrete elements of the production system. This measure could be used to evaluate the man-machine system also. What kinds of errors or mistakes can be made on a piece of equipment that can lead to injury or property damage? A measurement technique that provides an index of man-machine accident potential would be extremely useful. An example of one such measure is the fault-tree analysis. Potential error points and unsafe conditions in the man-machine system can be identified through the application of this technique.

PITFALLS OF STATISTICAL MEASUREMENT TECHNIQUES

Several pitfalls exist in the use of statistical tools as measures of safety performance. One of these is overreacting or reacting prematurely to a run of events. As a measure of safety performance is applied in a given situation there is likely to occur by chance a run of events that show up and are detected by the measurement technique. If a manager or safety professional is not aware of this statistical phenomenon, he is frequently going to "jump on" these chance runs and put a lot of pressure on the decision makers to change the procedures or modify the system drastically. It is possible that instead of producing corrective results the exact opposite effect is achieved and the situation may actually worsen. Simple statistical tools are available for use in determining the likelihood that a run of negative events could have occurred by chance, or resulted from a change in the causal system.

In attempting to establish a "cause-and-effect" relationship between the introduction of a countermeasure and a subsequent reduction in the accident loss criterion, we often fail to consider the "regression-to-the-mean" phenomena. The "regression-to-the-mean" concept describes the fact that when successive samples are selected from a population and certain variables measured, there is a tendency for these variables to return from an extreme to an average condition as repeated measures are taken. For example, if we select 10,000 samples of workers with similar environmental exposure with respect to accident potential, we would find that some samples would have a no-accident record, some samples would have around an average record, and some would have a bad record. If we subsequently sampled the same groups in a second time period, we would find that some of the groups with a bad accident record the first time would continue to have a bad record, some would get better, and some would get worse. In general it would be seen that the samples exhibiting the extreme characteristics in the first instance would tend to exhibit less extreme characteristics in the second instance. Workers who tended to perform poorly in the first observation would tend to improve in the second, while workers who performed exceptionally well in the first observation would tend to get worse in the second. We can conclude from this phenomenon that so-called "unsafe workers" identified in one time period will tend to show an apparent improved record during a second time period, regardless of whether a countermeasure or program activity was introduced or not. Thus an apparent improvement during the period following the introduction of a countermeasure may be falsely attributed to the effects of the countermeasure when, in fact, the natural tendency for sample measures to regress toward the mean may be producing the shift in criterion values. If we select from a similar environment a second or "control" group which is equally as poor but is not exposed to the countermeasure, it may show the same improvement, even though nothing at all was done to influence this change.

Other pitfalls include inadequate sample size, unrepresentative (biased) sample, assumption of causation from correlation, and extrapolation of results beyond the limits of the data.

Chapter 15

THE APPLICATION OF INFERENTIAL STATISTICS TO THE APPRAISAL OF SAFETY PERFORMANCE

In modern management practice the use of statistics as a tool for decision making is becoming increasingly important. Statistics, by means of their power to reduce data to manageable forms and their power to allow the study and analysis of variances, enable managers to make maximum use of available information in arriving at a decision concerning the best course of action in selecting one of two or more alternatives. It is important that the data upon which decisions are based have maximum relevance to the problem area as well as adequate reliability. This demands the development of measurement techniques that will produce the most and best information possible within the limits of available time and budget.

The safety professional has an important responsibility to develop good information so that managers can make sound decisions regarding the safety aspects of their operations. Alternative solutions to accident problems must be tested and evaluated in terms of their injury and loss prevention reliability, cost of implementation, and estimated return on investment. Recommended solutions and their accompanying substantiating evidence should be based on factual, unbiased, and objective information about accident-producing problems and their consequences. This requires more than the presently used problem appraisal procedures with their emphasis on after-the-fact analysis and their sensitivity to fortuitous severity oscillations. Unfortunately, today the safety professional concentrates most efforts on solving problems, that is, providing answers, when the emphasis should be placed on looking ahead and finding the right questions.

Modern statistical tools have been developed that will increase the safety professional's ability to measure, predict, and control the *loss-potential* factors within an organization. The application of these more objective techniques of systems performance appraisal will make the safety professional a vital member of the management staff as he teams with others to stress the scientific, economic, and management aspects of loss prevention programs and their contributions to the critical functions of the organization.

The term "statistics" is used as a general name for a large group of mathematical tools based on the laws of probability that are used to collect, analyze, and interpret numerical data. Statistics has been defined more formally as "the theory, discipline, and method of studying quantitative data gathered from samples of observations in order to study and compare sources of variance of phenomena, to help make decisions to accept or reject hypothesized relations between the phenomena so studied, and to aid in making reliable inferences from observations" (Kerlinger, 1973). This rather complex definition of statistics can be reduced to one major purpose: to aid in inference making. The application of inferential statistics enables the safety professional to attach probability estimates to the conclusions drawn from data analysis. Statistics say, in effect, "The inference you have drawn is correct at a particular level of significance. One may act as though your hypothesis were true, remembering that there is a certain probability that it is untrue." Statisticians often call statistics the discipline of decision making under conditions of uncertainty.

This modern conception of the subject is considerably different from that usually held by people outside the field. To the layperson, the term "statistics" usually carries only a nebulous (and too often distasteful) connotation of cold facts and boring figures. In modern usage, however, statistics has become a body of methods for obtaining and analyzing data in order to help decide questions of practical action.

The techniques of statistics serve in at least two capacities. First, they provide a means of organizing, summarizing, and communicating information. Second, they provide methods for making inferences beyond the observations actually made to statements about large classes of potential observations. The set of methods serving the first of these functions is called *descriptive statistics*—the body of techniques for effective organization and communication of data. When most safety professionals speak of "statistics," they usually mean data organized by these methods. The body of methods for arriving at conclusions extending beyond the immediate data is called *inferential statistics.* When one must make predictions and decisions on the basis of partial knowledge—and in accident and loss prevention work this is usually the case—statistics provide a means for performing these tasks well and wisely. Most of the interesting and important applications of statistics in the safety field involve problems of inference. Thus, it is of major importance that we establish a foundation and develop a framework for thoroughly understanding inferential methods and their application to accident data.

STATISTICS AND THE SCIENTIFIC METHOD

The techniques of inferential statistics are inexorably interwoven with the methods of scientific research. As such, they should be viewed against the background of general methods of obtaining knowledge. Assuming we have completed the planning phase, including problem identification and analysis, we are ready to apply the scientific method in seeking a solution to our research problem. In general, there are four steps involved in using the scientific method to solve research problems (see Figure 15-1, p. 268):

1. *Observation.* The researcher observes what happens and collects and studies facts relevant to his problem.
2. *Hypothesis.* In order to explain the facts observed, he formulates his conjectures into a hypothesis or tentative proposition expressing the relationships he thinks he has detected in the data.
3. *Prediction.* From the hypothesis or theory he makes deductions concerning the consequences of the hypothesis that he has formulated.
4. *Verification.* The researcher collects new data to test the predictions made from the hypothesis.

It is the prediction stage where experience, knowledge, and perspicuity are important. Predictive reasoning can help lead to more basic problems as well as provide operational or testable implications of the original hypothesis. At this stage one is anticipating or predicting what will be seen if certain observations not yet made are made. Then, as noted, one verifies the prediction. The essence of verification is to test the relation between the variables identified by the hypothesis. On the basis of research evidence the hypothesis is finally accepted or rejected.

What is important in this procedure for the safety professional is the overall fundamental idea of scientific problem solving as a controlled rational process of reflective inquiry. The safety professional is continually engaged in research activity, whether he calls it by this title or some other. He should be able to understand the scientific research process when others use it and be able to apply it himself in action research designed to improve his own performance capability.

Statistics plays an important role in the research process. During observation it suggests what can most advantageously be observed and how the

Figure 15-1
A Model for Problem Solution.

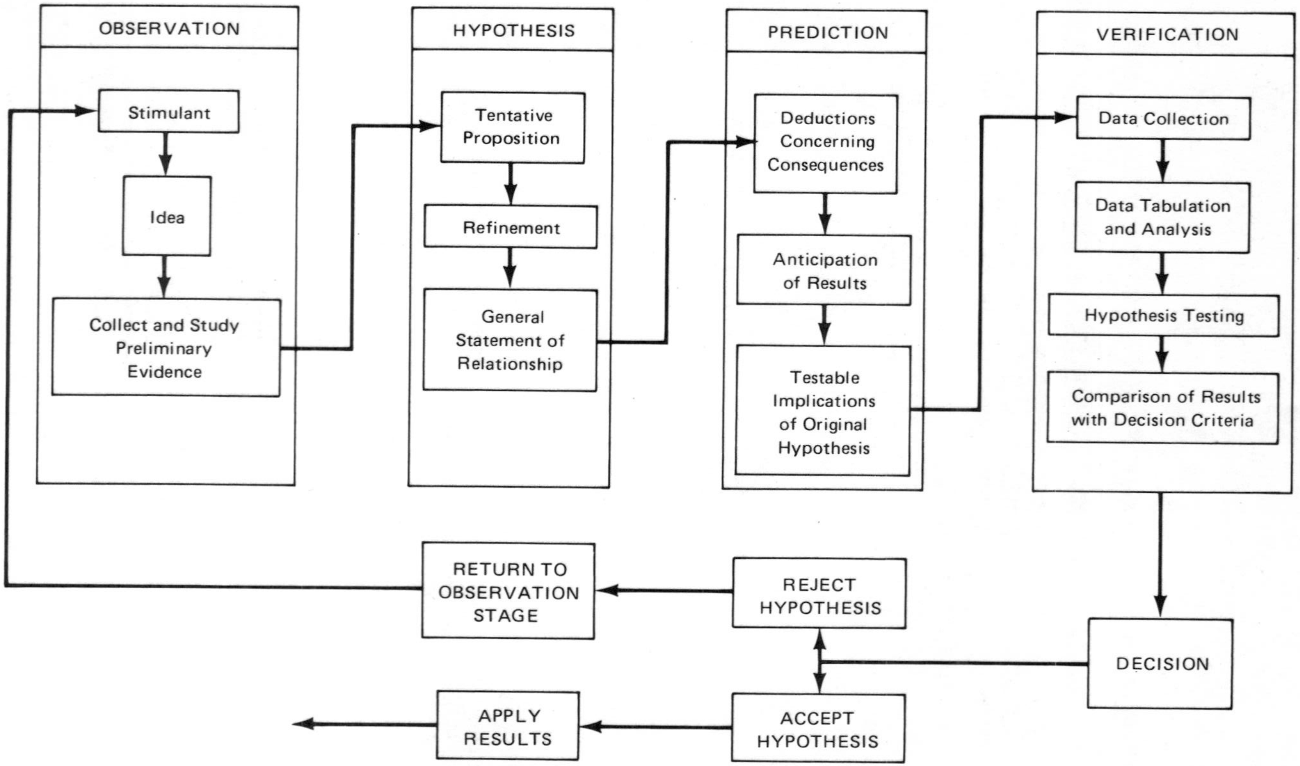

results of observations can be interpreted. At the second stage it helps to classify, summarize, and present the results of observations in forms that are comprehensible and likely to suggest fruitful hypotheses. At the final stage of the scientific method of problem solving hypotheses are considered verified to the extent that predictions deduced from them are borne out by later events. A hypothesis is verified or "tested" to the extent that the influence of chance on the evidence has been correctly interpreted. Statistical procedures have been evolved for measuring the risk of incorrect interpretation objectively in terms of numerical probabilities, or, to state it differently, for measuring the risks of erroneous conclusions.

In formulating hypotheses prime consideration must be given to the criterion or measurement standard to be used in conducting the test. The problem of what to measure when appraising safety performance is a serious one for the safety professional. By tradition, the measures most often available are based on the ex post facto occurrence of injuries with a certain degree of severity and the accumulation of a considerable quantity of exposure. These measures have rather severe limitations for purposes of testing hypotheses and measuring the internal safety effectiveness of an organization.

As a result of the limitations on currently used criteria, several new measures of safety effectiveness have been proposed. At least two of these measures have been tested experimentally and found to be more sensitive, valid, and reliable as indicators of accident loss potential. Both involve the use of statistical sampling techniques.

HYPOTHESIS TESTING

In hypothesis testing we are interested in measuring the risk of incorrect interpretation objectively in terms of numerical probabilities. Inferential statistics provide tools that formalize and standardize our procedure for drawing conclusions. The procedures of statistical hypothesis testing enable us to determine, in terms of probability, whether the observed difference is within a range that could easily occur by chance or whether it is so large that we can be confident that the differences are actually meaningful.

Inferential statistics applied to accident control problems enables the safety professional to gain knowledge through the use of experiments as opposed to intuitive speculation and the application of "trial-and-error" prevention methods. An experiment is simply a means of testing an idea under controlled conditions. A safety director, in his haste to combat the causes of today's disabling injuries, presents a program of accident countermeasures that are rarely supported by any objective evidence of probable success. He rarely, if ever, prescribes programs based on experimental evidence; he merely speculates that certain actions will prevent future accident losses.

This procedure involves the use of a "shotgun with hope" when what is really needed is a rifle that is "zeroed in" on the target problems. A statistically oriented safety professional, perhaps using some of the same program elements, will combine these elements according to a preconceived plan to obtain empirical confirmation or rejection of a hypothesis concerning their usefulness.

A test of a hypothesis, then, is a means of determining the validity of some prediction or expectation. Establishing a working hypothesis is an important step in control planning. It suggests the kinds of data needed and how they should be arranged and classified. It also establishes the basis for making decisions. A safety professional, for example, might wonder about the effectiveness of his safety program "package" in reducing unsafe worker behavior. He might, of course, base his judgment on prior experience and the recommendations printed in safety textbooks and manuals and simply take a chance that the program he recommends will effectively solve the problems at hand. If he is a little more perceptive, and has perhaps armed himself with knowledge of inferential statistics, he is likely to want some preliminary indication of the likelihood of success for whatever countermeasures he recommends before committing himself and his company to a particular course of action. Even if he has tentatively decided not to change his program "package," he has, in fact, set up the hypothesis that a change would not improve the safety effectiveness of the system. By preparing an appropriate sampling experiment, he can provide an objective basis for confirming or rejecting a hypothesis concerning success and proceed accordingly. He can thus base his decisions on the information he obtains from his "research" and on the risk he is willing to take that his decisions with respect to the hypotheses are incorrect.

Let us apply this technique. Suppose we would like to determine whether a particular safety training program for workers is effective in changing behavior. We select a fairly large group of workers and randomly assign these workers to an experimental group and a control group. The experimental group is given a safety training program while the control group is given no special safety training. Other relevant factors that may influence unsafe behavior are held constant for both groups. We test the null hypothesis that there is no significant difference in the behavior of the two groups at the conclusion of the training program; that is, hypothesize that the observed differences are due to chance factors alone. If the trained group is demonstrated to be significantly better than the nontrained group at an acceptable level of confidence (that is, unsafe behavior significantly improved), the null hypothesis must be rejected and the alternative hypothesis of a meaningful change accepted.

The steps in hypothesis testing for the general case are as follows:

1. State the null hypothesis (H_o) and the alternative hypothesis (H_1). For example:

$$H_0: P(A) = P(B) = \tfrac{1}{2}$$
$$H_1: P(A) < P(B)$$

2. Choose a statistical test (with its associated statistical model) for testing the null hypothesis. From among the several tests that might be used with a given research design, choose a test whose model most closely approximates the conditions of the experiment and whose measurement requirement is met by the measures used.
3. Specify a significance level (alpha or α) and a sample size (N).
4. Find or assume a sampling distribution of the statistical test assuming H_0. The sampling distribution is the distribution obtained if all possible samples of the same size from the same population, drawing each randomly, were taken. The sampling distribution of a statistic shows the probabilities assuming H_0 associated with the various possible numerical values of the statistic. Sampling distributions for frequently used test statistics appear in most statistics textbooks.
5. On the basis of 2, 3, and 4, above, define the region of rejection. The region of rejection consists of a set of possible values that are so extreme that when H_0 is true, the probability is very small (that is, the probability is α) that the sample actually observed will yield a value which is among them—for example,

Figure 15-2
Region of Rejection for the Null Hypothesis.

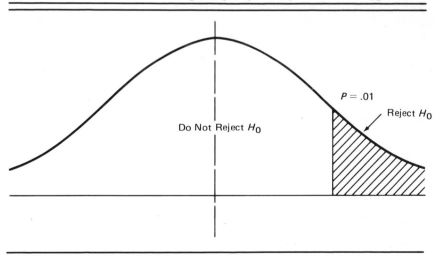

the grid area under the curve in Figure 15-2 is the one-tail region of rejection when $\alpha = .01$. The probability associated with any value in the region of rejection is equal to or less than α.

6. Compute the value of the statistical test using the data obtained from the samples.
7. Compare the computed test results with the tabled value for the sampling distribution of the test statistic at the specified significance level (α).
8. Make a decision concerning the findings. If the value of the test is in the region of rejection, we reject the null hypothesis (H_0). The null hypothesis is rejected whenever a "significant" result occurs. A "significant" value is one whose associated probability of occurrence under H_0 as shown by the sampling distribution is equal to or less than alpha (α). In symbolic form—using the t-distribution, a one-tail test, $\alpha = .01$, and degrees of freedom $(df) = n_1 + n_2 - 2$—

If t-test $<$ t-table (.01), we do not reject H_0
If t-test $\geq$ t-table (.01), we reject H_0

With these steps in mind, let us restructure the example described into a form suitable for hypothesis testing.

Problem

A large group of workers exposed to approximately the same types of hazards are randomly assigned to an experimental and a control group, with an equal number assigned to each. The experimental group (Group A) is given one hour of safety training per person each week while the control group (Group B) receives no safety training. Other relevant factors that may influence accidents are held constant for both groups.

ASSUMPTIONS.   In this problem we will make five assumptions:

1. The type of work and amount of exposure is equivalent for each group.
2. Worker past accident experience is equivalent for each group.
3. The only safety activity conducted for either group is the safety training program conducted for Group A.
4. There is independence among the sample observations of behavior.
5. The population approximates a normal distribution.

MEASURE.   The percent of workers involved in unsafe behavior within each group is used as the unit of measure. Standard behavior sampling techniques are applied during each week following the presentation of the safety training program to the experimental group. The measures are taken during 20 consecutive weeks. The data collected are shown in Table 15-1.

HYPOTHESIS.   The difference between the unsafe behavior of workers in the trained and untrained groups is due to chance.
  The steps in testing this hypothesis are as follows:

1. The null hypothesis (H_0) states that there is no significant difference between the percent of workers involved in unsafe behavior in the trained and untrained worker groups.
    The alternative hypothesis (H_1) states that the percent of workers involved in unsafe behavior in the trained worker group is significantly lower than in the untrained group. In symbolic form

$$H_0:\ P(A) = P(B) = \tfrac{1}{2}$$
$$H_1:\ P(A) < P(B)$$

Table 15-1
Percent of Workers Involved in Unsafe Behavior
in the Experimental and Control Groups

Week	Percent of Workers Involved in Unsafe Behavior Experimental (Group A)	Control (Group B)
1	36	37
2	16	18
3	19	46
4	25	18
5	34	53
6	30	26
7	29	28
8	5	33
9	29	30
10	18	20
11	30	35
12	15	25
13	19	19
14	12	18
15	16	14
16	30	40
17	19	19
18	35	37
19	21	21
20	16	21

2. The test statistic chosen is the t-test since this statistic most closely meets the requirements of the assumptions and is suited to the relatively small sample of 20 observations of behavior within each of the two groups. The statistical model for this design may be expressed as

$$t = \sqrt{\frac{n_1 n_2 (n_1 + n_2 - 2)(\bar{X}_1 - \bar{X}_2)^2}{(n_1 + n_2)(\Sigma x_1^2 + \Sigma x_2^2)}}$$

where:

n_1 = size of first sample (Group A)

n_2 = size of second sample (Group B)

$$\Sigma x_1^2 = \Sigma X_1^2 - \frac{(\Sigma X_1)^2}{n_1}$$

$$\Sigma x_2^2 = \Sigma X_2^2 - \frac{(\Sigma X_2)^2}{n_2}$$

df = degrees of freedom = $n_1 + n_2 - 2$

In the above model degrees of freedom refers to the number of restrictions placed on the data or the freedom left for the data to vary when all restrictions are imposed.

The foregoing formula yields a t-value that indicates, by comparison with a t-table, the probability that both sample groups could have resulted by random sampling from a single homogeneous population.

3. A significance level of 1 percent ($\alpha = .01$) is chosen. The sample size (N) in this case is 20 weekly random applications of the behavior sampling technique for the experimental group and for the control group.

4. The sampling distribution under the null hypothesis is given by the t-distribution with $df = n_1 + n_2 - 2$. Critical values of t from the sampling distribution, together with their associated probabilities of occurrence under H_0, are presented in the tabled appendix of most statistics textbooks.

5. The probability associated with any value in the region of rejection is equal to or less than $\alpha = .01$. In this case, since the alternate hypothesis (H_1) indicates the predicted direction of the difference between the two groups, a one-tailed test is called for. One-tailed and two-tailed tests differ in the location (but not in the size) of the region of rejection. In a one-tailed test the region of rejection is entirely at one end (or tail) of the sampling distribution. The size of the region of rejection is expressed by α, the level of significance. If $\alpha = .01$, then the size of the region of rejection is 1 percent of the entire space included under the curve in the sampling distribution (see Figure 15-3).

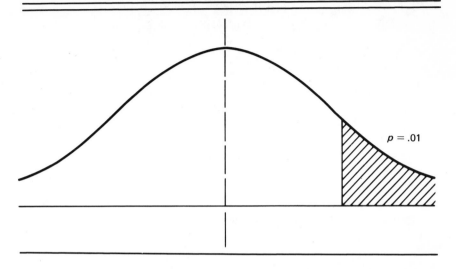

$p = .01$

The grid area of Figure 15-3 shows the one-tailed region of rejection when $\alpha = .01$.

6. Using the data obtained from the samples, we compute the value of the statistical test as follows:

$$t = \sqrt{\frac{n_1 n_2 (n_1 + n_2 - 2)(\bar{X}_1 - \bar{X}_2)^2}{(n_1 + n_2)(\Sigma x_1^2 + \Sigma x_2^2)}}$$

where:

$$n_1 = n_2 = 20$$

$$\bar{X}_1 = 27.9$$

$$\bar{X}_2 = 22.7$$

$$\Sigma X_1 = 558$$

$$\Sigma X_2 = 454$$

$$\Sigma X_1^2 = 17,734$$

$$\Sigma X_2^2 = 11,694$$

$$\Sigma x_1^2 = \Sigma X_1^2 - \frac{(\Sigma X_1)^2}{n_1} = 2165.8$$

$$\Sigma x_2^2 = \Sigma X_2^2 - \frac{(\Sigma X_2)^2}{n_2} = 1388.2$$

$$df = n_1 + n_2 - 2 = 38$$

$$t = \sqrt{\frac{(20)(20)(40-2)(27.9-22.7)^2}{40(2165.8+1388.2)}}$$

$$t = 1.70$$

7. The computed t-value is compared with the tabled value for the sampling distribution of the test statistic at the .01 level of significance (see Figure 15-4). At .01 significance level for $df = 38$, t-table is 2.42.
8. *Decision:* Since the value of the test is smaller than that demanded for significance at the 1 percent level as indicated by the tabled value, the test results fall in the region in which the null hypothesis is not rejected. Thus, these findings indicate that there is no significant difference between the percent of workers involved in unsafe behavior in the trained and untrained groups. In symbolic form

$$t\text{-test} = 1.70 < t\text{-table}(.01) = 2.42$$

Thus we do not reject H_0.

In this example the interpretation may be made that, with two samples of 20 observations each, the usefulness of the safety training program has not

Figure 15-4

t-Test Compared with t-Table at the .01 Level of Significance.

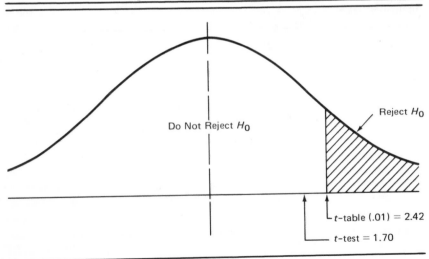

been demonstrated for an infinite hypothetical worker population from which the 40 observations, on which the analysis has been based, might be assumed to represent a random sample. This simply means that the 5.2 percentage points by which the untrained group exceeded the trained group in unsafe behavior resulted from chance in choosing at random groups of observations as small as 20 each.

Thus, this difference is not significant at $\alpha = .01$ for the sample size. It should be noted that if a significance level of $\alpha = .05$ had been chosen, the one-tail tabled critical value would be 1.68. At the .05 level the resulting computed t-value would fall in the region of rejection for H_0—since t-test = $1.70 > t$-table $(.05) = 1.68$—and the difference in unsafe behavior between the two groups would be barely significant.

EXPERIMENTAL VERSUS EX POST FACTO RESEARCH

There are numerous experimental designs in inferential statistics, each having its strengths and weaknesses in relation to the need for decision-making information. There is no perfect design that fits all situations. Some statisticians make a distinction between the experimental and the so-called ex post facto approaches to information gathering.

An experiment is a scientific form of inquiry in which the experimenter manipulates and controls one or more independent variables and observes one or more dependent variables for evidence of change. The independent variable is sometimes referred to as the task variable and the dependent variable as the criterion variable. In an experimental design the investigator has direct control over and is able to manipulate at least one independent variable. Thus in our safety training program example the program would be identified as the independent variable and the change in unsafe behavior as measured by behavior sampling techniques would be the dependent or criterion variable. In the experiment we attempted to change the group of workers by means of a training program, and we measured the effect of this program by observing the subsequent change in behavior.

In ex post facto research something is done or occurs after an event with a retroactive effect on the event. In this type of study the independent variable or variables have already occurred. The investigator starts with observations of the criterion and retrospectively studies the independent variables for their possible effects on the criterion or dependent variable. The safety professional who investigates an accident is, in effect, conducting ex post facto research. He examines certain evidence and retrospectively studies the independent variable or behavioral and conditional factors that he believes were causally related to the accident result. Most current accident prevention activity is based on similar after-the-fact appraisals of losses that have already occurred. It should be clear that the desired model of safety research is the controlled experiment since the safety professional can have more confidence that the relationships he discovers are really the ones he thinks they are.

SEQUENTIAL COMPARISON RESEARCH

One way to find out whether a planned program really produced a desirable change is to conduct a sequential comparison study. In this method the final status of a group is compared with its earlier status and judgments are made based on the amount of change. The technique can also be used to explore the potency of a number of different programs by comparing the amount of change resulting from each. Even their interaction can be measured.

For example, the effectiveness of three different safety posters in changing behavior can be tested using three different but matched worker groups. By comparing the proportion of workers engaged in unsafe behavior in each group *before* the introduction of the safety posters with the proportion of workers similarly engaged *after* the posters are presented, a more precise measure of poster effectiveness can be obtained. However, it is necessary to be careful in the interpretation of the results. Where time is an essential force, this design can introduce error by imputing change to the posters when change was the result of time alone.

THE USE OF CONTROL GROUPS

One of the preferred techniques of data collection includes the use of a control group. Comparison groups are necessary for the internal validity of any research program. The purpose of a control group is to rule out variables that are possible "causes" of the effects being studied other than the variables hypothesized to be the "causes." In other words, the objective is to control extraneous sources of influence. The control of extraneous variables means that the influences of independent variables extraneous to the purpose of the study are minimized or nullified. One way of controlling extraneous variables is by randomization. By randomly assigning subjects to the experimental group and the control group, it is possible to assume the preexperimental approximate equality of these two groups in all possible independent variables.

A major weakness in much of the accident research conducted in the past has been the lack of adequate controls. We frequently encounter so-called "research" evidence that supports conclusions based on uncontrolled observations of certain characteristics of persons involved in accidents. For example, in an investigation into one aspect of the motor vehicle problem— violations by drivers in fatal accidents—it was shown that nearly one-third of those involved were exceeding the speed limit at the time of the accident.

Another study concluded that in over 50 percent of all fatal accidents occurring in a particular region the driver had been drinking. Still another study concluded that the majority of drivers involved in one-car accidents were smoking a cigarette. Such statements contribute very little to understanding the problems unless it is known whether the cited characteristics of the accident-involved drivers are present to a different extent among those who are not involved in accidents. For example, it may well have been that an equal proportion of the *noninvolved* drivers operating at the times and places of the fatal accidents were also exceeding the speed limit or had been drinking or were smoking a cigarette and that this characteristic, therefore, did not discriminate between those involved and those not involved in the accidents. Such data are also frequently biased because of the tendency of those investigating accidents to conclude that the occurrence of an accident is sufficient evidence that such violations and characteristics were present. These conclusions are then used in a circular fashion to support preexisting biases concerning accident causes.

It is important to compare the characteristics of accident cases with the characteristics of the corresponding populations and situations from which they are derived. The informed safety professional, armed with the knowledge of inferential statistics and research design, can readily identify these sources of bias, or at least suspect their presence, when no mention is made of the frequency distribution of the same characteristics among appropriate control populations.

MISUSE OF STATISTICS

One added benefit derived from understanding inferential statistics and research design is the ability to interpret data. We must also remain as objective as possible. We must be wary that the statistics cited are not deliberately biased in favor of the presenter's interests, or that data are not being interpreted or processed carelessly. In fact, the deluge of statistically bathed information that confronts us in all phases of commercial advertising almost forces us to acquire a certain amount of statistical insight if we are to make intelligent decisions.

In our TV-oriented society all have no doubt been exposed to the very attractive young lady on television who smilingly attests to the fact that "her group had 34 percent fewer cavities" as a result of using a certain brand of toothpaste. To make the presentation sound more "scientific," an announcer then talks about "control groups." It may be that this toothpaste is quite good, but we cannot necessarily draw this conclusion from the evidence presented. What the advertiser would like us to believe is that his toothpaste is so much better than other toothpastes in reducing cavities that we will be motivated to buy his product and ignore all the others. Several questions are left unanswered by his presentation: What was the sample size used in the experimental and control groups? How many cavities are involved? What is the standard error of the mean proportion of improvements when the experiment is replicated? Is the product used for comparison representative of other toothpastes on the market or was it especially selected for this study? How many times was the experiment conducted with a finding of no significant difference? In fact, it could be that the teeth of some who used the product all rotted out! It is not necessarily suggested here that this particular brand of toothpaste is not good. It *is* suggested that, on the basis of the evidence presented, there is not enough information on which to base a sound conclusion. Statistical data can be extremely useful when carefully collected and critically interpreted. Unless handled with care, skill, and, above all, objectivity, statistical data may seem to prove things which are not so at all.

A television interviewer asks the question, "Which pile contains the whiter wash?" After the selection is made, we discover that, sure enough, the pile selected was washed with the sponsor's product! Would they show the film if the "typical housewife" had picked the wrong pile? What about the before-and-after photograph that is a familiar stunt in advertising? A pair of pictures shows what happened when a girl began to use a certain hair rinse. By golly, she does look better afterward at that. But wait! If you look closely, you note that most of her new-found attractiveness can be traced to a bright smile, combed hair, a spot light turned on her face, and a back light thrown on her hair. The typical viewer attributes all of these changes to the hair rinse! In addition, the advertiser leads the viewer to believe that this magic hair rinse will change one's personality, provide a full social life, and even produce the husband of one's dreams.

We do not have to leave the field of accident statistics to find evidence of logical fallacies. A midwestern manufacturing plant with a fairly stable injury frequency rate decided to hire a full-time safety engineer to see if this rate could be cut down. Within a short time after the safety engineer was hired, the injury frequency rate nearly doubled. Should we conclude that the safety engineer caused these accidents and quickly fire him? A closer look revealed that the safety engineer instituted a new accident investigation and reporting system that produced more reports of disabling injuries, thus increasing the frequency rate.

A not so easily-detected example of the fallacy of assuming causation from correlation is the situation where a new safety program is introduced and the frequency rate goes down. We may assume that the new safety program caused the rate reduction, when in reality any number of intervening variables, such as a few remarks by a plant manager, the influence of seasonal change, or a huge layoff program may have been the true influencing factor.

During World War II about 375,000 persons were killed in the United States by accidents and about 408,000 were killed by war action. From these figures it has been argued that it was not much more dangerous to be overseas in the armed forces than to be at home. A more meaningful comparison, however, would consider rates, not numbers, and would also consider the same age groups. This comparison would reflect adversely on the safety of the armed forces during the war—in fact, the armed forces death rate (about 12 per thousand men per year) was 15 to 20 times as high per person per year as the overall civilian death rate from accidents (about .7 per thousand per year).

Peacetime versions of the same fallacy are also common. We often hear that "off-the-job activities are more dangerous than places of work since more accidents occur off the job than on the job" or that "the bedroom is the most hazardous room in the home since more injuries occur in the bedroom than in any other room." Here again the originators fail to consider differences in quantity of exposure, type of exposure, age, and other influencing factors.

W. Allan Wallis and Harry V. Roberts (1956) in their book *Statistics: A New Approach* have listed several categories of statistics misuses, including those due to shifting definitions, inaccurate measurement or classification of cases, methods of selecting cases, inappropriate comparisons, shifting composition of groups, misinterpretation of association or correlation, disregard of dispersion, technical errors, misleading statements, and misleading charts. A rather interesting treatment of this subject can also be found in a book by Darrell Huff (1954) called *How to Lie with Statistics.*

Misuses, unfortunately, are probably as common as valid uses of statistics. The lack of ability to discriminate between a valid and an invalid statistical application is a very real handicap for the safety professional as well as for others who use statistical information as a basis for defining problems and measuring performance.

BENEFITS OF INFERENTIAL STATISTICS

What have we learned from these remarks about the subject of statistics that will help the safety professional do a better job? In a very practical sense, application of the techniques of inferential statistics can help the safety professional in the following ways:

1. In summarizing a mass of data or observations into a concise and understandable form that will be most useful for the problem at hand.
2. In determining limits of precision for conclusions reached from the

analysis of sample data; in other words, ascertaining the degree of confidence one may have in information obtained by sampling.

3. In determining the number of observations that must be made or the amount of information that must be gathered and summarized to give conclusions with the required degree of accuracy.
4. In extracting the maximum amount of useful information from available data and observations.
5. In appraising in specific terms the uncertainty or variability that is inherent in most procedures, processes, materials, activities, and situations.
6. In planning studies, planning for the collection of data, and reaching conclusions in such a way as to avoid the effects of known sources of bias and to minimize the possible effects of these biases.
7. In giving a more complete and meaningful analysis and interpretation of data.
8. In computing the probability of some specific event happening, thus allowing a more precise appraisal of the uncertainty of a situation than would be possible from a hunch or guess.
9. In giving evidence of relationships among factors in a process or system that might go unnoticed if a statistical approach were not used.
10. In placing emphasis on planning for a study and for the collection of data so that only necessary information will be gathered and so that the information used as the basis for decisions will be representative of the entire population for which accident loss control is desired.

References

Huff, D. *How to Lie with Statistics.* New York: Norton, 1954.
Kerlinger, F. N. *Foundations of Behavioral Research,* 2nd ed. New York: Holt, Rinehart and Winston, 1973.
Wallis, W. A., and H. V. Roberts. *Statistics: A New Approach.* New York: Free Press, 1956.

Chapter 16

BEHAVIOR SAMPLING

Work sampling has long been regarded as an industrial engineering tool for establishing time standards, setting delay allowances, and surveying areas for improvements in methods. It has been tested in experimental research and applied as a measurement tool in actual industrial operations to determine its usefulness as a means of measuring unsafe behavior. (Jewel, 1958; Rockwell, 1959; Schreiber, 1956; Meyer, 1963; Pollina, 1962). Work sampling has been proven to be a valid, stable, sensitive, and practical technique for directly measuring unsafe behavior. It is now being used by several industrial companies as a means of appraising safety performance on a continuing basis.

Behavior (work) sampling is based on the statistical principle of random sampling. By observing a portion of the total, a prediction can be made concerning the makeup of the whole. Behavior sampling can be defined as the evaluation of an activity performed in an area using a statistical measurement technique based on a series of instantaneous random observations.

Perhaps an analogy can be drawn between behavior sampling and a lapsed-time camera. An interesting use of the lapsed-time camera can be seen in the color nature movies where buds burst into blossoms and caterpillars turn into butterflies in a matter of seconds. These pictures are produced by shooting a few frames of the film periodically until the picture is complete. The actual length of time required by nature to complete a process is compressed in order to allow the entire process to be viewed on a relatively short length of movie film. By substituting an individual for the camera and permitting him to observe behavior several times during the day, an activity check using behavior sampling techniques has been achieved.

Work sampling was developed by L.C.H. Tippett in Great Britain and used in the English textile mills in the late 1930s to determine the percent of running time and idle time of loom operations. Since then it has been adopted as a standard industrial engineering technique for measuring work done by either man or machine, or a combination of both, in factories, offices, hospitals, warehouses, and many other places where people do work. It is used to determine the percent of time that people are productive and the percent of time that they are idle, to measure the percent of machine down-time, and also to establish a standard of performance for a manufacturing operation. The behavior sampling application to the measurement of safety performance through its quantitative units of measure will alert the safety professional to areas needing improvement. Furthermore, it can be used on a continuing basis to determine the amount of improvement made in unsafe behavior over time.

Behavior sampling employs the same principles used in industrial quality control. When a lot of material is received, a representative random sample is taken; based on this sample, a determination is made as to the quality level of the entire lot. The basis of this sampling is that the quality of the lot is assumed to be the same as the quality of the sample (within a certain preselected level of confidence).

In applying this principle to behavior sampling, we assume that it is necessary to determine the percent of time a worker is behaving safely and the percent of time he or she is behaving unsafely. One way to determine this is to watch the worker all day. The same determination can be made, however, by conducting a behavior sampling study, where the worker is observed at certain times and a record made whether he or she is working safely or unsafely. The results of the behavior sampling study will be the same as the all-day study within predictable limits, depending on the number of readings.

The behavior sampling study may be conducted as follows: (assuming the computed sample size $N = 100$ will give the desired degree of accuracy):

Total Observations

Safe behavior	72
Unsafe behavior	28
	$N = 100$

$$\text{Percent safe behavior} = \frac{72}{100} = 72 \text{ percent}$$

$$\text{Percent unsafe behavior} = \frac{28}{100} = 28 \text{ percent}$$

Thus, the worker was behaving safely 72 percent of the time and unsafely, 28 percent.

FUNDAMENTALS OF BEHAVIOR SAMPLING

Behavior sampling is based on three fundamentals: (1) the laws of probability, (2) the concept of the normal distribution, and (3) randomness. When conducting a behavior sampling study, one is concerned with the confidence level, accuracy, and the number of observations required.

Probability

Most games of chance are based on the laws of probability. One of the simplest chance games is the tossing of coins. When a coin is tossed, the chance of a head coming up is one out of two or one-half (assuming a fair coin). The same is true for a tail. However, everytime the coin is tossed twice, the result will not be one head and one tail. It is possible that in the course of tossing coins, two or three heads may come up in succession. If a coin were tossed 10 times, the possibility of 50 percent heads and 50 percent tails coming up is not very great. If, however, the coin is tossed 10,000 times, the chance of a 50/50 split would be much greater. This is also true of behavior sampling. A sufficiently large sample must be taken in order to obtain valid results. The larger the sample, the more confident we can be of the results. The principle of confidence is related to the normal curve.

The Normal Curve

Assuming we have an underlying normal distribution within our population for the characteristic we are measuring, as we take a larger number of similar readings, the results approach the shape of a normal curve.

One of the characteristics of a normal distribution is that the curve is symmetrical around the mean (symbolized by $\bar{X}$). This point is also the median and mode. As indicated in Figure 16-1 (p. 286), 68.27 percent of all readings fall within what is called the ± 1-sigma limits, 95.45 percent fall within ± 2-sigma limits, and 99.73 percent of all readings fall within ± 3-sigma limits.

Behavior sampling is based on the principle that the readings taken will conform to the characteristics of the normal distribution. The normal distribution may be assumed only when large numbers are involved. When a sample is taken from a population, a determination must be made as to the adequacy of the sample in terms of the level of confidence desired.

Confidence Level

The greater the number of readings taken, the closer the plot of them will approach the normal curve, and the more confidence one can have that the sample readings are representative of the population. In behavior sampling it is necessary to take only the number of readings required to satisfy the desired confidence level. The confidence level is related to the sigma limits established for the normal curve. A confidence level of 95 percent, or within ± 2-sigma limits, is considered adequate for most behavior sampling studies. A confidence level of 95 percent means that the conclusions will be representative of the true population 95 percent of the time and 5 percent of the time they will not.

Figure 16-1
Characteristics of a Normal Distribution.

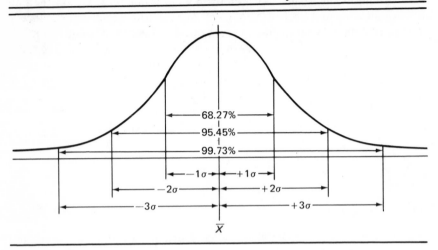

Accuracy

In addition to confidence level, a behavior sampling study has another characteristic called accuracy. It should be kept in mind that confidence level and accuracy are *not* synonymous. Accuracy is the tolerance limit of the readings that fall within a desired confidence level. Accuracy is a function of the number of readings taken. The tolerance becomes smaller as the number of readings increases. A plus or minus 5 percent accuracy with a 95 percent confidence level (a combination often used in behavior sampling) would mean that 95 percent of the time within a ± 5 percent accuracy limit the conclusions will be representative of the actual population and 5 percent of the time they will not.

The formula for determining accuracy at the 95% level of confidence (± 2-sigma limits) is as follows:

$$Sp = 2 \sqrt{\frac{p\,(1-p)}{N}}$$

where:

S = accuracy expressed as a decimal
p = percent occurrence of the activity (unsafe acts expressed as a decimal)
N = number of observations

Example

Assume we calculate the percent of time a group of workers behave unsafely to be equal to 20 percent and the number of readings taken equals 4000. What degree of accuracy is present? We have

$$p = .20$$
$$N = 4000$$
$$S = \text{accuracy (unknown)}$$

$$Sp = 2 \sqrt{\frac{p(1-p)}{N}}$$

$$S(.20) = 2 \sqrt{\frac{.2(1-.2)}{4000}}$$

$$S(.20) = 2 \sqrt{.00004}$$

$$S = \frac{2(.00632)}{.20}$$

$$S = .063$$

Therefore, accuracy expressed as a percent equals ± 6.3 percent. It can then be stated that 95 percent of the time the conclusions will be accurate to ± 6.3 percent of the true value and 5 percent of the time they will not.

Number of Observations Required

The number of observations to be made depends on both the confidence level and the desired accuracy. The formula for the number of readings required for a 95 percent confidence level is as follows:

$$N = \frac{4(1-p)}{S^2 p}$$

where:

$N = $ number of readings
$S = $ desired accuracy expressed as a decimal
$p = $ percent unsafe behavior expressed as a decimal

Example

Suppose a trial study revealed that 40 percent of the time a group of workers was behaving unsafely and that we agreed upon a desired accuracy of ± 10 percent with a 95 percent confidence level. How many readings are required to achieve this accuracy at the 95 percent level of confidence?

$$N = \frac{4(1 - .40)}{(.10)^2 \, (.40)}$$

$$N = \frac{2.4}{.004}$$

$N = 600 =$ number of readings required

Randomness

Behavior sampling requires randomness in the selection of times for making the observations. Randomness is achieved when each period of the work day has an equal chance of being selected as the time for making the observations. A predetermined system of selecting the various times for making the observations is necessary so that the observer does not bias the study by his own conscious or unconscious time preferences. The selection of observation times can best be accomplished by the use of a table of random numbers found in Appendix C. In using a table of random numbers, the observer merely determines how many readings he wishes to take and then consults the table and selects the times for those readings. (Times falling within scheduled nonwork periods are discarded.) For example, suppose six readings are needed for the day. The workday is from 8:00 to 12:00 A.M. and from 1:00 to 5:00 P.M. (no regularly scheduled breaks in between).

Time Selected from
Random Table

12:15	lunch—discard
1:25	selected
12:30	lunch—discard
11:15	selected
2:00	selected
4:20	selected
9:15	selected
12:40	lunch—discard
11:40	selected
7:30	not valid—discard

Thus the times selected for the observations are:

9:15	1:25
11:15	2:00
11:40	4:20

STRATIFIED RANDOM SAMPLING. Under the random sampling principle of time selection each portion of the workday has an equal chance of being selected for observation. This method may produce several readings within one hour and none within another hour. In order to distribute the readings more evenly throughout the workday and thus eliminate the possibility of taking too many readings during normally slow periods, we may select our sample observation times on a stratified basis within each working hour. Stratified random sampling simply means that the readings are distributed equally over the hours each day so that times within each hour are assured of being included as observation periods. In order to assure randomness, a table of random numbers is used to select the exact times within each hourly period when the observations will be made. For example, if we plan to make eight observations during a workday which runs from 8:00 A.M. to 5:00 P.M. with one hour for lunch from 12:00 to 1:00 P.M., one reading would be made at a random time between 8:00 and 9:00, another at a random time between 9:00 and 10:00, and so on, until eight readings have been made. No reading would be taken during the lunch hour.

When a behavior sampling study is made over only a few days, stratified random sampling will insure that the readings are spread out over the entire day. If the study is conducted during one week or longer, then sufficient readings will be taken so that adequate coverage throughout the day is obtained. Of course, if one is interested in determining hourly differences in unsafe behavior, then the hourly stratification plan should be followed in making the random selections.

Random number tables containing five-minute time intervals, such as the table included in Appendix C, can be used for selecting times for random stratified sampling. Let us assume that eight readings are to be made during an 8:00 A.M. to 5:00 P.M. workday, with a one-hour lunch period from 12:00 to 1:00 P.M. The procedure for this is as follows:

1. Record the hours on a work sheet (8, 9, 10, 11, 1, 2, 3, 4).
2. Select the minutes (one for each hour) from the random table. In using the random table, discard the hour and record only the minute digits next to the successive hours listed above. For example, assume that the random number table produces readings as follows:

1:25	8:05
1:55	3:10
6:15	12:10
2:35	7:15

The selected random readings stratified according to work hour are designated as follows:

Hour	*Reading Time*
8–9 A.M.	8:25
9–10 A.M.	9:55
10–11 A.M.	10:15
11–12 A.M.	11:35
1–2 P.M.	1:05
2–3 P.M.	2:10
3–4 P.M.	3:10
4–5 P.M.	4:15

3. Simply drop the hours from the selected readings and place the minutes next to each hour in succession.

This is a random selection process since both the hour and minutes are random in the random number table.

STEPS IN APPLYING BEHAVIOR SAMPLING TO UNSAFE BEHAVIOR ANALYSIS

1. Prepare a list of unsafe acts that have the potential of existing within the departments chosen for study. Previous records of all accidents (disabling injury, recordable injury, first aid, and so forth) and/or lists of unsafe acts contained in the National Safety Council's *Accident Prevention Manual for Industrial Operations* or other sources may be used as a guide in preparing the list. One study used 17 unsafe acts. In another plant 65 unsafe acts were included on the infraction list. The critical incident technique (see chapter 17) may be applied to identify unsafe acts associated with particular industrial operations.
2. Select trial observation periods in time by a random process. These are the times at which observations of worker behavior will be made. The number of trial observation periods needed depends upon the number of persons to be observed. A sufficient number of observation periods should be selected so that the total sample (N) is at least 100. For example, if 20 persons are to be observed, then at least five observation periods are needed to produce an N of 100. The trial observations will be used to determine the sample size of observations needed in the full-scale study to obtain the desired accuracy at the selected confidence level.
3. Dichotomize behavior as being either safe or unsafe as defined by the criterion of behaviors included on the unsafe act list.
4. Make the trial observations and instantaneously decide whether the observed behavior is safe or unsafe.

5. Two observation procedures are possible:
    a. Observe workers within a selected group from a single observation point and determine whether the behavior of each is safe or unsafe at the time of each observation.
    b. Walk through the department or plant, observe all workers who are actually working, and determine the proportion of these workers behaving unsafely.
6. Determine the number of observations required. The number of observations is based on data collected in the preliminary survey, the degree of accuracy desired, and the selected level of confidence.
    a. Two items are recorded in the preliminary survey:
        (1) Total number of observations made (N_1)
        (2) Number of these observations in which unsafe behavior was involved (N_2)
    b. Calculate the percent unsafe behavior found in the preliminary survey.
    c. Using this percentage (p), the desired accuracy ($\pm$ 10 percent), and a selected level of confidence, calculate the total number of observations required (N) by using the following formula:

$$N = \frac{4(1 - p)}{S^2(p)}$$

where:

N = total number of observations required
P = percent unsafe operations or observations
S = desired accuracy = (.10)
4 = constant factor relating to level of confidence of 95 percent.

Example

If the preliminary survey produced the following results:

Total observations = 126 = N_1
Unsafe observations = 32 = N_2

The percent of unsafe observations to total observations would be:

$$p = \frac{N_2}{N_1} = \frac{32}{126} = 25.4 \text{ or } 25 \text{ percent}$$

$A \pm 10$ percent desired accuracy at the .05 confidence level means that 95 percent of the time the correct number falls between 22.5 percent and 27.5 percent of the total.

$$N = \frac{4(1 - p)}{S^2 (p)}$$

$$N = \frac{4(1 - .25)}{(.10)^2 (.25)}$$

$$N = \frac{3}{.0025}$$

$$N = 1200$$

Thus, for these data, the study must have a minimum of 1200 observations to give satisfactory results ($\pm$ 10 percent at .05 confidence level).

7. Select the random time periods when the observations are to be made by the use of a table of random numbers. Either the normal random sampling process or a stratified random sampling procedure may be used, depending on the requirements and objectives of the study.

8. Conduct the actual study by making the number of random observations required and recording the observed behavior according to the safe or unsafe classifications.

9. a. If method 5a is used:
        (1) Compute the percent of time each worker is involved in unsafe behavior.
        (2) Compute the mean percent unsafe behavior in time for the entire group.
    b. If method 5b is used:
        (1) Compute the percent of workers involved in unsafe behavior.

10. Repeat steps one through nine once each week for a series of five to ten weeks, depending on the number of persons observed and the computed sample size requirements (N).

11. Construct a behavior control chart based on a mean ($\bar{p}$) percent of time each worker is involved in unsafe acts or the mean percent unsafe behavior for the entire group during the observation period.

    Use a 95 percent level of confidence (5 percent control limits). Compute the upper control limit (UCL) and lower control limit (LCL)

    where:

$$UCL = \bar{p} + 1.96 \sqrt{\frac{\bar{p}(1 - \bar{p})}{n}}$$

$$LCL = \bar{p} - 1.96 \sqrt{\frac{\bar{p}(1 - \bar{p})}{n}}$$

12. Determine if stability exists in terms of percent unsafe behavior for the test period, that is, determine if all points fall within the upper and lower control limits.

13. If the situation is not stable, introduce a minor change in the accident prevention program for the population studied and repeat the sampling study until stability in unsafe behavior is achieved.
14. Having achieved stability, introduce a major program change (such as a new inspection system, a new poster program, a new film, a safety training program for supervisors, worker training program, a lecture series) and repeat the behavior sampling study on a weekly basis during and after the initiation of the program.
15. Plot the percent unsafe behavior on the control chart for each period following the initiation of the program and determine if a significant improvement in unsafe behavior has been achieved.
16. Continue to modify the safety programming components until unsafe behavior has been reduced to a minimum. Repeat the behavior sampling study each week and continue to plot data on the control chart to assure that unsafe behavior remains stable at the minimum level.

SAMPLE PROBLEM

Define the Problem

The purpose of the study for our example is to measure the effectiveness of a worker safety training program presented to members of the automatic lathe department. We are interested in determining whether or not the safety training program produces a significant reduction in the percent of time the automatic lathe operators are involved in unsafe behavior. We would also like to determine if any significant reduction achieved is stable over time.

Plan the Study

1. The following is a list of unsafe acts developed from past records of first-aid and recordable injuries and from a previous application of the critical incident technique to determine the unsafe acts associated with noninjurious accidents.
   a. Not wearing safety goggles
   b. Adjusting the machine while it is in operation
   c. Running the machine with the plastic guard removed from the cutting point
   d. Handling rough metal stock with bare hands (no hand protection)
   e. Reaching into running machine to remove part in process
   f. Running machine at excessive speed
   g. Failure to place "speedy-dry" oil absorbing compound on oil spills
   h. Wearing rings, loose clothing, long sleeves, and the like, around machines in operation
   i. Misuse of air hose (blowing chips, cleaning clothing, and so on)

   j. Horseplay with fellow employee
   k. Handling chips with bare hands
   l. Lifting improperly (stock pans, raw materials, and so forth)
   m. Calipering or gauging the job while the machine is in operation
   n. Cleaning chips while machine is in operation
   o. Using the power of the machine to start the face plate or chuck onto the spindle
   p. Reaching into the turret area in close proximity of the tool (possibly permitting hand to strike against tool)
2. Behavior is dichotomized as being either safe or unsafe, depending on whether or not one of the above unsafe acts is observed.
3. All 20 workers in the automatic lathe department will be observed. Only those who are actually working will be included during each observation period.
4. A trial observation period of $N = 100$ will be conducted to determine the number of observations needed to obtain the desired accuracy.
5. A 95 percent confidence level will be used.
6. Observations in the full-scale study will be taken each week for a period of 10 weeks, and a behavior control chart will be plotted based on the mean and standard deviation of these data.
7. A 10-hour safety training program will be presented to the automatic lathe operators.
8. The weekly behavior sampling studies will be continued to determine what influence the training program has on the percent of time workers are involved in unsafe behavior.

Determine the Accuracy

1. During the trial study it was found that the operators were behaving unsafely 40 percent of the time (mean).
2. The number of readings required for a ± 10 percent accuracy at the 95 percent level of confidence is computed as follows:

$$N = \frac{4(1-p)}{S^2 p}$$

where:

N = number of readings
S = desired accuracy = .10
p = percent of unsafe acts = .40

$$N = \frac{4(1 - .40)}{(.10)^2 (.40)}$$

$$N = \frac{2.4}{.004}$$

$N = 600$ readings required

3. With 20 operators being observed and 600 observations required for the desired accuracy at .95 confidence level, we can obtain the necessary readings by making 10 trips per day for 3 days (10 trips $\times$ 20 operators = 200 observations $\times$ 3 days = 600 readings).

Gather the Observations

1. Design an observation sheet. In this case, four columns are needed. The observation sheet is used for tabulating the frequencies and computing the percentages (see Figure 16-2).

Figure 16-2
Behavior Sampling Observation Sheet

BEHAVIOR SAMPLING OBSERVATION SHEET		Sheet _____ of _____	
Study: AUTOMATIC LATHE DEPARTMENT		Date:	
Random Observation Times	No. Persons Behaving Safely	No. Persons Engaged in Unsafe Acts	Total No. Persons Observed
Total			
Calculations:			

2. Select the times for the observation using a table of random numbers. For this study normal random sampling is used. Times are selected for the 10 trips per day and recorded on the observation sheet. Different times are selected for each day's readings. The random times selected within the workday of 8:00 A.M. to 5:00 P.M., with 12:00 to 1:00 P.M. lunch hour, are as follows:

First Day	Second Day	Third Day
8:50	8:20	8.00
9:05	8:55	10:10
9:30	9:30	10:50
9:45	9:50	10:55
11:10	9:55	1:55
11:25	11:45	2:45
11:50	11:55	3:55
3:50	1:00	4:15
4:30	2:05	4:50
4:45	3:35	4:50

3. Conduct the actual study by walking through the area and recording the number of persons engaged in any of the unsafe acts included on the criterion list. Also note the number of operators working safely on the automatic lathes. These numbers can easily be recorded by the use of two hand counters. Record the number of persons observed working safely on a counter held in one hand and the number of persons observed working unsafely on the second counter held in the other hand. The counters can be kept out of sight by holding them inside the pockets of trousers or skirts. The observations should be made by someone *not* associated with the safety department, since the presence of a safety engineer or safety specialist tends to alter the operators' behavior. Enter these totals on the observation sheet. Figure 16-3 shows the observation sheet at the end of the first day. Calculate the percent unsafe behavior for each day.

Summarize the Results

1. Summarize the data entered on the daily observation sheets for the week (three-day period). Set it up as shown in Figure 16-4.
2. Determine the percent unsafe behavior for the total number of observations made during the week.

Figure 16-3
Observation Sheet for the First Day of the Study

BEHAVIOR SAMPLING OBSERVATION SHEET		Sheet __1__ of __3__	
Study: AUTOMATIC LATHE DEPARTMENT		*Date:* January 13	
Random Observation Times	**No. Persons Behaving Safely**	**No. Persons Engaged in Unsafe Acts**	**Total No. Persons Observed**
8:50	12	8	20
9:05	13	7	20
9:30	10	10	20
9:45	14	6	20
11:10	12	8	20
11:25	12	8	20
11:50	11	9	20
3:50	10	10	20
4:30	14	6	20
4:45	15	5	20
Total	123	77	200

Calculations:

$$\% \text{ unsafe acts} = \frac{77}{200} = 38.5\%$$

Figure 16-4
Daily Observation Sheet for a Three-day Period.

UNSAFE BEHAVIOR SUMMARY						
Study: AUTOMATIC LATHE DEPARTMENT				*Dates:* January 13–15		
Date	**No. Persons Behaving Safely**	**No. Persons Engaged in Unsafe Acts**	**% Unsafe Acts**	**Cum. No. of Daily Obs.**	**Cum. No. of Unsafe Acts**	**Cum. % Unsafe Acts**
1/13	123	77	38.5	200	77	38.50
1/14	118	82	41.0	400	159	39.75
1/15	120	80	40.0	600	239	39.83

Repeat the Study

1. Repeat the study each week for 10 consecutive weeks, using the same group of operators.
2. Construct a behavior control chart based on a mean percent ($\bar{p}$) of time the group is involved in unsafe behavior. Use a 95 percent confidence level. Compute the UCL and the LCL.
3. Determine if stability in percent of unsafe behavior exists for the group of operators.

Introduce the Training Program

1. Having achieved stability, introduce the safety training program for the operators in the automatic lathe department. Make certain all other prevention actions affecting unsafe acts are held constant throughout the entire study.
2. At the conclusion of the training program, repeat the entire study. Begin by making the trial observations needed in order to establish a new observation sample size (N) for the ± 10 percent accuracy desired.
3. Evaluate any downward shift in the mean percent of unsafe behavior which may follow the training program to determine if a significant (other than chance) improvement in the safety behavior of the department has been achieved.

References

Barnes, R. M. *Work Sampling,* 2nd ed. New York: Wiley, 1957.

Hansen, R. F. *Work Sampling for Modern Management.* Englewood Cliffs, N.J.: Prentice-Hall, 1960.

Heiland, R. F., and W. J. Richardson. *Work Sampling.* New York: McGraw-Hill, 1957.

Jewel, A. J. An investigation and evaluation of the application of work sampling techniques to the safety behavior measurement problem. Unpublished master's thesis, The Ohio State University, 1958.

Lambrow, F. H. *Guide to Work Sampling.* New York: John F. Rider Publishing Co., 1962.

Maynard, T. H. *Industrial Engineering Handbook,* 2nd ed. New York: McGraw-Hill, 1964.

Meyer, Joseph J. Statistical sampling and control for safety. *Industrial Quality Control,* June, 1963.

Pollina, Vincent. Safety sampling. *ASSE Journal,* 7(8):19–22, August 1962.

Rockwell, Thomas H. Safety performance measurement. *The Journal of Industrial Engineering, 10*(1), January–February 1959.

Schreiber, Robert. The development of engineering techniques for the evaluation of safety programs. *Transactions of the New York Academy of Sciences.* Series II, *18*(3), January 1956.

Chapter 17

THE CRITICAL INCIDENT TECHNIQUE AS A METHOD FOR IDENTIFYING POTENTIAL ACCIDENT CAUSES

The problem of indentifying accident causal factors has long plagued the safety specialist and the safety researcher who are interested in obtaining factual information for use in developing safety programs and measuring safety performance. Like other phases of modern business management, accident prevention must be based on facts that clearly identify the problem. Unfortunately, safety practitioners and researchers alike are forced to define their accident problems by using vague, unreliable, and insensitive criteria, such as data obtained from lost-time accident reports and first-aid injury reports. As a consequence, present attempts to control accidents and their resulting losses can well be described as trial and error chiefly because accurate information about accident-producing problems does not exist.

Since lost-time accidents are rare events and first-aid cases are subject to serious reporting inaccuracies, the safety professional is faced with only an intuitive notion of his accident experience. Consequently, he has only a limited expectation of success when he applies his various accident countermeasure programs. The researcher in this field faces the same criterion problem.

As was previously stated, most present-day safety efforts are based on after-the-fact appraisals of loss-producing causes that happen by chance to produce an accident of sufficient severity to be included within the limits of the reporting criterion. Under the present method of safety program operation accident causal factors are identified primarily through the analysis of accident data obtained from investigations of injury-producing, illness-producing, and property-damaging accidents. Some loss must have occurred before accident problems can be identified. As noted in Part I, according to the method of recording and measuring work injury experience established by the American National Standards Institute (1967, r. 1973), an injurious accident loss must involve a minimum of 24 hours away from the job before it is included in the statistical evaluation system. The BLS-OSHA (U.S. Department of Labor, 1975) measurement system expanded the scope of recordable injuries and illnesses to include every work-related injury or illness that involves loss of consciousness, requires medical treatment, or prevents an employee from carrying out all of his regularly assigned duties. Both of these measurement systems are based on the use of some type of loss event as the evaluative criterion.

Even under the more frequently occurring measurement criteria contained in the BLS-OSHA reporting system, the recordable cases are a loss-producing, after-the-fact means of identifying accident problems. These less severe injuries, unfortunately, are rarely given the attention needed to probe fully into their causes. Safety professionals and managers need to accept the necessity for modifying current methods of accident problem appraisal, with their extreme sensitivity to fortuitous severity oscillations, and to seek new measures that will improve their accident problem identification and control capability.

Because lost-time accidents, disabling injuries, and BLS recordable cases are statistically rare events and first-aid cases and serious injury frequency rates are subject to large-scale reporting inaccuracies, the safety specialist is faced with only an intuitive notion of the effectiveness of various accident prevention methods. It is preferable to appraise the *internal effectiveness* of an accident prevention program by directly measuring its influence on an acceptable criterion of safety performance as it fluctuates over time.

An accident causal factor identification technique is needed that will identify noninjurious accidents as well as those involving injuries. One should be able to identify unsafe conditions or defects in the environment that have an accident-producing potential. The inclusion of noninjurious accidents within the scope of a safety performance appraisal system obviates many of the difficulties associated with current measuring techniques. Since noninjurious accidents generally occur much more frequently than disabling injury or property-damaging accidents, even small organizations can collect sizeable amounts of causal data within a relatively short time. Moreover, studies have shown that people are more willing to talk about "close calls" than about injurious accidents in which they were personally involved (Chapanis, 1959), the implication being that since no loss ensued, no blame for the accident would be forthcoming.

The general justification for the collection of noninjurious or nonhealth impairment accident data is that the severity of accident results appears to be largely a matter of luck. As Brody and Dunbar (1959), Chapanis (1959), Suchman and Scherzer (1960), and others have indicated, accidents from the same causes can recur with high frequency without a resulting injury. Which accident does produce an injury appears to be determined largely by chance. This suggests that the real importance of any accident is that it identifies a situation that could potentially result in injury or property damage. Whether an accident does or does not result in such losses is less significant.

If one accepts the position that the severity of accident consequences is largely a fortuitous or chance occurrence, then a measuring technique that would identify the relatively high-frequency noninjurious accident could be used to identify loss-potential problems at the *no-loss* stage. This information could then be used as a basis for a prevention program designed to remove or control these problems before more severe accidents occur. Moreover, if the same technique could be used to identify causal factors involved in injurious accidents of varying severity, then the entire range of the accident severity spectrum could be represented by a single method of problem identification.

A relatively new procedure known as the *critical incident technique* (CIT) has been tested and found to meet these requirements. This technique, originally developed by Flanagan (1954), is essentially an outgrowth of studies in the aviation psychology program of the United States Air Force. One of the early studies using the technique, conducted by Fitts and Jones (1947), surveyed psychological and man-machine systems problems involved in the use and operation of aircraft equipment. The investigators asked a large number of pilots if they had ever made, or seen anyone else make, "an error in reading or interpreting an aircraft instrument, detecting a signal, or understanding instructions." As an example, one pilot responded as follows:

> "It was an extremely dark night. My copilot was at the controls. I gave him instructions to take the ship into the traffic pattern and land. He began letting down from an altitude of 4000 feet. At 1000 feet above the ground I expected him to level off. Instead, he kept right on letting down until I finally had to take over. His trouble was that he had misread the altimeter by 1000 feet. The incident might seem extremely stupid, but it was not the first time that I had seen it happen. Pilots are pushing up plenty of daisies today because they read their altimeter wrong while letting down on a dark night."

A total of 270 "pilot-error" incidents were collected during the study and many similar reports were found. The classification of errors suggested some remedies. For example, the fact that so many pilots reported serious errors in reading multirevolution indicators suggested that perhaps such an instrument is too hard to read. This led to experiments on the design of long-scale indicators and recommendations for a less confusing altimeter than the conventional one.

The critical incident technique has been used successfully by Flanagan (1954) and others in such areas as job analysis, the analysis of performance requirements, and proficiency testing. The use of the technique in educational research has been suggested by Corbally (1956) and Mayhew (1956).

THE CRITICAL INCIDENT TECHNIQUE DEFINED

The critical incident technique (CIT) is a method of identifying errors and unsafe conditions that contribute to both potential and actual injurious accidents within a given population by means of a stratified random sample of participant-observers selected from within this population. These participant-observers are selected from the major plant departments so that a representative sample of operations existing within different hazard categories can be obtained.

In applying the technique an interviewer questions a number of persons who have performed particular jobs within certain environments and asks them to recall and describe unsafe errors that they have made or observed, or unsafe conditions that have come to their attention in connection with plant operations. The participant-observer is encouraged to describe as many specific "critical incidents" as he or she can recall, regardless of whether or not they resulted in injury or property damage.

The incidents described by a number of participant-observers are transcribed and classified into hazard categories from which accident problem areas are defined. The incident information is analyzed in the same way any accident data are analyzed. Normally, this involves applying a modification of the American National Standards Institute Z-16.2 (1962, r. 1969), *Method of Recording Basic Facts Relating to the Nature and Occurrence of Work Injuries* or some other causal classification system (examples of the description and analysis of typical critical incidents can be found in Appendix F). When the potential accident causes are identified, a decision is made concerning a priority system for allocating the available resources, and an accident prevention program is organized and directed toward solving these problems. A reapplication of the technique using a new stratified random sample is then conducted periodically in order to detect new problem areas or for use as a measure of the effectiveness of a prevention program in attacking the previously discovered problems.

A careful selection of a representative sample of participant-observers each time, based on valid stratification criteria, should enable the safety professional to make an inference concerning the accident potential state of the entire system at that particular time. The frequency of reapplication would be a function of the degree of variance in the types of critical incidents over time. A reapplication should follow any major alteration in size, function, or mission of the system itself. The objective of the technique is to discover causal factors that are critical—that is, factors that have contributed to an actual or potential loss-producing accident or occupational health problem.

The critical incident technique has been tested several times in industry. Tarrants (1963) conducted a series of studies to determine if the CIT meets the requirements of an improved method for identifying the causes of industrial accidents. The research was designed to determine the feasiblity of using the technique for revealing errors and/or unsafe conditions that lead to industrial accidents regardless of whether or not losses occur. The positive findings of these studies have resulted in an expansion of safety performance measurement, thus enabling the safety specialist to broaden loss prevention activities by gathering information more representative of the true state of the entire system.

THE WESTERN ELECTRIC COMPANY STUDY: AN APPLICATION OF THE CIT

An early CIT study was conducted by Tarrants at the Kearny, New Jersey, plant of the Western Electric Company. The following discussion summarizes that study and its findings.

The Problem

The purpose of the Western Electric study was to evaluate the critical incident technique to determine its usefulness as a means of identifying causal factors in industrial accidents. In seeking a solution to the main problem, two specific subordinate problems were examined:

Subproblem 1.   Does the critical incident technique dependably reveal causal factors in terms of errors and/or unsafe conditions that lead to industrial accidents?
Subproblem 2.   How does the critical incident technique compare with other methods of accident study, such as reports of disabling injuries and first-aid cases, as a means of identifying accident causal factors?

Definition of Terms

The following are terms used in applying the critical incident technique:

Accident.   An unplanned, not necessarily injurious or damaging event, that interrupts the completion of an activity and is invariably preceded by an error and/or an unsafe condition or some combination of errors and/or unsafe conditions.

Causal Factors.   Errors and/or unsafe conditions that are associated with an accident.

Error.   A departure from an accepted, normal, or correct procedure; an unnecessary exposure to a hazard; or conduct reducing the degree of safety normally present. An error may be either an act of commission or an act of omission.

Participant-Observers.   Persons exposed to potential accident situations who are asked to recall errors and/or unsafe conditions that they have experienced and/or seen within a particular work environment.

Unsafe Act.   A human error that in the past has produced a disabling or lost-time injury and thus has the potential for producing a similar loss each time it occurs.

Unsafe Conditions.   Accident factors that are present due to physical defects, errors in design, faulty planning, or omission of recognized safety requirements for maintaining a relatively hazard-free environment.

Delimitations

The study was delimited in the following manner:

1. The subjects were industrial workers engaged in operations involving similar task and environment conditions. This selection system was used to limit the differential in exposure to risk of accidents among the subjects.
2. The subjects in the full-scale study were chosen from a single industrial plant that had a carefully maintained accident reporting system in operation for a minimum of two years. The Western Electric Kearny plant was chosen to avoid wide variations in accident reporting procedures that may exist among different plants. The two-year period was selected so that a sufficient number of reported accidents could be collected to provide data for comparison purposes.
3. No attempt was made to probe at the reasons for a person's behavior or reasons for the unsafe conditions.

Basic Hypotheses

The study was based on two hypotheses: (1) There is no significant difference between the accident causal factors derived from the independent application of the critical incident technique to the same subjects by two trained investigators. (2) There is no significant difference between the accident causal factors revealed by the critical incident technique and the causal factors revealed by other methods of accident study.

Method

The critical incident technique was pretested in a pilot study conducted in order to develop a sound interview plan and to establish and refine the procedures to be followed in the full-scale study. The pilot study was conducted in a general manufacturing plant engaged in operations similar to those involved in the full-scale study.

In the full-scale study 30 participant-observers (plant workers) were chosen from five departments within the Kearny plant of the Western Electric Company. The departments were engaged in general manufacturing operations involving such machinery as drill presses, milling machines, engine lathes, and the like. Other requirements for selection were that a system of accident reporting must have been in existence for the previous two years and that a sufficient quantity of accident reports must have been available for comparison purposes.

The 30 participant-observers were selected from a total population of approximately 300 on a stratified random sampling basis. The two stratifications of department size and proportion of male versus female operators within each department were used in the selection process. There were 7 female and 23 male observers who participated in the study.

Causal factor data were collected from lost-time and minor injury accident records for all workers in the five departments over a two-year period. A comparison was made between the unsafe acts and unsafe conditions contained in the accident records and similar information revealed by the observers during the critical incident interviews.

Two trained investigators independently administered the critical incident technique using the same group of subjects, as a means of securing an estimate of reliability. The interview sequence consisted of a preliminary interview followed after a time lapse of approximately 24 hours by a data-collection interview. The objective of the preliminary interview was to inform the observers (workers) concerning the purpose of the study, explain the type of information desired, and explain the procedures to be followed in collecting the data. Approximately one week after the first data-collection interview, a second interview was conducted by the second investigator. The two investigators conducted their interviews independently, with each having no prior knowledge of the results obtained by the other. The observers were assigned at random to the two trained investigators for the first data-collection interview by means of a table of random numbers. The order of assignment of the observers was reversed for the second interview. All interviews were recorded on magnetic tape.

Special care was taken to eliminate the threat of punishment or making a value judgment on the individual observers as a result of the information they divulged during the interviews. The research goals of the study were clearly stated at the outset and the anonymity of the observers in relation to the data was guaranteed.

The questions presented to the observers were essentially "open ended" in that the observer was free to express himself or herself as he or she chose and to discuss as many incidents as could be recalled. There were two primary data-collection questions:

1. Would you describe as fully as you can the last time you saw an unsafe human error or unsafe condition in your department?
2. Would you please describe as fully as you can other unsafe human errors or unsafe conditions you have seen in your department during the past two years?

After the observer stated that he or she could not remember any additional incidents, the interviewer began to probe systematically for other incidents that may have been forgotten. The interviewer asked such questions as the following:

1. Have you ever seen anyone cleaning a machine or removing a part while the machine was in motion? Tell me about it.
2. Have you ever seen any speeding or other improper handling of a power truck? Tell me about it.
3. Have you ever seen anyone "beating" or "cheating" the machine guard? Would you describe exactly what you saw?
4. Have you ever seen anyone operating equipment such as a milling machine or a grinding machine without eye protection? Tell me about it.

A total of 45 probe questions of a similar nature were used during each interview for the purpose of stimulating the observer's recall processes (See Sample Preliminary Interview questions in Appendix E).

Data Analysis Procedures

The information derived from the observer's descriptions of critical incidents was analyzed for the purpose of identifying specific unsafe acts and/or unsafe conditions involved in the incidents. These causal factors were then classified according to type of unsafe act and type of unsafe condition using the American National Standards Institute Z-16.2 (1956) standard for classifying accident causes. Examples of the procedure followed in this analysis are given in Appendix F. This analysis procedure consisted of the following steps:

1. Listening to the magnetic tape transcription of a critical incident described by an observer
2. Preparing a written summary of the critical incident description
3. Identifying the unsafe acts and unsafe conditions contained in the critical incident and classifying them according to the categories defined in the ANSI Z-16.2 standard
4. Reporting this analysis procedure for all critical incidents described by the observers
5. Summarizing the results in tabular form

The accident records for the five departments from which the observers were chosen were reviewed. The accident causal factors contained in the records were identified and classified according to type of error or unsafe act and type of unsafe condition using the ANSI classification system.

It was found that the analysis of the critical incidents and the records using the standard ANSI Z-16.2 classification system resulted in broad, general categories that excluded from the incidents information considered useful for isolating more specific problem areas. In order to make more complete use of the available causal information, a modified form of the ANSI standard was developed, and data were reclassified according to this system. The basic framework of the original ANSI standard was retained in the modified system, and the descriptions of the unsafe acts and conditions contained in the standard were expanded in order to identify more fully the problem areas revealed by the critical incident descriptions as applied to the particular operations and environments involved in the study. Examples of the modified accident causal classification system are included in Appendix D.

Comparisons were made between the types of unsafe acts and unsafe conditions revealed by the two interviewers. Comparisons were also made between the accident causal factors obtained by the critical incident technique and the causal factors contained in the accident records collected over a two-year period. Data summarizing the unsafe acts and conditions classified according to the modified classification system provided the basis for developing the solutions to the two subproblems and for testing the hypotheses identified in the study.

Results

The 30 observers (workers) participating in the study described a total of 102 different incidents to the two investigators. Of these incidents, 91 were revealed by both investigators during their separate and independent interviews with the same observers. Seven incidents were revealed by investigator

A and not by investigator B, while four incidents were revealed by investigator B and not by investigator A. Thus investigator A obtained a total of 97 different critical incidents while investigator B obtained 95. The mean number of incidents per observer obtained by investigators A and B was 25.63 and 24.33, respectively.

The length of the interviews varied from 20 minutes to over 2 hours, with an average (mean) length of around 45 minutes. Approximately 45 hours of taped incidents were recorded during the study. Examples of typical incidents described by the observers are included in Appendix F.

As noted previously, details concerning unsafe acts and conditions useful for precise identification of accident problems could not be presented within the framework of the existing ANSI classification system. A modification of this system was made by introducing subheadings within the original standard that would provide a more refined categorization of the accident causal factors. When the incidents described by the observers were classified into unsafe acts and unsafe conditions according to the modified classification system, it was found that the 30 observers identified 187 different acts and conditions during interviews conducted by the two investigators. A total of 171 of these acts and conditions were revealed by both investigators during their independent interviews with the same observers. Investigator A revealed 9 unsafe acts and conditions that were not noted by investigator B, while 7 were revealed by investigator B and not by investigator A. Thus, investigator A obtained a total of 180 unsafe acts and conditions while investigator B obtained a total of 178.

Under the modified classification system 90.1 percent of the unsafe acts and conditions revealed by the records were also revealed by investigator A using the critical incident technique ($N = 111$). Investigator A revealed 80 unsafe acts and conditions not contained in accident records while missing only 11. A total of 180 acts and conditions was revealed by investigator A, while the records contained 111. Thus, under the modified classification system investigator A revealed 62 percent more unsafe acts and conditions than the records while missing only 9.9 percent, for a net gain of 52.1 percent. Similar relationships were found between investigator B and the accident records.

Subproblem 1

Does the critical incident technique dependably reveal errors and/or unsafe conditions that lead to industrial accidents?

The types of causal factors revealed by the data collected fom the application of the critical incident technique by investigator A were compared with types of causal factors derived from the data independently obtained by investigator B. Numbers were used to identify the types of unsafe acts and unsafe conditions elicited from the observers by each of the two investigators.

Hypothesis 1—there is no significant difference between the number of accident causal factors identified by investigator A using the critical incident technique and the number of accident causal factors identified by investigator B using the critical incident technique—was then tested.

STATISTICAL TEST OF NULL HYPOTHESIS 1. In this case the same individuals were measured on two different occasions and a comparison of the number of accident causal factors identified by the two investigators was made. The statistical test selected is an adaptation of chi-square for use with noninependent groups developed by McNemar (1955)

$$\chi^2 = \frac{(A-D)^2}{A+D} \text{ with 1 degree of freedom}$$

where:

A = number of casual factors identified by the first interviewer but not by the second

D = number of causal factors identified by the second interviewer but not by the first

This statistic considers the frequency of causal factors that were identified by one investigator and not the other (that is, investigator A not B and investigator B not A), since a net change in identified causal factors between the two investigators must necessarily involve the difference between these two frequencies. The null hypothesis is that the universe frequencies are not different—that is, for a given sample A and D would differ only as a result of chance sampling. Since $A + D$ represents the total number of causal factors that changed between the two applications of the critical incident technique, in setting up the null hypothesis concerning the net change it would seem appropriate to say that if $A + D$ causal factors changed, $(A + D)/2$ would change in one direction and $(A + D)/2$ in the other direction. Thus $(A + D)/2$ would become the expected frequency; then $A - (A + D)/2$ and $D - (A + D)/2$ would become the discrepancy between observed and expected frequencies (on the basis of the null hypothesis). If $A = D$, both discrepancies would become zero. As indicated by McNemar (1955), squaring each discrepancy and dividing by the expected frequency and then summing the two quotients will give a χ^2 that is based on one degree of freedom.

$$\chi^2 = \frac{(A-D)^2}{A+D}$$

where:

$$A = 9$$
$$D = 7$$
$$\alpha = .01$$
$$df = 1$$

$$\chi^2 = \frac{(9-7)^2}{9+7} = \frac{(2)^2}{16} = \frac{4}{16}$$

$$\chi^2 = 0.25$$

χ^2 table = 6.6 at α = .01 for df = *1*
(Reference Table D in McNemar, 1955, p. 386)

CONCLUSION.   The difference between the number of accident causal factors identified by investigators A and B, both using the critical incident technique, is not significant at the .01 level and can be attributed to chance since

$$\chi^2 \text{ computed} = 0.25 < \chi^2 \text{ table} = 6.6$$

χ^2 computed = 0.25 for df = 1 has a probability of occurrence between p = .50 and p = .70, that is, .50 < p < .70. Inasmuch as that probability is larger than the previously set level of significance, α = .01, the null hypothesis cannot be rejected at this significance level.

Subproblem 2

How does the critical incident technique compare with other methods of accident study as a means of identifying accident causal factors?
    Causal factors revealed by the application of the CIT to the randomly selected observers were compared with causal factors derived from reportable and nonreportable (lost-time and minor injury) accident data collected for the workers in the departments studied. In conducting the analysis, numbers were used to identify the types of unsafe acts and unsafe conditions elicited from the observers by each of the two investigators. Hypothesis 2—there is no significant difference between the number of accident causal factors revealed by the critical incident technique and the number of accident causal factors revealed by existing accident records—was then tested.

STATISTICAL TEST OF NULL HYPOTHESIS 2.   In this case the same causal factors were measured by two different instruments, and a comparison of the number of different causal factors revealed by each method was made. The statistical test selected is an adaptation of chi-square for use with nonindependent groups developed by McNemar.

$$\chi^2 = \frac{(A - D)^2}{(A + D)} \text{ with 1 degree of freedom}$$

where:

A = number of accident causal factors identified by the critical incident technique but not by the records

D = number of causal factors identified by the records but not by the critical incident technique

Since the results of testing hypothesis 1 indicated that the two investigators identified the same causal factors, the information derived from the application of the critical incident technique by one of the investigators (investigator A) was used in testing hypothesis 2.

$$\chi^2 = \frac{(A - D)^2}{A + D}$$

$A = 80$
$D = 11$
$df = 1$
$\alpha = .01$

$$\chi^2 = \frac{(80 - 11)^2}{80 + 11} = \frac{(69)^2}{91} = \frac{4761}{91}$$

$$\chi^2 = 52.32$$

χ^2 table = 6.6 at $\alpha = .01$ for $df = 1$
(Reference Table D in McNemar, 1955, p. 386)

CONCLUSION.   The difference between the number of accident causal factors identified by the critical incident technique and the number of accident causal factors identified by the accident records is significant at the .01 level and cannot be attributed to chance since

$$\chi^2 \text{ computed} = 52.32 > \chi^2 \text{ table} = 6.6$$

χ^2 computed = 52.32 has a probability of occurrence beyond $p = .001$, that is, $p < .001$. Inasmuch as that probability is smaller than the previously set level of significance, $\alpha = .01$, the null hypothesis is rejected. The conclusion reached from an examination of the data is that significantly more accident causal factors were revealed by the CIT than were revealed by the accident records.

Other Findings Revealed by the CIT

The ability of the critical incident technique to identify accident causal factors existing along the range of the accident severity spectrum from no-injury results to lost-time injuries is another consideration related to the sensitivity of the technique. The published literature supports the underlying assumption that a range of severity from no-injury to major injury exists for the same type of accident causal factors. The assumption that the severity of accident results appears to be largely fortuitous serves as justification for the collection of noninjurious data for use in predicting and controlling future potential accident losses.

A comparison between noninjurious and injurious accidents using the critical incident technique as the source of comparative information indicated that all injury-producing causal factors revealed by the critical incident technique also appeared as causes of noninjurious accidents. In addition, 19 noninjurious accident causal factor classifications appeared which were not identified with injurious causal factors. Thus, it is seen that all injurious accident causal factors were included in the noninjurious causal factor classifications and that noninjurious accident sources produced new causal factors that were not revealed by the injurious acts and conditions. A comparison between the critical incident technique and the accident records revealed that all causes of lost-time injuries and causes of 100 out of 111 (90.1 percent) of the minor injuries contained in the accident records for the departments studied were also revealed by the critical incident technique. Consequently, the technique is able to identify accident causes not resulting in injuries as well as those resulting in injurious losses with varying degrees of severity. This finding supports the conclusion of other writers in the field who state that injurious and noninjurious accidents emanate from the same causes (Blake, 1953; Chapanis, 1959; Heinrich, 1959; Suchman and Scherzer, 1960).

Of special interest was the proportion of incidents that were estimated by the observers to have occurred *every day* during the two-year period studied. It was found that 76 out of the 102 different incidents reported, or 74.5 percent, were estimated by at least one observer to have occurred *every day* during the two-year period. Apparently within the population studied there is daily exposure to a considerable number of potential injury-producing situations as a result of the daily repetition of numerous unsafe acts and conditions.

An approximation of the adequacy of the sample size chosen was obtained from a backward glance at the number of new incidents identified by each successive observer interviewed. For the first interviews the observers were randomly assigned to the two investigators by means of a table of random numbers. The order of assignment was reversed for the second interview to reduce the influence of the time lag between interviews on the incidents obtained during the second interview period. Analysis of the cumulative frequency of new incidents indicated that the first 22 observers revealed approximately 97 percent of the total number of incidents revealed by all 30 observers interviewed by each investigator. Plotting a cumulative frequency distribution of new incidents revealed by each successive observer interviewed was found to assist in determining when a sufficient number of interviews had been conducted.

THE WESTINGHOUSE CIT STUDY

Another critical incident technique study was conducted at the Underseas Division of the Westinghouse Electric Corporation plant at Baltimore, Maryland. One purpose of the Baltimore study was similar to the Western Electric study, namely, to evaluate the usefulness of this technique as a method for identifying potential accident causes. A second purpose was to develop procedures for its practical application by in-plant personnel. In earlier studies trained interviewers were utilized. For the Westinghouse study it was decided to use the resident safety specialists as interviewers. The study was conducted under the direction of the Division of Accident Research of the Bureau of Labor Statistics in Washington, D.C. The BLS was interested in further exploring the value of this technique as a research tool. The study was initially proposed by the Accident Prevention Administrator of Westinghouse. He was interested in examining the usefulness of the critical incident technique as a method for identifying potential accident problems in the corporation's plants.

The Problem

The answers to three questions were sought:

1. Does the critical incident technique reveal information about accident causal factors in terms of human errors and unsafe conditions that lead to potential industrial accidents?
2. Does the technique reveal a greater amount of information about accident causes than presently used methods of accident study?
3. Is the technique useful for Westinghouse plants?

Method

The population selected for study included approximately 200 employees in the Underseas Division of the Baltimore plant. After eliminating those who had been on the job less than one year, those who had left Westinghouse, and others who were not available for other reasons, such as transfer, the list was reduced to 155 employees. This group was chosen because the operations in which they were engaged are typical of general manufacturing establishments and the equipment and processes involved in these operations are found in most industrial plants. Employees from two work shifts and two plant locations participated in the study. Both male and female workers were included as subjects. These factors determined the major stratifications in the sample selection process. The selection of additional stratifications was based on the "budget center unit" and the type of equipment or operation involved. The criteria for selecting the various subject stratifications were determined by the number of factors judged to have an influence on the nature of the exposure to potential accidents. These factors included the work shift, the plant location, the male-female differential, the budget center unit (which, in turn, produced a supervisor variable), and the type of equipment involved or the particular job performed by the worker.

An approximate 10 percent sample of 20 workers was selected as subjects for the study. These subjects, called participant-observers, were chosen by a stratified random process using a table of random numbers and the stratifications previously defined. Since participation in the study was on a voluntary basis, it was anticipated that a few persons would elect not to participate. The representative integrity of the sample was preserved by selecting additional individuals from within each stratification by the same random process. Then, when a person chose to decline, he or she was replaced by the next individual on the randomly selected list within the same stratification. Only one subject out of the original 20 selected at random elected not to participate.

Preliminary interviews approximately 15 minutes in length were conducted with each subject. During these interviews a prepared statement describing the study and its objectives was read and any questions asked about the study were answered (see "Critical Incident Technique Sample Preliminary Interview," in Appendix E). Each person was given the opportunity to decline at this point. None elected to do so. Later, after apparently thinking it over, one subject withdrew.

Each subject was advised that both management and the union were in agreement concerning the conduct of the study and supported its objectives. At the end of each preliminary interview the subject was presented with a copy of the prepared descriptive statement along with a list of typical incidents that had occurred in similar operations within other plants (see Appendix E). Unsafe practices and conditions prohibited by Westinghouse company safety rules were also included. The purpose of this list was to stimulate the recall process and to identify specifically the types of information sought. The subjects were told that they would remain anonymous as individuals in relation to the information they provided, that they would not be penalized for participating in the study, and that no blame would be imposed as a result of the information they revealed. A minimum of 24 hours was allowed between the preliminary interviews and the data-collection interviews in order to provide sufficient time for recalling incidents.

The subjects were first requested to recall the last time they had observed or participated in an unsafe error or an unsafe condition or operation in the plant. No distinction was made between observation and participation. Subjects were then requested to think back over the past year and to recall and describe fully any incidents occurring during this period of which they were aware, regardless of whether or not the incident resulted in injury or property damage. The typical incidents included on the previously presented list (see Appendix E) were then converted into probe questions, and each participant-observer was asked if he or she could recall observing any of them. This procedure usually resulted in obtaining a considerable number of incidents in addition to those revealed by the initial open-ended questions. The interviews were recorded on magnetic tape, and the participants were informed in advance that this method of data recording would be used.

The data-collection interviews were conducted by two members of the safety staff at the Westinghouse Baltimore plant. The investigators were given a brief orientation consisting of a discussion of the objectives of the critical incident technique, a review of the procedures for its application, and general instructions concerning interviewing methods.

The investigators questioned the subjects on each incident they described until sufficient information was provided to identify the human errors and unsafe conditions involved. (In all of our critical incident studies error and unsafe conditions were defined as given on page 306.)

The length of the open-ended interviews varied from 25 minutes to 1 hour and 40 minutes, with an average length of about 47 minutes. The observers were very frank and candid in their incident descriptions. For the most part, they spoke freely and were quite willing to tell every incident they could remember. Occasionally someone would get "carried away" and begin talking about personal problems or gripes. As soon as it was practical to do so, the investigator would attempt to guide the worker's thinking back into a "critical incident" channel.

Results

Approximately 14 hours and 10 minutes of taped incidents were collected from the Baltimore study. A total of 389 incidents were identified by the 20 participant-observers during the interviews. Analysis of these data revealed 117 different types of incidents occurring in these operations during the year studied. The number of different incidents revealed by each subject ranged from 4 to 41 with a mean of approximately 19 and a standard deviation of 8.7. Four subjects disclosed 30 or more incidents each, four revealed between 20 and 30, and the remainder revealed 12 or more incidents each, with the exception of one female who could think of only four.

The cumulative frequency distribution of new incidents revealed by each successive person interviewed was plotted. The graph indicated that 12 subjects provided 73.5 percent of the different types of incidents revealed; 14 subjects provided 86.3 percent; 16 subjects, 88.1 percent; 17 subjects, 94.1 percent; and 18 subjects, 97.4 percent of the total information obtained in the study. Thus, the study could have stopped with 17 subjects and still have obtained well over 90 percent of the total information received from all 20.

During the one-year period for which critical incidents were collected, a total of *206* minor and *6* serious but nondisabling injuries occurred within the population studied. Serious injuries included lacerations requiring at least one suture, fractures, and foreign bodies in the eye requiring a physician's attention. There were no disabling or lost-time injuries during this period. In all cases where cause information was contained in the injury report, these same causal factors were also identified by a participant-observer during the critical incident interviews. This supports the finding of the previous study that most of the causes of recorded minor injuries that occurred during the period studied were revealed by the *critical incident technique*. In addition the technique revealed numerous potential accident causes not identified by existing records. In the Western Electric study, where injury cause data were available for comparison, it was found that 52.1 percent more unsafe acts and conditions were revealed by the critical incident technique than were identified by the accident records during the two-year period studied. Thus, it was found that this new technique is capable of *identifying causes* of accidents at the no-injury stage *before* they result in losses of sufficient magnitude to appear in any of the current "reportable" categories.

Similar to the Western Electric finding, it was discovered in the Westinghouse study that 67.52 percent of the different incidents reported were estimated by at least one participant-observer to have occurred *every day* during the year studied. This means that there was a tremendous exposure to potential injury-producing accidents as a result of the daily repetition of numerous unsafe acts and conditions. Under the present accident appraisal system, these potential loss-producing situations would normally not be revealed until losses with a certain degree of severity actually occurred. Using the CIT, most causal factors can be identified well in advance of injury losses.

WHAT CAN BE EXPECTED FROM A CIT STUDY

The results of the Westinghouse study and other studies proved that the CIT is a valuable safety measurement tool. These studies showed that

1.   The critical incident technique dependably reveals. causal factors in terms of errors and unsafe conditions that lead to industrial accidents.

2. The technique is able to identify causal factors associated with both injurious and noninjurious accidents.
3. The technique reveals a greater amount of information about accident causes than presently available methods of accident study and provides a more sensitive measure of total accident performance.
4. The causes of noninjurious accidents as identified by the critical incident technique can be used to identify the sources of potentially injurious accidents.
5. Use of the critical incident technique to identify accident causes is feasible.

The discovery that the critical incident technique is sensitive to accident problems that have a *potential* for accident loss but have not yet produced a loss provides a significant advancement in safety performance measurement capability. The existence of an injury or property damage loss is no longer a necessary condition for appraising accident performance. It is now possible to identify and examine accident problems "before the fact" instead of "after the fact" in terms of their injury-producing or property-damaging consequences. This allows the safety professional to concentrate on the measurement of *loss potential* or near-misses and removes the necessity of relying on measurement techniques based on the probabilistic, fortuitous, rare-event injurious accident.

Future Applications

What can be expected from future applications of the critical incident technique? Measuring and analyzing the circumstances of accidents according to critical incident procedures can produce the following results:

1. Identify and locate the principal sources of accident and health problems by determining, from actual incidents, the materials, machines, and tools most frequently involved in accidents and the jobs or environments most likely to produce injuries or health problems.
2. Disclose the nature and size of all safety and health problems in departments and among occupations.
3. Provide an insight into accident causes and health problems without waiting for losses in terms of injuries and/or property damage.
4. Provide before-the-fact information about accident and health problem areas for use as a basis for accident prevention programming. (This results in a progression beyond the "fire-fighting" stage and leads to true prevention work.)
5. Indicate the need for engineering revision by identifying the principal unsafe and unhealthful conditions of various types of equipment, materials, and environments.

6. Reveal inefficiencies in man-machine-environment systems, operating processes and procedures where, for example, poor layout contributes to accidents or where stress-producing methods or procedures overtax the physiological or psychological limitations of the operator.

7. Uncover unsafe practices that need special attention in the training of employees.

8. Disclose instances where inabilities or physical handicaps contribute to accidents.

9. Enable supervisors to use the time available for safety work to the greatest advantage by providing them with information about the principal hazards and unsafe practices in their department.

10. Permit a relatively objective evaluation of the progress of a safety program by noting in continuing analysis the effect of different safety procedures, educational techniques, engineering improvements, and other methods adopted to prevent accidents.

11. Provide a more sensitive measurement technique by including no-injury or near-miss accident information in the causal identification system. Lost-time accidents are relatively rare events and are often not sensitive to the influence of safety programming improvements.

12. Circumvent the formal, written report that presently serves as a limited communications instrument for accident problem information. The technique provides a direct approach to causal information in the search for accident problem areas.

13. De-emphasize the placement of blame on individuals and give increased emphasis to the discovery of the problems themselves.

14. Provide an improved criterion of safety performance for use by researchers who study the accident phenomenon.

PRECAUTIONS IN APPLYING THE CIT

It is assumed that the observers are capable of perceiving certain acts and conditions occurring within a particular setting as being unsafe. Therefore, care must be taken to ensure that the unsafe acts and conditions observed are clearly recognized as such by the observers. Both written and oral instructions should be given to the observers so that they will become fully aware of the crucial aspects of the acts and conditions they have seen. Appendix E contains a "partial list" of typical incidents which will assist the participant observer in recalling specific incidents and assist the investigator in developing probe questions for use during the data collection interview. This "partial list" is presented as an example applicable to general manufacturing operations. In actual use a "List of typical incidents" should be prepared which is applicable to the specific operations being studied.

There is danger that the frequency of mention of a critical incident will be confused with the "degree of criticalness." As Corbally (1956) suggests, some incidents may occur more often than others in the course of the job because of the nature of the job itself. In addition, the estimates of the frequencies of incident occurrences as reported by the observers are not considered meaningful per se, because of the necessity for making general approximations.

The isolation and categorization of unsafe acts and unsafe conditions may be difficult in certain instances where the meanings of descriptions supplied by the observers may be more obscure. The interviewer should be alert to these obscurities and introduce probe questions designed to elucidate the causal factors involved.

Instructions should clearly convey to the observers that they are to describe only errors they have made or observed and/or unsafe conditions they observed. Instructions should warn against evaluating the behavior or condition.

The critical incident technique only determines that certain unsafe acts and/or unsafe conditions occurred. It does not probe at the reasons for a person's behavior or reasons for the unsafe condition. Furthermore, it does not tell what actions should be taken to correct the unsafe act or unsafe condition once it has been identified.

The unsafe acts and unsafe conditions revealed are limited to those that have actually been seen by the observers included in the selected sample. If an inference is to be made concerning the unsafe acts and unsafe conditions that have occurred within a particular department or throughout an entire plant, great care should be taken to assure that the observers selected constitute a representative sample of the larger population.

If personal penalties are invoked upon the observers or other persons in the plant as a result of the information they reveal, the workers may become reluctant to participate in future critical incident interviews. It is important that the identity of the observers be kept anonymous in relation to the incidents they describe and that no one is penalized as a result of the critical incident study.

Increased knowledge of all accident causal factors will enhance the safety professional's ability to define his accident problems. Subsequently, with better information, managers of industrial plants will be able to make better decisions regarding the allocation of accident prevention resources. The critical incident technique has the potential for providing this needed knowledge, thus allowing industry to progress toward the objective of maximizing its accident problem identification capability.

CONCLUSIONS

The objective in applying the *critical incident technique* is to improve internal measurement capability. We know that measurement is an absolute prerequisite for control, whether this be the control of production or accidents.

Accurate measurement teamed with proper control provides prediction of future performance.

We know that attempts to control accidents and their consequences can, at best, be described as trial and error chiefly because adequate measures of the effectiveness of this control do not now exist. New measuring instruments such as the *critical incident technique* hold much promise as improved methods of measuring safety effectiveness.

An issue that often arises when the CIT is discussed is whether or not critical incidents or near-accidents come from the same population as loss-type accidents (disabling injuries, recordable accident cases, property damage accidents, and so forth). If the CIT clearly produced data that are not from the same population as injury or loss accidents, then its value might be questioned. The question to be asked is, Are we obtaining information useful for accident prevention purposes or not? This issue can be settled once and for all by examining the basic causal factors associated with both the critical incidents and the disabling injury, recordable incident, fatality, and property damage types of accidents. The causal factors of unsafe behavior (errors of omission or errors of commission) and unsafe conditions (equipment, environments, and so forth) are by definition *identical* in both the CIT and the accident loss events. Unsafe behaviors, equipment, machinery, and environments identified by the CIT process are *defined* as those behaviors and conditions that, in the past, have actually *produced* losses in similar situations. If these behaviors and conditions are permitted to continue, then at some point in time, a re-occurrence of these events will produce future losses. Which one of these events will produce a loss-type accident is not detectable by the CIT and criticism of this technique for not predicting which future event will produce a loss is unwarranted.

The CIT is intended to provide information about *loss-potential* behaviors and conditions. It is not intended to predict which exposure will produce a loss in the future. Furthermore, by definition, the unsafe behaviors and conditions identified by the technique are potentially loss producing because only those errors and conditions that have actually resulted in accident losses in the past are included in the incident descriptions. These losses may not have occurred at a particular location during a particular time period. Instead, we are opening up for consideration a vast number of similar operations by including in our definition the actual loss data from a broad universe of similar operations. For example, we know that placing the hands between the dies of a punch press has produced loss in the past. We do not need to wait for loss to occur again before we are able to define this action (incident) as having an unsafe behavior component and perhaps an unsafe condition component as well. The CIT will reveal that this unsafe behavior or condition has occurred. An investigation of a disabling injury accident resulting from the same situation will also disclose that an unsafe behavior or unsafe condition has occurred. By *definition* the underlying causal factors are *identical* in the CIT and in the loss-type accident investigation. No doubt the reader can recall his or her own "incidents" where exposure to unsafe conditions or commission or omission of certain behaviors has produced disabling or recordable accidents in some instances and no measurable loss in others. Yet the potential for loss is present with *each exposure*. True loss prevention and cost control occur when action is taken to correct a problem that has been revealed by the CIT without waiting for that problem to produce a loss with a measurable cost.

No matter what measure evolves, there will be those who misapply it or criticize it for not doing something it was not intended to do or for not solving all safety measurement problems. For example, the critical incident technique does not tell which countermeasures to select or how best to solve the problems it detects. Other techniques are available for this purpose, such as hypothesis testing with various experimental designs that provide objective measures of which countermeasure or combinations of countermeasures are most effective. At one level we need a measurement system that will reveal what the problem is—a *problem identification* technique. At another level we need a measurement system that will tell what to do about the problem—a *problem solution* technique. Very rarely, if ever, will a single measuring instrument perform both functions. Because of the role of chance, the consequences of accidents need not be fatal, they need not be injurious, they need not even be property damaging. But the *potential* for producing loss is always present. As long as there is such potential, our primary concern lies with environmental conditions that are not right and with human conditions that are not right, regardless of whether or not they demonstrate any statistical correlation with injury involvement within any fixed time period.

Studying injury accidents, therefore, is less profitable than studying work conditions and worker behavior in a broad sense. Fortunately most of these conditions, human and environmental, are modifiable or compensable. Correct or adjust these conditions and accident losses will inevitably be reduced.

In the evaluation of safety programs as well as in research generally, injury frequency rates, injury severity rates, BLS incidence rates, and other currently used measures of loss-type accidents are not sufficiently sensitive, stable, or representative to serve as criteria for safety effectiveness. What is needed are measures of safety performance that do not depend upon injury involvement.

Since there is increasing evidence that unsafe acts—regardless of the occurrence of injury—are inconsistent with desirable production or service, the measurement of improper or inefficient work performance and near-misses will enable us to broaden our loss prevention activity by gathering information more representative of the true state of our system. The *critical incident technique* provides this needed knowledge, thus allowing us to improve significantly our accident problem identification and control capability.

References

American National Standards Institute. *Classifying Accident Causes: Z-16.2.* New York: The Institute, 1956.

American National Standards Institute. *Method of Recording Basic Facts Relating to the Nature and Occurrence of Work Injuries: Z-l6.2.* New York: The Institute, 1962, r. 1969.

American National Standards Institute. *Method of Recording and Measuring Work Injury Experience: Z-16.1.* New York: The Institute, 1967, r. 1973.

Blake, R. P. *Industrial Safety.* Englewood Cliffs, N.J.: Prentice-Hall, 1953, pp. 134–135.

Brody, L., and F. Dunbar. *Basic Aspects and Applications of the Psychology of Safety,* New York: Center for Safety Education, New York University, 1959, p. 9.

Chapanis, A. *Research Techniques in Human Engineering.* Baltimore: John Hopkins Press, 1959, p. 85.

Corbally, J. E., Jr. The critical incident technique and educational research. *Educational Research Bulletin, 35*: 57–62, 1956.

Fitts, P. M., and R. E. Jones. *Psychological aspects of instrument display, I. Analysis of 270 "pilot error" experience in reading and interpreting aircraft instruments.* Memorandum Report TSEAA-694-12A. Dayton, Ohio: U.S. Air Force, 1947.

Flanagan, J. C. The critical incident technique. *Psychological Bulletin, 51*:327–358, 1954.

Heinrich, H. W. *Industrial Accident Prevention.* New York: McGraw-Hill, 1959, pp. 26–32.

Mayhew, L. B. The critical incident technique in evaluation. *Journal of Educational Research, 49*:591–598, 1956.

McNemar, Q. *Psychological Statistics.* New York: Wiley, 1955, pp. 228–231.

Suchman, E. A., and A. L. Scherzer. *Current Research in Childhood Accidents.* New York: Association for the Aid to Crippled Children, 1960, p. 2.

Tarrants, W. E. *An evaluation of the critical incident technique as a method for identifying industrial accident causal factors.* Unpublished doctoral dissertation, New York University, 1963. (Also available from University Microfilms, Inc., Ann Arbor, Michigan.)

U.S. Department of Labor. *Recordkeeping Requirements Under the Williams-Steiger Occupational Safety and Health Act of 1970,* rev. With amendments October 1972 and January 1973. Washington, D.C.: Occupational Safety and Health Administration, 1975.

APPENDICES

SOME FURTHER INADEQUACIES OF INCIDENCE RATIOS

The theory of measurement is concerned with the use of numbers to represent certain characteristic properties of systems and their components. Sometimes it seems that in order to apply the theory the only problem usually considered is how to obtain the numbers, that is, how to take the measurements. But numbers are, after all, only symbols. Unless they are relevant to the system's properties and are manipulated meaningfully, they are apt to be useless as measures. The need for relevance plus the necessity that the measured values be capable of intelligible handling are requirements that existing measures of safety performance do not seem to satisfy. Consideration of these weaknesses appears to be important to the consideration of suggested improvements.

With numbers one can do as one pleases, but with measurements one can do only what reality permits. This is a complication that familiarity with common measurements may obscure from the general user of standard safety performance measures. For example, weight measurements can be manipulated with comfort, allowing several evaluative comparisons. It can be said, for instance, that a 10-pound bundle of feathers is twice as heavy as 5 pounds of lead or that the feathers weigh 5 pounds more than the lead and that the two weights total 15 pounds. Such an expressive manipulation is not possible with existing safety measures, yet such relative evaluations of safety performance mistakenly occur. Many experienced administrators are inclined to say that a disabling injury frequency rate half that of another indicates a doubly effective safety performance. It seems important for the sake of utility that a measure should allow meaningful comparisons. At least it should discourage easy misinterpretation.

Weight measures (and frequency rates) are examples of *ratio scale*[1] applications. Algebraically, ratio scales have the form $Y = bx$ and may be contrasted with *interval scales,* which algebraically are $Y = a + cx$. Although several scale types are commonly used, and others are possible, the two given will enable the illustration of certain limitations of existing safety measures.

Measurements need not employ scaled values. It is possible, of course, to review a performance—as an expert analyst for example—and give an opinion of its quality without ranking or relating the performance to either its own record or that of others. Certain auditors and consultants make careers of such reviewing and reporting. Administrators, however, generally want to rate levels of progress, and this usually requires the employment of a measurement scale. The scale performs its function when appropriate distinction can be made between the values it depicts.

The interval scale, contrasted to the ratio scale, illustrates the significance of differences between the interpretations permitted by the scale designs. The Fahrenheit temperature scale is an example of an interval scale. If an open-door refrigerator at 34° F is considered to be in a room at 68° F, it would be reasonable to refer to their average temperature as $(34 + 68)/2 = 51°$ F. It would not seem correct to say one is twice as hot as the other, and it is difficult to understand what is meant if it were said one is 34° hotter. For the ratio scale there is a *natural zero,* which exists when there is a satisfactory answer to the question: Is there real meaning if there is nothing or none of the quantity measured? The concept of having no, or zero, weight has a clear meaning: zero degrees Fahrenheit does not allow an interpretation of no temperature.

A basic principle is involved here; namely, no operation can be permitted that changes the *real characteristics* of the system measured as a result of the selection of the units of measurement. This may be explained algebraically in the following example:

Assume two measurements, X_1 and X_2, and that an arithmetic operation is done with them to give a resulting value V. If the same measurements were taken with different units, giving Y_1 and Y_2, and the same arithmetic is done with them yielding Z, then it is required that V have the same relationship to Z that X has to Y and $V = Z$. This is the situation when the ratio of a circle's circumference is expressed with respect to its diameter. Regardless of the units used in measuring the diameter, $C/D = 3.14$ or the principle is violated.

Now consider the inappropriateness of relating the safety performance of an operation with a disabling injury frequency rate of 1.12 to another whose rate is 2.24. Is an operation with a rate of 1.12 twice as safe as an operation with a rate of 2.24? Would a zero rate indicate perfection? These distinctions are beyond the capability of existing safety measurements. Without knowing the quality of the hazard control effort involved, such interpretive differentiations clearly are improper.

Notes

1. The concept of measurement scales has been discussed in Chapter 1. Measurement concepts have been reviewed critically by W. S. Torgerson, *Theory and Methods of Scaling,* (New York: Wiley, (1958), and B. Ellis, *Basic Concepts of Measurement,* Cambridge, Eng.: University Press, (1966), and the theory has been detailed by C. W. Churchman and P. Ratoosh, *Measurement: Definition and Theories,* (New York: Wiley, 1959) among others.

THE FEASIBILITY OF USING COST-ACCOUNTING DATA TO MEASURE ACCIDENT CONTROL EFFECTIVENESS

John V. Grimaldi

At present there is no valid indicator of an operating department's effectiveness in controlling accidents. Managers now are alerted to a deficient performance usually by

1. The rate of injury occurrence
2. Plant safety audits

A low injury rate, however, does not necessarily imply effective accident control. Chance may be largely responsible. Also, measuring accident control according to the rate of accident occurrence is like appraising a barn door lock by counting horses that are stolen. A measure that would indicate effectiveness in controlling accidents and that would be independent of the occurrence of injuries would have merit. If successful, it would forecast severe accidents. To some extent, plant audits provide this intelligence, but their value is a function of many factors, for example, the auditor's knowledge, attitude, available time, and so on, and thus they have questionable reliability.

A complication in appraising safety effectiveness has been the logical question, How much control is practical? It is apparent, but unreasonable, that plant injuries can be eliminated by stopping work—so could scrap, rework, and other costs that are incidental to production but accidental in their occurrence. The measure sought, therefore, should reflect management's ability to control unwanted losses while striving to reach an optimum profit objective.

In order to reach this goal, it was considered that

1. Accidents are the result of either an improper (or inadequate) operational procedure or failure of personnel to carry out a proper procedure. In either case the control rests with the plant management. (In this concept there is as much interest in the machine tool that is damaged, the spoilage of the product, and destruction by fire as in the injury to personnel.)
2. Certain cost-accounting factors, when used in a suitable relationship, might indicate manufacturing management's effectiveness in controlling accidents in general and broadly might warn of a growing need for tighter control before a severe accidental loss occurs.

It should be noted that although this study was based on data collected in the middle to late 1950s, the principles, methodology, and conclusions remain equally applicable today. This study was and still is the definitive work on this subject.

The Problem

In order to evolve a measure that answered management's need, it was necessary to investigate the feasibility of developing a measure that would enable an operating component to do three things:

1. Appraise its effectiveness in controlling accident losses in general.
2. Obtain warning of its slipping out of control with respect to its keeping accidental losses to a minimum.
3. Apply the measure by using data that are customarily collected and already available. There should be no significant additional expense or inconvenience associated with the application.

The Hypothesis

Disabling injuries (the conventional target for attack by safety programs) are only one type of accident. Management oversights, which precede their occurrence, are also associated with such unwanted (accidental) operational results as scrap, customer complaints, and increased shop costs. Hence, there may exist a statistical intercorrelation.

The Method

A number of cost-accounting elements were selected as *management variables.*

1. Scrap cost (includes overhead)
2. Rework cost (excludes overhead)
3. Direct labor cost
4. Shop cost of product produced (includes overhead but not engineering and development)
5. Workers' compensation[1]
6. Maintenance (buildings)
7. Maintenance (equipment)
8. Sales value
9. Complaint costs

Several *safety variables*[2] were considered:

1. Absenteeism
2. Days lost and charged due to disabling injuries
3. Number of disabling injuries
4. Total of all injuries (disabling and minor)
5. Total man-hours

The safety variables customarily are reported as rates (that is, injuries/man-hours) in order to place large and small operations on an equivalent basis. The management variables, on the other hand, were not given in such a way. Therefore, the study analyzed the management and safety variables as given as well as on an equivalent man-hour basis.

The data for each variable were obtained from the following 17 businesses (chosen at random from the 108 diverse enterprises of a multibusiness, decentralized corporation) for a three-year period.

1. Large motor and generator
2. Locomotive and car equipment
3. Industry control
4. Capacitor
5. Distribution transformer
6. Large steam turbine
7. TV receiver
8. Automatic blanket and fan
9. Home laundry
10. Metallurgical products
11. General-purpose motor

12. Hermetic motor
13. Conduit products
14. Instrument
15. Heavy military electronic equipment
16. Receiving tube
17. Semiconductor

It will be seen that a substantial variety of industrial businesses is represented.

The method of investigation required summarizing the data in each group of variables (that is, management and safety) into a smaller and more compact set. The data first were converted to logarithms as a preliminary to conducting a factor analysis. The factor analysis method converts the variables into factors that are expressible as linear combinations of the raw variables. The mathematics of such an analysis are represented in Addendum A to Appendix B. The use of logarithms in the performance of the factor analysis was desirable since some of the variables, when provided, were expressed as ratios (for example, scrap cost/shop cost). In the application of the data, the numerator and denominator of certain ratios were considered separately (for instance, scrap and shop cost). It was believed that this and other ratios were significant expressions of performance since the businesses' accountants must have good reason for reporting the values in such a relationship. In using the logarithms of the variables (management and safety) a difference of logarithms corresponded mathematically to a ratio of the original variables and a difference is one form of linear combination. Another reason for performing factor analysis of the logarithms of the data was that it was felt the total man-hour exposure effect would enter as a ratio and thus could be removed, allowing all data to be considered on an equivalent man-hour basis. Also, some variables were expressible in terms of others, as ratios. For example, the cost accountants gave "scrap" in dollars, "rework" in dollars, "percent of scrap losses to direct labor costs," and "percent of rework losses to direct labor." The variable "direct labor costs" was computed from these data. When scrap and direct labor costs are expressed in logarithms and the pertinent variable is "percent of scrap losses to direct labor costs," the regression equation, in logarithms, will have approximately the same regression coefficients for the variables "scrap in dollars" and "direct labor costs" but of opposite sign. Conversely, if in the regression equation "scrap in dollars" and "direct labor costs" occur with the same numerical coefficients but with opposite signs, the implication is that the pertinent variable is "percent of scrap losses to direct labor costs."

The factor analysis summarization reduced the twelve *safety variables* (four variables for three years) to four *safety* factors, with little loss of information, since the criterion used was that the factors should retain 90 percent of the original information when tested statistically. Similarly, the three man-hour variables (one variable for three years) were reduced to one factor and the management variables were reduced from twenty-seven variables (nine variables for three years) to six management factors. These variables, summarized into a compact set of data, then required analysis. The analysis should show all relationships that might be present among the variables. To facilitate this analysis, a stepwise linear regression technique was employed. This procedure yields statistical equations indicating the relationship between managerial and safety factors.

In computing the regression analysis, the following equations resulted:

With man-hours

$$\hat{Z}_1 = -.46215 \ Z_7 + .74929 \ Z_5 - .0055121$$
$$\hat{Z}_2 = \text{(no significant relation)}$$
$$\hat{Z}_3 = .46623 \ Z_9 + .69084 \ Z_{10} - .1518250$$
$$\hat{Z}_4 = .54184 \ Z_9 - .0623008$$

Without man-hours

$$\hat{Z}_1 = .72425 \ Z_6 - .50519 \ Z_7 - .22044 \ Z_{11} - .0091230$$

The equations for $\hat{Z}_2$, $\hat{Z}_3$, $\hat{Z}_4$ are the same with and without man-hours as can be seen from the fact that they are independent of Z_5 when man-hours are included. The "caret" over the Zs ($\hat{Z}$) represents the estimated Zs as predicted by the analysis.

See Addendum B of this appendix for the regression coefficients, standard deviations, and the degrees of freedom. The correlations matrix is included in Addendum A. The diagonals are not unity because the means have been removed in the factor analysis process. Using the regression equations, the hypothesis that a relationship exists between managerial and safety factors can be tested and is substantiated. In applying the equations the coefficients of the regression lines are multiplied by the component factor scores (Table B-1) of the corresponding factor. That is, in the case of $\hat{Z}_4$ we would have

$$\hat{Z}_4 = .54184 \left\{ \begin{array}{r} -122 \\ 924 \\ -272 \\ 301 \\ 692 \\ -132 \\ 136 \\ -260 \\ 654 \end{array} \right\} - .0623008 =$$

The results are tabulated in Table B-2 (p. 336).

Table B-1
Average Factor Scores (Three-year Period—14 Variables)

		Z_1*	Safety Factors			Man-Hours	Management Factors					
			Z_2	Z_3	Z_4	Z_5	Z_6	Z_7	Z_8	Z_9	Z_{10}	Z_{11}
Safety												
Absenteeism	x_1	172	91	35	247							
Days lost	x_2	1329	635	456	−538							
No. disabling injury	x_3	863	−340	−102	993							
Total no. injuries	x_4	322	577	−742	−1014							
Total man-hours	x_5					1730						
Managerial												
Scrap	x_6						610	1280	385	−122	283	770
Rework	x_7						618	−239	330	924	−775	47
Direct labor	x_8						522	67	71	−272	57	−440
Shop	x_9						450	−304	−166	301	−176	1330
Workers' comp.	x_{10}						501	−489	−711	692	−307	803
Building maint.	x_{11}						617	163	−1030	−132	671	−445
Equipment maint.	x_{12}						370	22	179	136	321	184
Sales	x_{13}						451	−264	−133	−260	−303	83
Complaints	x_{14}						802	627	1008	654	447	122

*Z_1 is the general safety factor.

335

Components of Management Variables on $\hat{Z}_1$ (Without Man-Hours)

$$\hat{Z}_1 = .72425 \ Z_6 - .50519 \ Z_7$$
$$- .22044 \ Z_{11} - .0091230$$

Component of	on $\hat{Z}_1$	
x_6	−384	scrap costs
x_7	553	rework costs
x_8	433	direct labor costs
x_9	178	shop costs
x_{10}	420	workers' compensation
x_{11}	456	building maintenance
x_{12}	407	equipment maintenance
x_{13}	433	sales
x_{14}	228	complaint costs

Component of Management Variables on $\hat{Z}_3$

$$\hat{Z}_3 = .46623 \ Z_9 + .69084 \ Z_{10} - .1518250$$

Component of	on $\hat{Z}_3$	
x_6	− 13	scrap costs
x_7	−256	rework costs
x_8	−240	direct labor costs
x_9	−135	shop costs
x_{10}	− 42	workers' compensation
x_{11}	250	building maintenance
x_{12}	7	equipment maintenance
x_{13}	−482	sales
x_{14}	462	complaint costs

Component of Management Variables on $\hat{Z}_4$

$$\hat{Z}_4 = .54184 \ Z_9 - .0623008$$

Component of	on $\hat{Z}_4$	
x_6	−128	scrap costs
x_7	438	rework costs
x_8	−209	direct labor costs
x_9	101	shop costs
x_{10}	312	workers' compensation
x_{11}	−134	building maintenance
x_{12}	8	equipment maintenance
x_{13}	203	sales
x_{14}	291	complaint costs

The regression analysis of the six management factors indicated that three factors (Z_6, Z_7, and Z_{11}) could be combined to give a composite that is highly significantly related to the general safety factor Z_1, which is essentially a measure of severity—see Table B-1 for the factor score. (The first and most prominent safety factor that has emerged is considered "the general safety factor.") This composite managerial factor shows a strong relationship to rework costs and, to a slightly lesser extent, all of the other variables except shop and complaint costs. The complaint cost is only slightly related, and a puzzling aspect is the negative relation to scrap costs. Perhaps this comes out of the relatively small sample of businesses studied (17 out of 108). This may be explained by variables that probably exist in the businesses' calculations of scrap costs. Also since scrap costs and complaint costs are inversely related quite strongly, the results may suggest that an effort to reduce one may adversely affect the other. For example, it might be concluded that a drive to reduce scrap costs may result in a poor quality product unless effective controls are in place.

The second safety factor was not significantly related to the managerial factors, but the third (Z_3), essentially a measure of average severity (that is, the ratio of severity to frequency rates), is significantly related to two of the managerial factors. The importance of this factor to the management variables is largest for complaint costs, and it shows a strong negative relation to sales, a slight positive association with building maintenance, and a negative relation to rework, direct labor, and shop costs.

The correlation of management factors and the four safety factors, with man-hours included, are shown in Table B-3. The four safety factors were then correlated with the management factors after removing the effect of man-hours. These correlations are given in Table B-4. The resultant multiple correlations of the safety factors as predicted from the management factors

Table B-3
Intercorrelation of Management and Safety Factors with Man-Hours

Management Factor	Safety Factor 1	2	3	4	5
5	.771	.147	−.339	.131	1.000
6	.720	.098	−.300	.191	.889
7	−.495	.315	−.201	.197	−.056
8	−.068	.386	−.024	.136	−.117
9	−.159	−.339	.006	.578	−.162
10	.007	.103	.461	−.259	−.061
11	−.222	−.082	.079	−.326	−.216

Management Factor	\	\	\	Safety Factor	\	\	\	\	\	\
	6	7	8	9	10	11	1	2	3	4
6	.458									
7	.1056	.998								
8	.2958	−.0086	.993							
9	.4130	−.0092	−.0214	.987						
10	.1534	−.0034	−.0075	−.0124	.998					
11	.5563	−.0123	−.0260	−.0363	−.0156	.976				
1	.1530	−.7119	.0350	−.0533	.0852	−.0902	.637			
2	−.0929	.3281	.4100	−.3226	.1133	−.0525	−.1817	.989		
3	.0032	−.2339	.0682	−.0523	.4690	.0060	.4361	.0522	.941	
4	.2135	.2086	.1554	.6186	−.2566	.3107	−.1618	−.0199	.0462	.981

were left effectively unchanged by the removal except for the general safety factor. The multiple correlation of the first safety factor is increased when the data are considered on a per man-hour basis, and the composite managerial factor was most affected by this consideration. (The effect of man-hours was "held constant" by using a mathematical technique.)

In general, it may be said that when the data were placed on an equivalent man-hour basis, little effect was noticed in the multiple correlations of the safety factors as predicted from the management factors.

The general safety factor is highly correlated with the composite managerial factor, and it has certain relationships with the management variables. These relationships are determined from the preceding tables (notice the minus in front of 384 in the components of management variables on $\hat{Z}_1$ in Table B-2 indicating the negative relation of scrap costs). One of the conclusions then is that if all of the variables are kept at a fairly low level, that is, building maintenance costs, shop costs, and so on, while a tolerable rise in scrap costs is present, a generally good safety profile will prevail.

The ability of the data to predict short-range safety performance cannot be evaluated properly by this study. However, some notion of whether this possibility exists can be gained from an examination of Z_1 (the general safety factor) and Z_6 (the first managerial factor)—see Figure B-1. The plot indicates that 7 businesses of the 17 in the example, on the basis of 1955–1957 data, could be expected to improve in 1958 and 5 would not. The following were the predictions and the 1957–1958 severity rate comparisons:

Businesses Expected to Improve in 1958	Severity Rates	
	1957	1958
Large motor and generator	458	1406
Locomotive and car equipment	1226	107
Distribution transformer	777	98
Large steam turbine	290	129
TV receiver	235	135
Home laundry	1328	8
Receiving tube	40	93
Businesses Expected to Retrogress in 1958		
Industry control	116	409
Metallurgical products	13	15
Hermetic motor	44	68
Conduit products	104	0
Semiconductor	84	4

It will be seen that of the 12 businesses for which predictions were made two in the better group (large motor and generator and receiving tube) did not do better. In the second group two (conduit products and semiconductor) did not perform as predicted. One business, metallurgical products, went as predicted, but the difference is slight. Thus it is seen that the predictions were correct in 66.6 percent of the cases. The predictability of the information was studied further by plotting the regression analysis for $\hat{Z}_1$ versus the composite managerial factor (see Figure B-2). This analysis gave 15 businesses for which predictions might be made as follows:

Businesses Expected to Improve in 1958	Severity Rates	
	1957	1958
TV receiver	235	135
Large steam turbine	290	129
Locomotive and car equipment	1226	107
Home laundry	1328	8
Large motor and generator	458	1406
Distribution transformer	777	98
Receiving tube	40	93
Heavy military electronic equipment	508	46
Instrument	184	60

Continued

Businesses Expected to Retrogress in 1958	Severity Rates	
	1957	1958
Automatic blanket and fan	363	598
General-purpose motor	248	83
Semiconductor	84	4
Metallurgical products	116	409
Hermetic motor	44	68
Conduit products	104	0

Figure B-1

Scatter Diagram—General Safety Factor versus First Managerial
(Z^1 versus Z^6).

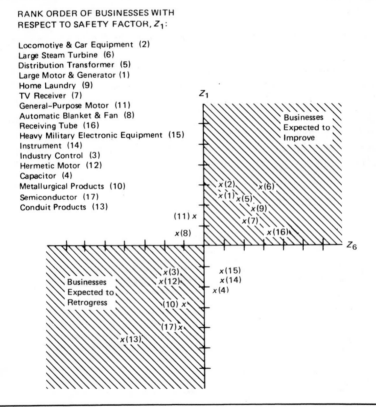

RANK ORDER OF BUSINESSES WITH
RESPECT TO SAFETY FACTOR, Z_1:

Locomotive & Car Equipment (2)
Large Steam Turbine (6)
Distribution Transformer (5)
Large Motor & Generator (1)
Home Laundry (9)
TV Receiver (7)
General-Purpose Motor (11)
Automatic Blanket & Fan (8)
Receiving Tube (16)
Heavy Military Electronic Equipment (15)
Instrument (14)
Industry Control (3)
Hermetic Motor (12)
Capacitor (4)
Metallurgical Products (10)
Semiconductor (17)
Conduit Products (13)

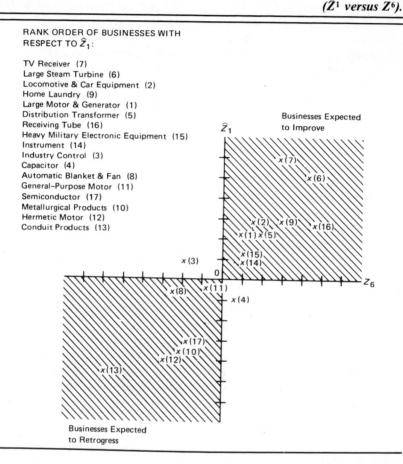

It will be noticed that of the 9 businesses predicted to improve in 1958, all but 2 (large motor and generator and receiving tube) did; of the 6 predicted to retrogress, 3 did not (general-purpose motor, semiconductor, and conduit products). Thus, 10 of the 15 or again 66.6 percent of the businesses performed as predicted. This percentage is greater than would be expected through chance. Since it is so high in both analyses, the possibility of predicting safety performance with cost-accounting data is considered quite good.

In general, the following can be said:

1. Two of the four computed safety factors were found to have a relatively close relationship with several of the management factors. Of the two

safety factors, the first (Z_1) is inherently a measure of severity and the second (Z_2) essentially measures "average severity" (that is, the ratio of the severity rate to the frequency rate). The most prominent safety factor was Z_1 and this may be considered the "general safety factor."

2. Differences in the businesses' comparative profiles of safety performance appear to be explained by the differences in their severity rates and the ratio of the severity rate to the frequency rate (that is, average severity) rather than the generally accepted measure: frequency rate. It is interesting that the severity rate and the average severity rate are also closely related to injury costs.

3. The regression analysis indicated several complex relationships that were less descriptive than the relationship between the general safety factor and the composite managerial factor. These are interesting, but they do not help to clarify the management-safety relationship. They are reported below:

   a. The second safety factor $(\hat{Z}_2)$ was not significantly related to the managerial factor.

   b. The third safety factor $(\hat{Z}_3)$, essentially a measure of average severity (that is, the ratio of severity to frequency rates), indicated a relatively strong negative relation to sales, lesser negative relation to rework, direct labor and shop costs. It was comparatively strongly related positively to complaint costs and less strongly related to building maintenance expense.

   c. The fourth safety factor $(\hat{Z}_4)$, which is not inherently descriptive of any one of the usual work injury experience measures, is relatively most strongly related positively to rework costs and to a lesser degree to complaint costs and sales. Negatively it is comparatively most strongly related to direct labor costs and to a lesser degree to building maintenance and scrap.

DISCUSSION AND CONCLUSIONS

The results broadly indicated that in an operating component where there is fairly good control of all contingency costs, a generally good safety profile will prevail. It appears that the better operating businesses, from a safety point of view, keep rework costs, direct labor costs, building and equipment maintenance costs, complaint and shop costs on a fairly low level while scrap costs may be tolerably high.

The inverse relationship of scrap costs to other costs, particularly complaint costs, suggests that an attempt to secure improvement in one may have an opposite effect on the other. This curious possibility may be interesting to study further.

The tendency for building and equipment maintenance costs to be low, where a good safety experience occurs, appears to be contrary to the theory of the case at first glance. Usually it is expected that a generous attitude toward maintenance expense exists in locations where good safety records are found. However, the well-managed operation (also the safer one) receives watchful preventive maintenance and thus keeps expenditures to a minimum.

The results indicate that sales are low in those businesses where a better safety (in general, severity) record occurs. This would be expected when the injury and sales data are expressed only as totals for a time interval. But the data were examined on an equivalent man-hour basis. Also, the safety factor is dominated by the severity variable that is influenced most heavily by the penalty charges and days lost for each injury rather than the man-hours of exposure. The comparison, therefore, is considered significant.

The counterrelationship between sales and safety effectiveness could be explained by the changing demand for better managerial performance that occurs as sales get better or worsen. For example, with the exception of scrap, other managerial variables included in the study tended to be lower when a good safety profile exists. It would seem that managers apply themselves more assiduously when faced with declining sales and that the resultant tightening of control affects contingency losses broadly. If this reasoning is sound, the study's hypothesis is supported.

One could conclude that safety success is independent—to an extent that may be considerable—of the efforts of the safety specialist. The principal factor seems to be the degree to which managers, by applying the knowledge they have, maintain a good plant and persuade subordinates to meet accepted work standards.

Inasmuch as the "general managerial factor" of the study predicts the "general safety factor" fairly well, it would be possible to compute from management variables a good predictor of overall accident control effectiveness.

Notes

1. Workers' compensation and absenteeism were included in the management and safety variables respectively because these data were so easily obtainable and it was considered worthwhile to explore their value with respect to the study's objective. Neither proved to have any significance and would not be considered in further exploration of the problem.

2. Although these five independent variables are labeled "safety variables," absenteeism and total man-hours, in truth, are not safety variables. They were included in this category for convenience but do not color the results because of such an arrangement.

ADDENDUM A TO APPENDIX B

Introduction

In order to "explain" the data of the study a mathematical model was constructed. Certain parameters were assumed, and a value was computed corresponding to each cell of the data matrix. If these values are in general close to the experimentally obtained or observed values, the model may be considered a good one.

In deciding which of two models is the better for representing a particular matrix, a quantitative measure of the goodness of fit of the competing models is necessary. A least-squares criterion—sum of squares of deviations of the model's predictions from experimentally obtained values—is appropriate.

Factors and Factor Analysis

The primary assumption made here is that there are such things as factors, that there are few enough of these to make the approach feasible, and that these act in the manner specified in equation (1) below. Factors are underlying variables, the effect of which is exerted only indirectly in the data. This is not to assume that such a factor might not be estimable directly by other means.

General Problems of Reducibility and Identifiability

Like most statistical parameters, uncertain estimates are made of the factor values for some particular representation. Owing to the fact that many different representations are possible in this case, there is a lack of identifiability which must be resolved by arbitrary but not essential decisions on the form of the representation. These problems will be treated more fully in the next three sections.

In the remainder of this section the specification of the statistical model to be used will be completed. The assumption of the existence of quantitative factors presupposes that, for some representation assumed given, they have numerical magnitudes for the elements of the rows and columns separately. The magnitude of one of these factors is called a "factor loading." For definiteness, call the loadings of the i-th row and the j-th column on the k-th factor h_{ik} and a_{jk}, respectively and assume that the factors are independent so that effects add. Failure to assume this merely necessitates a more redundant notation and contributes further to the already difficult problem of the identification of the factors. The contribution of the k-th factor is then $h_{ik}a_{jk}$. The total effect, under the above assumption of independence, is then

$$x_{ij} = \sum_k h_{ik}a_{jk} \tag{1}$$

Now the nature of the model is such that any set of data could be represented exactly if sufficiently many factors were assumed. Such a use of the model would be worth nothing for purposes of prediction or analysis since any value of the missing effect would be compatible with the remaining data. On the other hand, even if only a few factors were present, the existence of errors of measurement would make it unlikely that an exact representation could be obtained with less than the full number of factors required to represent exactly in all cases. This number of factors is known to be the smaller of the two numbers m and n where m is the number of rows and n the number of columns (see Table B-1). Thus a more reasonable assumption is that in equation (1) the number of factors is small and in (2) a random "error" is added to obtain the observed result. Measurement error, of necessity, will provide a lower limit for the error of the extrapolation to be made. The revised equation then is

$$ x_{ij} = \sum_{k=1}^{p} h_{ik} a_{jk} + \epsilon_{ij} \tag{2} $$

In this equation:

$p = \min (m, n)$; $i = 1, 2, \ldots, m$; $j = 1, 2, \ldots, n$; where the errors ϵ_{ij} are independently and identically distributed (that is, random).

Problems of Identification and Representation

If the model specified in the first section held exactly without the introduction of random errors of measurement, there would still be problems of representation and identifiability. Let us consider equation (2) and translate it into matrix form.

$$ X = H A' \tag{3} $$

where X is the table or matrix of effects, H is the matrix of factor loadings for the rows, and A for the columns. The expression HA' indicates that matrix H is "postmultiplied" by matrix A', the transpose of A. However, if one takes any nonsingular matrix U of the appropriate size (square with as many rows and columns as there are independent factors), one can obtain a different representation—$X = (HU^{-1})(UA')$ since $U^{-1}U = 1$ and $HIA' = HA'$. But by postmultiplying H by a properly chosen U^{-1} one can convert it into a matrix (we still call it H for convenience since we will assume it done hereafter) such that $(HU^{-1})'(HU^{-1}) = A_1$, where A_1 is a positive diagonal matrix. Once this is done, it is possible to choose an orthogonal matrix Q (that is, a matrix such that $QQ' = Q'Q = 1$) with the property that $Q'A'AQ = A_2$ where A_2 is a positive matrix diagonal (positive if the number of factors is chosen to be minimal, otherwise just non-negative). By further converting H and A as follows:

$$HA_1^{-1/2}, \; AA_2^{-1/2} \text{ with } A = A_1^{1/2} - A_2^{-1/2}$$

one has

$$X = HAA' \tag{4}$$

where

$$H'H = 1$$

$$A'A = 1$$

A = a non-negative diagonal matrix

It can be shown that A is just the matrix of the non-zero characteristic roots of $X'X$ and XX'.

These matrices $X'X$ and XX' are just the matrices of the sums of the cross-products of rows and columns that are computed in many multivariate statistical problems.

Before we conclude this topic, it is perhaps important to leave the above generalization of the "method of principal components" and consider a more easily computed method of representing X, one related to the so-called "centroid method." It follows from the fact that if H is properly chosen in (3) then $X = HA'$, $H'X = H'HA'$, and $A' = (H'H)^{-1} H'X$ since $H'H$ is non-singular, if only the number of factors necessary for the representation are used. But this is a special case of the more general result that the rank of the matrix $X - H(H'H)^{-1}H'X$ is reduced from that of X by the amount of the rank of $H'X$. This may be forced to be the smaller of the two numbers, the rank of X and the number of columns of H.

Two observations are appropriate at this point. The above formulation is a basis for a stepwise procedure for finding a representation as in (1), and the above indicates that the number of independent factors is the rank of the matrix X when there are no errors.

The Estimation of Factor Loadings

In the preceding section it was assumed that the data are of the sort that, using (4), the differences

$$x_{ij} = \sum_{k=1}^{p} \lambda_k h_{ik} a_{jk} \qquad i = 1, 2, \ldots, m \qquad j = 1, 2, \ldots, n \tag{5}$$

are independently-distributed errors. We shall further assume them to be normally distributed. Then their joint distribution is given by

$$\prod_{i=1}^{m} \prod_{j=1}^{n} \frac{1}{\sigma\sqrt{2\pi\sigma}} \; e \; -\frac{1}{2\sigma^2} \left(x_{ij} - \sum_{k=1}^{p} \lambda_k h_{ik} a_{jk}\right)^2 \tag{6}$$

If one proceeds to estimate the h_{ik} and a_{jk} by the method of maximum likelihood, one sees immediately that the appropriate matrices are those that minimize

$$\sum_{i=1}^{m} \sum_{j=1}^{n} \left(x_{ij} - \sum_{k=1}^{p} \lambda_k h_{ik} a_{jk}\right)^2 \tag{7}$$

That is to say, we use the least-squares criterion.

Before proceeding further, it should be noted that the following results hold: If X is an $m \times n$ matrix and v is a given (column) vector, then the vector u that provides a minimum of

$$\sum_{i=1}^{m} \sum_{j=1}^{n} \left(x_{ij} - u_i v_j\right)^2$$

is given by

$$u_i = \frac{\sum\limits_{j=1}^{n} x_{ij} v_j}{\sum\limits_{j=1}^{n} v_j^2} \qquad i = 1, 2, \ldots, m$$

This result expressed in matrix form is: The vector u that minimizes the trace

$$tr(X - uv')\,'(X - uv') = tr(X - uv')(X - uv')'$$

is given by

$$u = \frac{1}{v'v} Xv$$

The minimum value attained is

$$\sum_{i=1}^{m} \sum_{j=1}^{n} x_{ij}^2 \frac{\sum\limits_{i=1}^{m} \left(\sum\limits_{j=1}^{n} x_{ij} a_j\right)^2}{\sum\limits_{j=1}^{n} a_j^2} = tr(X'X) = \frac{v'X'Xv}{(v'v)^2}$$

If X is a given $m \times n$ matrix and V is a $n \times k$ matrix, then the $m \times k$ matrix U that minimizes

$$\sum_{i=1}^{m} \sum_{j=1}^{n} \left(x_{ij} - \sum_{k=1}^{p} u_{ik} v_{jk}\right)^2$$

is given by the solution of the equations

$$\sum_{k=1}^{p} u_{ik} \sum_{j=1}^{n} v_{jk}v_{jr} = \sum_{j=1}^{n} x_{ij}v_{jr} \qquad \begin{array}{l} r = 1, 2, \ldots, p \\ i = 1, 2, \ldots, m \end{array}$$

In matrix form this equation is

$$UV'V = XV$$

with the solution

$$U = X \, V(V'V)^{-1} \tag{8}$$

provided V is non-singular as we shall require. The matrix

$$X - UV'$$

is unaltered if V is replaced by VB where B has an inverse so that the solution depends only on the linear subspace determined by the columns of V. Since this is the case, it will cause no difficulty to require that $V'V = 1$, the identity matrix. This choice leads to the simple expression

$$U = XV$$

where

$$V'V = 1$$

The minimum value of the expression to be minimized is the sum of squares of the residual matrix $X - UV'$; that is,

$$tr(X - UV')\,'(X - UV') = tr(X - UV')(X - UV')'$$

which reduces to

$$tr \, X(1 - VV')X' = tr \, XX' - tr \, UU'$$

If now we return to (7) and apply the results obtained from (8), we find that necessary conditions for a minimum of (7) are

$$\hat{H}\hat{A} = X \, \hat{A} \, (\hat{A}'\hat{A})^{-1} \tag{9a}$$
$$\hat{A}\hat{A}' = (\hat{H}'\hat{H})^{-1} \, \hat{H}'X \tag{9b}$$

where the carat sign,^, indicates the estimates of H, A, and A' that give the minimum value. In order to obtain the above results, one considers also $X' - VU'$ takes U as given and finds the corresponding V.

In order to make use of this result, it is necessary to call attention to the result in the previous section in a more general framework. Namely, it is asserted that an arbitrary rectangular matrix X is representable in the form $X = R\Gamma S'$ where $R'R = I$, $S'S = I$, and Γ is a non-negative diagonal matrix whose rank is the same as that of X. If Γ, R, and S are partitioned into two parts as

$$\Gamma = \begin{pmatrix} \Gamma_1 & 0 \\ 0 & \Gamma_2 \end{pmatrix}, \quad R = (R_1, R_2), \quad S = (S_1, S_2),$$

then

$$R'_1 R_1 = I = S'_1 S_1, \ R'_1 R_2 = 0, \ S'_1 S_2 = 0, \text{ and } X = R_1 \Gamma_1 S'_1 + R_2 \Gamma_2 S'_2$$

It can then be shown, using the above relations, that

$$\sum_{i=1}^{m} \sum_{j=1}^{n} x_{ij}^2 = \sum_{k=1}^{r_1} \gamma^2 kk + \sum_{k=r_1+1}^{r} \gamma^2 kk,$$

where

$$\Gamma_1 = \begin{pmatrix} \gamma_{11} & & & 0 \\ & \gamma_{22} & & \\ & & \ddots & \\ 0 & & & \gamma^r_1{}^r_1 \end{pmatrix} \quad \text{and} \quad \Gamma_2 = \begin{pmatrix} \gamma r_1+1, r_1+1 & & 0 \\ & \ddots & \\ 0 & & \gamma_{rr} \end{pmatrix}$$

It is now easy to choose Λ by noting that if

$$\gamma_{11} \geqslant \gamma_{22} \geqslant \gamma_{33} \geqslant \dots \geqslant \gamma_{rr}$$

then

$$\Lambda = \begin{pmatrix} \gamma_{11} & & 0 \\ & \ddots & \\ 0 & & \gamma_{pp} \end{pmatrix}$$

H is the matrix consisting of the first p rows in R and A is the matrix consisting of the first p rows in S.

That is to say, since the above choice of H, Λ, and A satisfies the conditions in (9), and since the residual sum of squares (7) then reduces to $\gamma^2 p+1, p+1^+ \dots {}^+\gamma^2_{rr}$, and since this, because of the ordering $\gamma_{11} \geqslant \gamma_{22} \geqslant \dots \geqslant \gamma_{rr}$, is as small as possible, the required solution has been defined.

Computation of Factor Loadings

Two schemes for the computation of the factor loadings are discussed below. The first, based on the generalized centroid scheme, is easier to compute, but, presumably, it is not completely efficient in comparison with the second scheme, which provides maximum likelihood estimates.

In the first scheme, a vector $v_1' = (v_{11}, v_{21}, \dots, v_{ml})$ is chosen arbitrarily or by common sense considerations. The appropriate vector u, to go with it, is found from the equations developed in the section on estimation. This equation is repeated here:

$$u_1 = \frac{\sum_{j=1}^{n} x_{ij}v_j}{\sum_{j=1}^{n} v_j^2} \qquad i = 1, 2, \ldots, m \tag{10}$$

The sum of squares of the elements of X,

$$\sum_{i=1}^{m} \sum_{j=1}^{n} x_{ij}^2$$

is now reduced by

$$\left(\sum_{i=1}^{m} u_i^2 \right) \left(\sum_{j=1}^{n} v_j^2 \right)$$

The next step is to choose a v_2 orthogonal to v_1 and proceed as above, or to compute the residual matrix $X - u_1 v'_1$ and proceed as above with the residual matrix. In this case it is advisable but not necessary to make v_2 orthogonal to v_1. This can be easily done using

$$v_2^* - \frac{(v_1'v_2^*)v_1}{(v_1'v_1)} = v_2$$

where v_2^* is chosen arbitrarily.

The procedure of estimation using (9a, 9b) has two variants that are appropriate in different circumstances depending on m and n, the number of rows and columns of X. If m and n are approximately equal, the procedure should be to start with vector v_1 of unit length, compute a vector u_1 of unit length in the same direction as Xv_1, find v_2 as the unit vector in the same direction as $X'u_1$, and repeat the cycle until one finds u and v satisfying the relations

$$\lambda u = Xv \tag{11}$$
$$\lambda v = X'u$$

When this is done, the residual matrix $X - uv'$ has a sum of squares of its elements reduced by λ^2 from the same sum for X. One proceeds anew from $X - \lambda uv'$. The potential computational difficulties in this procedure will not be explored in detail. They arise when V is improperly chosen and when $\gamma_{11} = \gamma_{22}$ or nearly so.

In case X is such that m and n are quite different, say $n \gg m$, then $X'X$ is smaller in size than X by a large factor. Further, if

$$X = R \, \Gamma \, S'$$
$$X'X = S \, \Gamma^2 \, S'$$

and hence the above procedure can be applied to get Γ^2 and S. From this $R = XS\,\Gamma^{-1}$ is obtained.

Treatment of Matrices with Missing Elements

The presence of at least one missing element in the data gives the entire analysis its point. Yet so far we have assumed all data present in the analysis and computations. This must now be remedied. Let us recall expression (7) and examine it when there are missing elements. It becomes

$$\sum_{ij}{}^{*}(x_{ij} - \sum_{k=1}^{p} \lambda_k h_{ik} a_{jk})^2 \tag{12}$$

where the symbol $\sum_{ij}{}^{*}$ indicates summation over those pairs (i, j) where data are present. Let us proceed to estimate the missing elements as though they were parameters. The estimates are

$$\hat{x}_{ij} = \sum_{k=1}^{p} \lambda_k h_{ik} a_{jk} \tag{13}$$

using the maximum likelihood criterion. This provides the following scheme: Determine the missing values arbitrarily or by using common-sense procedures. Estimate the factor loading matrices H and A and the matrix a. Estimate the missing elements using (13). These values are better estimates than those used previously. Replace the missing elements by these estimates, and repeat the procedure. Repeat until the process converges.

ADDENDUM B TO APPENDIX B

Regression Coefficients

	Factor	With 5	Without 5
(1)	5*	.41	
	6	.40	.75*
	7*	−.55	−.55*
	8	.01	−.04
	9	.16	.11
	10	.08	.05
	11	−.13	−.22*
	a	−.0295	−.0222
	+s	.1289	.135
	++d/f	9	10
(2)	5	.73	
	6	−.59	.04
	7	.42	.41
	8	.46	.35
	9	−.13	−.21
	10	.14	.07
	11	.07	−.09
	a	−.0065	.0063
	s	.2604	.2614
	d/f	9	10
(3)	5	.09	
	6	−.16	−.08
	7	−.59	−.59
	8	.17	.16
	9	.76*	.75*
	10	.80*	.79*
	11	.12	.10
	a	−.2079	−.2063
	s	.1182	.1126
	d/f	9	10
(4)	5	.59	
	6	.73	.38
	7	−.11	−.11
	8	.20	.26
	9	−.67*	.71*
	10	−.12	−.08
	11	.25	−.29
	a	−.0792	−.0862
	s	.2172	.2114
	d/f	9	10

* indicates the regression coefficients
+ standard deviation from the line; ++ degrees of freedom

RANDOM NUMBER TABLE

5-Minute Interval

9:40	10:40	9:40	1:45	7:00	11:00	9:30	7:20
5:15	8:35	3:15	11:50	1:55	12:35	6:40	9:10
10:00	7:30	7:30	4:05	9:10	3:35	7:40	5:30
11:25	8:30	5:15	8:50	12:40	5:10	8:20	11:20
10:15	7:50	3:40	8:45	3:55	3:20	4:30	12:35
12:10	2:40	10:30	9:20	5:05	8:05	7:15	6:10
11:25	9:20	2:05	8:35	6:40	8:30	8:45	6:20
2:50	9:35	12:10	2:05	6:55	12:50	2:30	10:35
6:00	2:30	7:15	6:10	11:45	9:10	11:35	11:30
8:35	10:05	7:30	4:15	12:40	1:10	9:30	10:10
7:15	11:00	12:50	11:20	10:00	5:25	9:00	1:30
5:40	10:55	1:05	7:40	6:10	8:10	3:00	12:30
9:30	4:00	2:20	6:55	6:50	2:45	12:15	4:50
2:10	9:30	5:20	11:05	8:20	5:15	10:45	5:50
2:30	5:35	5:05	3:20	8:45	2:45	7:05	2:00
7:00	4:35	8:20	2:45	4:15	9:30	12:40	10:25
2:15	12:05	12:10	2:25	7:45	1:45	9:05	3:25
3:05	2:30	3:35	1:30	12:15	6:40	9:10	11:15
5:55	5:50	3:00	1:55	6:50	6:50	8:20	3:35
2:25	7:40	3:50	11:10	1:50	12:55	12:50	5:10

Source: F. H. Lambrow, *Guide to Work Sampling* (New York: Rider, 1962), p. 86. Used by permission.

5-Minute Interval

8:50	9:55	2:45	5:00	3:15	10:35	9:55	8:45
9:05	9:30	8:00	12:45	5:15	10:55	11:55	3:50
11:50	2:05	6:15	6:45	3:55	9:15	4:50	6:40
6:05	11:55	3:55	10:05	10:30	10:10	10:05	10:15
9:30	12:25	4:50	7:25	6:50	3:10	2:55	4:55
4:30	12:15	10:50	7:40	6:30	1:55	4:20	10:55
9:45	11:55	7:25	2:30	4:35	3:30	5:25	8:35
11:25	7:05	10:55	8:10	12:10	3:40	2:25	10:25
4:45	7:55	4:40	1:45	9:00	9:55	3:00	2:15
3:50	6:20	1:55	3:10	1:55	9:35	6:05	9:55
5:30	9:30	8:10	12:00	4:05	10:10	8:30	4:50
11:10	7:05	6:35	1:25	6:35	1:45	4:10	3:45
8:55	7:35	9:30	2:45	9:05	8:40	6:05	3:25
1:00	8:20	9:35	1:15	11:55	12:30	4:40	11:40
5:25	11:45	11:25	2:50	12:40	7:10	3:10	4:40
2:05	3:35	10:25	9:05	2:30	1:25	5:05	12:00
7:20	9:50	8:40	9:35	2:55	5:25	7:00	11:20
5:20	4:15	2:20	9:00	12:40	11:35	7:50	3:45
7:05	7:10	12:30	10:00	9:55	1:00	12:40	8:45
7:20	10:10	6:55	9:05	3:00	5:10	3:15	3:25

Source: F. H. Lambrow, *Guide to Work Sampling* (New York: Rider, 1962), p. 87. Used by permission.

ACCIDENT CAUSAL FACTORS OBTAINED FROM THE CRITICAL INCIDENT STUDY

This appendix includes the codes and classifications of causes developed from the Western Electric Study—see pages 305–315 in Chapter 17.

Accident Causal Factors Obtained from the Critical Incident Study Classified According to a Modified ANSI Classification System

General Code		Unsafe Acts
0		*Operating without authority, failure to secure or warn*
	00	Starting, stopping, using, operating, moving without authority
	00.1	Cleaning personal clothing in degreaser
	00.2	Using unauthorized solvent for cleaning parts, hands, etc. (carbon tetrachloride, etc.)
	01	Starting, stopping, using, operating, moving, etc., without giving proper signal
	01.1	Power truck operator failing to signal when passing through blind pedestrian doors
	01.2	Walking through two-way blind pedestrian door without giving indication to oncoming persons
	02	Failing to lock, block, or secure vehicles, switches, valves, press rams, other tools, materials, or equipment against unexpected motion, flow of electric current, steam, etc.
	02.1	Failure to secure material on flats, power trucks
	02.2	Changing cutter on milling machine while machine is running
	02.3	Holding part by hand while drilling (no jig)

General Code	Unsafe Acts

02.4 Failing to secure material in jig

02.5 Failure to secure jig or fixture on table—no backstop, loose fastenings, etc.

02.6 Failing to hold part securely while burring, using bench machine

02.7 Unintentionally tripping punch press

02.8 Failure to secure drill in chuck

02.9 Failure to secure vehicle, power truck, etc., against unexpected motion

03 Failing to shut off equipment not in use

03.1 Leaving machine unattended while it is running

05 Failure to place warning signs, signals, tags, etc.

05.1 Stringing air hose, cords, wire, etc., across aisle without marking or other indication of hazard

05.2 Failure to label hot parts removed from oven

1 *Operating or working at unsafe speed*

10 Running

10.1 Running in or around plant

11 Feeding or supplying too rapidly

11.1 Machine operated at excessive speed

12 Driving too rapidly

12.1 Speeding or improper handling of power truck

12.2 Speeding with automobile in parking lot (collisions)

2 *Making safety devices inoperative*

20 Removing safety devices

20.1 Guard on milling machine removed

20.2 Grinding without eye shield in place

20.3 Removing guard or making guard inoperative on punch press, welding machine, other machines

20.4 Guard (shield) removed from automatic screw machine

20.5 Guard removed from chuck motor

20.6 Guard removed from circular saw

21 Blocking, plugging, tying, etc., of safety devices

21.1 Reaching behind guard on punch press, welding machine ("cheating" the guard— circumventing the guard)

21.2 Circumventing the glass shield on grinding wheel

Accident Causal Factors (Continued)

General Code	Unsafe Acts

23   Misadjusting safety device
    23.1   Guard on milling machine not properly adjusted
    23.2   Work rest improperly adjusted on grinding wheel
    23.3   Guard on punch press, welding machine, or other machine improperly adjusted or inadequate (opening too wide, guard loose, etc.)

3   *Using unsafe equipment, hands instead of equipment, or equipment unsafely*
30   Using defective equipment
    30.1   Using makeshift fastener (wire) to fasten guard or shield in place
    30.2   Using improperly maintained tools and equipment
    30.3   Using weak or damaged flats
    30.4   Using worn nuts, bolts, threads, wrenches, etc., on jigs, fixtures, vises, machines, etc.
    30.5   Working with exposed, unguarded work light
    30.6   Using makeshift, unstable stand for holding stock pans

31   Unsafe use of equipment
    31.1   Using air hose to blow chips off machine, out of jig, etc.
    31.2   Using air hose to clean body, clothes, hair, etc.
    31.3   Wiring air hose in "on" position—fastening air hose down on drill press in open position
    31.4   Unsafe improvising—using improper or makeshift tools or equipment
    31.5   Cleaning under dies, removing parts from under dies, putting hands under punch press dies while machine is running
    31.6   Excessively high *psi* pressure on air hose
    31.7   Grinding material too heavy for size of grinding wheel

32   Using hands instead of hand tools, etc.
    32.1   Handling metal chips by hand
    32.2   Removing frozen drill from jig bushing by hand
    32.3   Holding part by hand while drilling, tapping (no jig)

33   Gripping objects insecurely, taking wrong hold of objects
    33.1   Hand slipped from part, striking belt sander, drill, etc.
    33.2   Dropping material, tools, pans, etc., onto floor, body
    33.3   Grabbing spindle of drill press by hand while it is rotating—attempting to stop spindle by hand with power on
    33.4   Burring part—hand or tool slipped, striking drill, tool

Accident Causal Factors (Continued)

General Code	Unsafe Acts

4   *Unsafe loading, placing, mixing, combining, etc.*
   40   Overloading
      40.1   Overloading flats, power trucks
      40.2   Overloading stock pans

   41   Crowding
      41.1   Machines too close together—inadequate operator work space

   42   Lifting or carrying too heavy loads

   43   Arranging or placing objects or materials unsafely
      43.1   Placing scrap material on floor around machine
      43.2   Unstable stacking of stock pans on makeshift stands
      43.3   Storing materials in aisles, workers trip, bump into material while walking
      43.4   Unstable stacking of stock pans on floor
      43.5   Material, equipment stored in designated aisles—struck by power trucks or other vehicles
      43.6   Tools kept on floor
      43.7   Stock rack for automatic screw machine placed in designated aisle
      43.8   Stringing air hose, cords, wire, etc., across aisle
      43.9   Holding part by hand while drilling, tapping—no jig
      43.10   Holding jig by hand while drilling, tapping—no backstop
      43.11   Plastic spoons thrown on floor in cafeteria and in work areas—slips, falls
      43.12   Chips retained in jig misaligns jig resulting in broken drill on drill press
      43.13   Misaligned material in punch press

   45   Introducing objects or materials unsafely
      45.1   Oil spilled on floor

5   *Taking unsafe position or posture*
   54   Lifting with bent back, while in awkward position, etc.
      54.1   Improper lifting procedure

   56   Exposure on vehicle right-of-way
      56.1   Power truck operated on route used by pedestrians—truck strikes pedestrians (entering aisle from restrooms, etc.)

   59   Miscellaneous unsafe positions or postures
      59.1   Touching hot light fixture

Accident Causal Factors (Continued)

General Code		Unsafe Acts
	59.2	Exposing parts of body to flying chips from milling, drilling, etc.
	59.3	Splash burn from welding—touching hot welding rod
	59.4	Assuming awkward position while tightening nut
	59.5	Reaching too far, lost balance, fell from chair
	59.6	Pinched finger in drill jig while closing jig
	59.7	Mashed finger or hands while moving flats, pans, etc. (not dropped)
	59.8	Stumbling over wooden platform supporting machinery

6   *Working on moving or dangerous equipment*
  61   Cleaning, oiling, adjusting, etc., of moving equipment
      61.1   Changing cutter on milling machine while machine is running
      61.2   Adjusting machine while it is in motion (gauging work, removing part, etc.)
      61.3   Cleaning chips from milling, drilling, and other machines while machines are in motion
      61.4   Cleaning under dies, removing parts, putting hands under dies of punch press with power on
      61.5   Reaching inside of automatic screw machine to remove part while machine is in motion
      61.6   Putting drill in drill press chuck while spindle is turning (standard chuck)

  63   Working on electrically charged equipment
      63.1   Electric shock from machine, faulty wiring, other causes (power-on maintenance, etc.)

7   *Distracting, teasing, abusing, startling, etc.*
  70   Calling, talking or making unnecessary noise
      70.1   Distraction from fellow worker or supervisor
      70.2   Horseplay—making loud, startling noise behind operator (drop pans, etc.)

  71   Throwing material
      71.1   Horseplay—throwing small pieces of stock at other operators

  72   Teasing, abusing, startling, etc.
      72.1   Tickling operator in ribs or under arms
      72.2   Pinching female operator
      72.3   Intentionally bumping operator using drill press, grinding wheel, etc.

General Code	Unsafe Acts

73 Practical joking, etc.

    73.1   Opening interlocked door on gear box of milling machine while operator is away—machine fails to start when operator returns

    73.2   Miscellaneous horseplay: (1) hooking can of oil behind worker, (2) putting chips on seat of chair, (3) making loud noises, (4) loosening a drill, (5) turning off machine while in use, (6) hiding jig, piece parts, etc.

8 *Failure to use safe attire or personal protective devices*

    80   Failing to wear goggles, gloves, masks, aprons, shoes, leggings, etc.

        80.1   Operating milling machine without safety glasses

        80.2   Operating a punch press without safety glasses

        80.3   Operating a screw machine or engine lathe without safety glasses

        80.4   Operating a drill press without safety glasses

        80.5   Operating a grinding machine, welding machine, other machine without safety glasses

        80.6   Outsiders walking through designated eye hazard area without safety glasses

        80.7   Oil or chemical splash in eye (no goggles or goggles unsuited for splash protection)

        80.8   Wearing defective or unsafe goggles or goggles unsuited for impact protection (glasses without side shields)

        80.9   Handling burred or sharp-edged stock without gloves

        80.10   Handling hot machined parts with unprotected hands

        80.11   Handling hot parts from oven without gloves

        80.12   Gloves worn while handling sheet steel or other material that are unsuited for protection from sharp edges of material

        80.13   Failure to wear hard-toe safety shoes while handling heavy material

        80.14   Failure to wear ear plugs in noise hazard areas

    81   Wearing high heels, loose hair, long sleeves, loose clothing, etc.

        81.1   Loose, long hair around revolving machinery

        81.2   Wearing loose clothes, long sleeves, rings, etc., around rotating machinery

        81.3   Wearing gloves while working with revolving machines (drills, milling machines, lathes, etc.)

        81.4   Wearing gloves while grinding, buffing, etc.

        81.5   Wearing high heels, tripping, etc., in areas containing holes, depressions, protrusions, etc., in floor

99 *Not elsewhere classified—jig heats up during drilling operation, burning operator (no coolant)*

Unsafe Mechanical or Physical Condition

General Code		Unsafe Condition
0		*Improperly guarded agencies*
	00	Unguarded
	00.1	Guard on milling machine removed
	00.2	Eye shield on grinder not in place
	00.3	Removing guard or making guard inoperative on punch press, welding machine, other machines
	00.4	Unguarded work light
	00.5	Guarded (shield) removed from automatic screw machine
	00.6	Guard removed from chuck motor
	00.7	Guard removed from saw
	01	Inadequately guarded
	01.1	Guard on milling machine not properly adjusted
	01.2	Makeshift fastener (wire) used to fasten guard in place
	01.3	Guard on punch press, welding machine, other machines not properly adjusted or inadequate
1		*Defects of agencies*
	11	Slippery
	11.1	Oil or coolant standing (spilled) on floor
	11.2	Plastic spoons from cafeteria discarded on floor
	12	Sharp-edged
	12.1	Stock with sharp burr, sharp-edged material
	12.2	Sharp edge on stock pan
	13	Poorly designed
	13.1	Blind, two-way swinging pedestrian doors
	15	Poorly constructed
	15.1	Makeshift stands for stock pans
	15.2	Equipment failure
	16	Inferior composition
	16.1	Faulty wiring
	16.2	Material failure
	17	Decayed, aged, worn, frayed, cracked, etc.
	17.1	Improperly maintained tools and equipment
	17.2	Makeshift or improper tools and equipment
	17.3	Worn nuts, bolts, threads, wrenches, etc., on jigs, fixtures, vises, machines, etc.
	17.4	Broken part of machine, mechanical failure

Unsafe Mechanical or Physical Condition (Continued)

General Code	Unsafe Condition

	17.5	Weak or damaged flats
	17.6	Holes, depressions, protrusions in floor
	17.7	Glass shield on grinder pitted, not transparent

2   *Hazardous arrangement, procedure, etc.*

20   Unsafely stored or piled tools, materials, etc.

     20.1   Materials stacked too high on flats, trucks—material not tied down on flats, power trucks

     20.2   Scrap material on floor around machine

     20.3   Stacking pans on makeshift standards (unstable)

     20.4   Unstable stacking of stock pans on floor

     20.5   Material stored in designated aisles—workers trip, bump into flats and material

     20.6   Material, equipment stored in designated aisles—struck by power trucks, other vehicles

     20.7   Tools kept on floor (in aisles, around machine)

     20.8   Stock rack for automatic screw machine extending into aisles

21   Congestion of working spaces

     21.1   Machines too close together, inadequate operator work space

22   Inadequate aisle space

     22.1   Flats and materials stored in aisles (workers trip, bump into material while walking in aisles)

     22.2   Material, equipment stored in aisles—struck by power trucks, other vehicles

     22.3   Stock rack for automatic screw machine extending into aisle

23   Unsafe planning and/or layout of traffic or process operations

     23.1   Restrooms, locker rooms exit directly into aisles—workers leaving restrooms struck by moving vehicles

     23.2   Blind door used by pedestrians and vehicles

24   Unsafe processes

     24.1   Heat up of jig held by operator during drilling operations

     24.2   Using air hose to blow chips off of machine, out of jig, etc.

     24.3   Using air hose to clean body, clothes, hair, etc.

     24.4   Wiring air hose in "on" position—fastening it down on drill press in open position

     24.5   Work rest improperly adjusted on grinding wheel—too far away from edge of wheel

     24.6   Jig or fixture not secured on table—no backstop, loose fastenings, etc.

Unsafe Mechanical or Physical Condition (Continued)

General Code		Unsafe Condition
	24.7	Machines operated at an excessive speed—speed increased to make rate, bonus, etc.
	24.8	Hot parts removed from oven not properly marked "hot"
	24.9	Flying chips from milling, drilling, etc., strike body
	24.10	Stringing air hose, cord, wire, etc., across aisle, floor
	24.11	Holding part by hand while drilling, tapping—no jig
	24.12	Unauthorized solvent used for cleaning parts, hands (carbon-tetrachloride, etc.)
	24.13	Excessively high *psi* pressure on air hose
	24.14	Grinding material too heavy for size of grinding wheel—grinding wheel too light for work
	24.15	Drill not secured in chuck
	24.16	Coolant produces rash on hands
25		Overloading
	25.1	Overloaded flats
	25.2	Overloaded stock pans
26		Misaligning
	26.1	Material not secured in jig—bushings misaligned
	26.2	Chips in or on jig, misaligning part or jig (breaking drill, etc.)
3		*Improper illumination*
30		Insufficient light
	30.1	Inadequate illumination of work area—(dirt on lights, oil covered lights, etc.)
31		Glare
	31.1	Glare produced by sunlight shining through unshaded window
4		*Improper ventilation*
41		Unsuitable capacity, location or arrangement of system
	41.1	Inadequate general or local ventilation system
5		*Unsafe dress or apparel*
50		No goggles
	50.1	Operating a milling machine without safety glasses
	50.2	Operating a punch press without safety glasses
	50.3	Operating a screw machine or engine lathe without safety glasses
	50.4	Operating a drill press without safety glasses
	50.5	Operating grinding machine, welding machine, other machines without safety glasses

Unsafe Mechanical or Physical Condition (Continued)

General Code	Unsafe Condition
50.6	Outsiders walking through designated eye hazard area without safety glasses
50.7	Oil or chemical splash in eye (no safety goggles or goggles unsuited for splash protection)
51	Goggles defective, unsafe, or unsuited for impact protection
51.1	Glasses without side shields—unsuited for work
52	No gloves
52.1	Handling burred or sharp-edged stock without gloves
52.2	Handling hot machine parts with unprotected hands
52.3	Handling hot parts from oven without gloves
53	Gloves defective, unsafe, or unsuited for work
53.1	Gloves worn while handling sheet steel, other material, unsuited for protection from edges of material
53.2	Foreign particles inside gloves—hands cut
54	No safety shoes
54.1	Failure to wear hard-toe safety shoes while handling heavy material
55	Loose hair
55.1	Loose, long hair around revolving machinery
56	Loose clothing
56.1	Wearing loose clothing, long sleeves, rings, etc., around rotating machinery
56.2	Wearing gloves while working with revolving machines (drills, milling machines, lathes, etc.)
56.3	Wearing gloves while grinding, buffing, etc.
57	No ear plugs
57.1	Failure to wear ear plugs in a high noise environment

CRITICAL INCIDENT TECHNIQUE SAMPLE PRELIMINARY INTERVIEW

You were selected at random from the persons in your department to help us discover and evaluate the causes of accidents before they result in disabling work injuries. We believe that because of your experience and close association with industrial operations you are especially well qualified as an observer of errors and unsafe conditions which have led to "near misses" or even to actual injurious accidents.

Within the next few days you will be asked to participate in an open-ended interview as part of our application of a method for identifying accident problems called the "critical incident technique." All statements you make will be kept strictly confidential. Neither management nor your co-workers will be given any information about your specific comments which can be identified with you.

During the interview you will be asked to describe as many unsafe human errors and unsafe conditions as you can recall which have occurred or existed in your department during the past year (or other designated time period). We are interested in hearing about all of these incidents you can remember, regardless of whether or not they resulted in an actual injury. We have defined a human error as an unsafe departure from an accepted, normal, or correct procedure; an unnecessary exposure to a hazard; or conduct reducing the degree of safety normally present in an operation. A human error may be either something a person failed to do or forgot to do, such as not wearing safety glasses, or something he actually did which is considered unsafe, such as blowing away metal chips with an air hose or lifting too heavy a load.

Unsafe conditions may include such things as physical defects in equipment, machinery, or in the general work area; errors in the design of machinery, equipment, or work areas; faulty planning; or omission of a safety requirement for maintaining a relatively hazard-free work area. Examples of unsafe conditions include such situations as a machine being operated without its guard in place or with a broken guard, aisleways cluttered with material, poor lighting, or a glass shield missing from a grinding wheel. The incidents may have been observed as a combination of human errors and unsafe conditions occurring at the same time.

During the interview you will be asked to describe fully each human error and unsafe condition you have observed in the plant. You will be requested to give an estimate of the number of times you have observed each type of error and unsafe condition during the past year (or other designated time period). In addition, you will be asked to estimate how many times you have observed the unsafe behaviors or conditions when there was no injury, and how many times you have seen someone actually get hurt from these causes. Let me again emphasize that the statements you make will be used for purposes of problem identification and analysis only and your identity will remain anonymous in relation to the information you give. We seek only information about incidents themselves and we will not associate these incidents with any particular person. With your permission, we will be recording your comments on tape to aid us in later classifying the incidents into a number of specific categories for analysis purposes.

In order to aid you in recalling details of the incidents, a partial list of accident causal factors frequently appearing within operations similar to those in your department has been prepared. Keep in mind that this list is by no means complete, and that its purpose is to help you recall the incidents you have actually seen.

PARTIAL LIST OF TYPICAL INCIDENTS

1. Adjusting and gauging (calipering) work while the machine is in operation.
2. Cleaning a machine or removing a part while the machine is in motion.
3. Using air hose to remove metal chips.
4. Using compressed air to blow dust or dirt off of clothing or out of hair.
5. Using excessive *psi* on air hose.
6. Operating machine tools (turning machines, knurling and grinding machines, drill presses, milling machines, punch presses, and so forth) without eye protection.
7. Not wearing safety glasses in a designated eye-hazard area.
8. Failing to use protective clothing or equipment (face shield, face mask, ear plugs, hard hat, cup goggles, and so on).
9. Failure to wear proper gloves or other hand protection when handling rough or sharp edged material.

10. Wearing gloves, ties, rings, long sleeves, or loose clothing around machine tools.
11. Wearing gloves while grinding, polishing or buffing.
12. Handling hot objects with unprotected hands.
13. No work rest or poorly adjusted work rest on a grinder.
14. Grinding without the protective shield in place.
15. Making safety devices inoperative (removing guards, tampering with the adjustment of a guard, "beating" or "cheating" the guard, failing to report safety device defects).
16. Not properly using a safety guard (for example, reaching behind a sweep guard on a punch press).
17. Placing any part of the body between the dies of a punch press (for example, using the fingers instead of tongs or mechanical ejection devices for removing parts).
18. Improperly designed safety guard (for example, a designed wide opening on a barrier guard which will allow the fingers to reach underneath the dies).
19. Accidentally tripping a punch press.
20. Failure to shut off power when adjusting, setting-up, or maintaining a machine.
21. Jig or fixture not secured on table. Work not secured in jig, fixture, or vise.
22. Handling metal chips by hand.
23. Worn nuts, bolts or wrenches used on jigs and fixtures.
24. Machines operated at excessive speeds.
25. Disruption of work by interference from another person.
26. Lifting heavy weight (such as one person lifting an object by hand which weighs over 50 pounds).
27. Lifting or carrying improperly (awkward position, failure to keep back straight and bend knees, twisting body, and so on).
28. Unintentionally dropping an object onto the body or onto the floor from a high level.
29. Not wearing safety shoes.
30. Using improper or unauthorized procedure.
31. Using a tool for purposes other than those for which it was designed (improvising).
32. Placing parts being processed or other materials in designated aisleways.
33. Scrap material, chips, or other wastes accumulated in aisleways.
34. Leaving tools on the floor.
35. Using poorly maintained tools and equipment (mushroomed chisel head, loose hammer head, ladder with cracked or broken rungs, and so forth).

36. Oil spills on floor (slips? falls?).
37. Holes, depressions, protrusions, or other defects in floors (slips? falls?).
38. Speeding or other improper handling of a power truck, tug, fork-lift, or similar vehicle.
39. Running inside the plant.
40. Horseplay.
41. Inadequate lighting.
42. Material stacked improperly (for example, unstable stacks of containers).
43. Using unauthorized solvent for cleaning.
44. Electrical shocks from machines or equipment.
45. Hazardous layout or arrangement (improper storage; inadequate aisle space; unmarked, locked, or inadequate number of exits; machines placed too close together; inadequate space for stock; and so forth).

You may prepare notes concerning these or other incidents and bring them with you to the interview if you wish. During the interview you will be asked to provide sufficient details about each incident to enable us to identify specific unsafe behaviors, unsafe conditions, or combinations of these which were causally related to the incident. Do you have any questions? If not, please return at (time) on (day and date) for your interview.

DESCRIPTION
AND ANALYSIS
OF TYPICAL
CRITICAL INCIDENTS

The following are examples of typical critical incidents described by observers during the interviews in the Western Electric study. These incidents were transcribed from the tape recordings made during the actual interviews. A summary and a modified ANSI Z-16.2 analysis follow each critical incident transcription. The codes referred to are found in Appendix D. In the examples *PO* represents the worker participant-observer and *I* the investigator.

INCIDENT NO. 1

Description

PO: The material handlers taking pans on flats from one place to another—usually they stack them too high without regard to tying them down with cord or anything. Usually they're supposed to tie them down with cord. I don't think they are supposed to stack them no higher than five pans on top of one another.

I: About how many pans do they include in one stack?

PO: Oh sometimes it varies—maybe six or seven high. I recall about 18 months ago, I guess—it happened right in our area—a guy had pans that weren't tied down on his power truck. He was going along at a little bit excessive speed and they tumbled off. It happened that there wasn't anyone there close by at the time, but he should have had them tied down.

I: Have you ever seen anyone get hurt from pans falling off of trucks?

PO: Once—it was last year—a guy had pans on the flats. There were two employees working right beside the driving area. They had their backs turned to the aisle. They weren't paying any attention to the guy. He came along—it happened when he came around the curve, the pans tumbled off. One of the guys was off work for approximately, I guess, four or five months. He still complains of his back. He got a little cut on his scalp up there. Most of the pans were the plastic type—had a few steel pans. It could have been a lot worse.

I: How many times have you seen anyone get hurt from pans falling off?

PO: Twice in the past two years.

Summary and Analysis

Pans were stacked too high on the power truck and they were not tied down. The power truck was operated at an excessive speed. According to the ANSI system for classifying accident factors, the following unsafe acts and unsafe mechanical and physical conditions were involved in this incident.

Code	Unsafe Mechanical or Physical Condition
20	Unsafely stored or piled tools, materials, etc.
25	Overloading

	Unsafe Acts
02	Failing to lock, block, or secure vehicles, switches, valves, press rams, other tools, materials, and equipment against unexpected motion, flow of electric current, steam, etc.
12	Driving too rapidly
40	Overloading

INCIDENT NO. 2

Description

PO: There's always this case where you will find when there is more than two or three people working in a group where they get the tendency sometimes to get playful. I've seen a case happen in the past two years where this particular employee—when the supervisor's eyes wasn't on him—he threw small pieces of material at people. He was just throwing them at random, not at no particular person. It happened at this particular time a woman was working across the aisle from where I was working. Just as she turned, a piece hit her glasses and knocked them off. She reported it to the supervisor but she didn't know who had thrown the material. I mean, it could have happened where it could have cut her or maybe it could have hit her in the eye or something.

I: About how many times have you seen this happen during the past two years—people throwing pieces around.

PO: Maybe four to five times during the past two years people have been hit by something. Fifteen or twenty times I've seen people throwing something.

Summary and Analysis

An operator was engaged in the horseplay act of throwing small pieces of material in the vicinity of other workers.

Code	Unsafe Act
71	Throwing material

INCIDENT NO. 3

Description

PO: And then that air hose—I've seen it happen where an operator, he's trying to make a rate, and rather than use the air hose by hand he wires it down on the machine where it continuously blows at all times. He figures that saves him the motion of reaching and grabbing the air hose, blowing out the jig, and putting it back. And sometimes the wire where he's got it wired down might come loose and the thing starts revolving around in the air before he can get it off. I mean, I have seen it happen.

I: About how many times during the past two years have you seen someone tie down the air hose to keep it on?

PO: That's a general practice. Practically every day you will see that.

I: About how many times do you think someone has gotten hurt from the air hose flying around?

PO: I've seen it come loose and fly around several times—maybe four or five times in the past two years. The operator might run a job maybe a week or two weeks and the wire he uses is a fine copper wire and it gets weak after awhile and starts breaking and when it happens it flies around.

Summary and Analysis

The operator used an air hose to blow chips out of a jig. He wired the air hose down in the open position so that chips could be removed continuously without extra time being required for this operation. The air hose frequently came loose and flew around.

Code	Unsafe Mechanical or Physical Condition
24	Hazardous arrangement, procedure, and so on—unsafe process

	Unsafe Act
31	Unsafe use of equipment

INCIDENT NO. 4

Description

PO: I've seen them operating machines without safety glasses. In the area you're supposed to have them at all times. Maybe the guy came back from lunch and he forgets about it and has to be reminded.

I: About how many times in the past two years have you seen anyone operating machinery without safety glasses?

PO: About four or five times.

I: Have you ever seen anyone get something in his eye when he wasn't wearing safety glasses?

PO: It's hard to say. Usually an operator won't admit that it happened when the glasses were down low on his nose bridge or something. But it has happened.

I: About how many times during the past two years?

PO: I would say two to three times. I mean, maybe sometimes the operator will say he gets tired of the glasses. If the supervisor's back is turned or maybe the setup man is out of the area, he'll take them off. Especially, I mean, people that don't normally wear glasses, they have a tendency more to put them down than people that usually wear glasses. This happens anywhere from six or seven or eight times during a year.

Summary and Analysis

Workers were operating machines without wearing safety glasses. In cases where the type of equipment was not mentioned by the observer, a check of available records was made to determine the department and the particular type of machine to which the observer was assigned. Unless it was otherwise indicated by the observer, it was assumed that the incident occurred on machines in his department that were similar to his own. This information was used in making the detailed entries in the expanded causal classification system.

Code	Unsafe Mechanical or Physical Condition
50	Unsafe dress or apparel—no goggles

	Unsafe Act
80	Failure to use safe attire or personal protective devices—failure to wear goggles, gloves, masks, aprons, shoes, leggings, etc.

INCIDENT NO. 5

Description

PO: Another unsafe act is wearing long sleeves on machines. I saw an accident where a fellow got his arm caught in a machine. The fellow went to reach for a piece part—it was an automatic—and the end of the tool grabbed onto his sleeve and pulled his arm right in and he got 35 stitches right across his arm here. He was lucky. The kid had sense enough to reach up and throw the stop button off, otherwise it would have took his arm off. This was an automatic welder. They couldn't get him out. I sawed the back of a plate off in order to get his arm out. There is a ruling in our shop—no long sleeves. If you have long sleeves, you have to roll them up. But sometimes you see guys—smart guys—who come in with long sleeves, especially in the winter.

I: During the past two years, how many times have you seen someone wearing long sleeves around these machines?

PO: I'd say about once a week there's someone in the department, you know, someone in the department. I've only seen one fellow get hurt.

Summary and Analysis

The operators were wearing long sleeves while working around moving machinery.

Code	Unsafe Mechanical or Physical Condition
63	Unsafe dress or apparel—loose clothing

	Unsafe Act
81	Failure to use safe attire or personal protective devices—wearing high heels, loose hair, long sleeves, loose clothing, etc.

INCIDENT NO. 6

Description

PO: I also saw a girl get her hair caught in a chuck motor. A chuck motor has a drill coming out to burr parts. Well, you see, they have guards that come right out to the end of your drill—just enough drill left to burr that part. Sometimes they leave these guards off. Sometimes it annoys them, so when the setup man puts it on, they take it off.

I: How often does this happen?

PO: That happens quite a bit. I'd say at least twice a week. So this drill was sticking out just that length. The operator dropped her piece part, bent down to pick up the piece part, and caught her hair in there and pulled it right out of her scalp. It wound right around because the guard wasn't on it. If the guard was on there, her hair would have brushed right along the guard. There's only that much sticking out, you know, to burr the part.

I: How many times have you seen someone get hurt because they left the guard off?

PO: Well, the hair is one instance—a sweater in another instance— the drill penetrated the girl's breast. She leaned over—it grabbed her. When she leaned over, it caught hold of her sweater and dragged her sweater then dragged her in. It just pierced her a little bit, but it could have been more serious.

Summary and Analysis

The guard was removed from the drill and left off while the machine was in operation. The operator's loose hair was allowed to come in contact with the exposed rotating drill. In another instance the operator's sweater was caught in the unguarded drill.

Code	Unsafe Mechanical or Physical Condition
00	Improperly guarded agencies—unguarded
62	Unsafe dress or apparel—loose hair

	Unsafe Act
20	Making safety devices inoperative—removing safety devices
81	Failure to use safe attire or personal protective devices—wearing high heels, loose hair, long sleeves, loose clothing, etc.

INCIDENT NO. 7

Description

PO: Lifting heavy stuff—that's another safety hazard. I see that every day. I mean, one guy lifting 85 or 90 pounds, that's too much. Especially when you have to do it all day long.

I: Do you know of anyone who has gotten hurt from heavy lifting?

PO: Yes, hernias.

I: About how many over the past two years?

PO: I'd say about four guys.

Summary and Analysis

The workers were engaged in manually lifting loads that were too heavy for one person to handle safely.

Code	Unsafe Act
32	Using unsafe equipment, hands instead of equipment, or equipment unsafely—using hands instead of tools
42	Lifting or carrying too heavy loads

INCIDENT NO. 8

Description

PO: Just this morning I saw an unsafe act. A fellow was holding a part in his hand and he was trying to drill it. The part slipped from the grasp of his hand and spun around. It scratched him but it didn't do any real damage. This is done constantly—holding a part by hand while drilling instead of holding it with a vise or a proper jig.

I: How often would you say this is done?

PO: It's done practically daily. It's a time-saving thing. Rather than clamping it or going to get a vise or something, you hold it by hand. You think you can hold it and the chips bind and it starts to spin. I've been guilty of it myself.

I: Could you give me an estimate of how many times you have seen someone get hurt from this?

PO: About six times in the last two years.

Summary and Analysis

The operator was drilling a part while holding it in his hand. Neither a vise nor a jig was used to secure the part during the drilling operation.

Code	Unsafe Mechanical or Physical Condition
24	Hazardous arrangement, procedure, and so on, in, on, or around the selected agency—unsafe process

	Unsafe Act
02	Failing to lock, block, or secure vehicles, switches, valves, press rams, other tools, materials and equipment against unexpected motion, flow of electric current, steam, and so on
32	Using unsafe equipment, hands instead of equipment, or equipment unsafely—using hands instead of tools

INCIDENT NO. 9

Description

PO: Using the air hose, that's done constantly where people will either use the hose to save the time sweeping the floor or blow the chips without regard as to where another person is standing. The air hose is used in summer especially, to cool off because it's fairly hot down there—like blowing down the back of the neck or in the pants or something. It's used to clean material off the shoes. You can see this happen every day if you look for it. It happens mostly about time to go home. They want to get cleaned up, and they figure blowing these chips will hurry things up. The pressure of the air has such great force that it could blow chips into the body and cause damage. . . . I've been hit with chips, but it didn't really go deep enough to do any damage. I just picked them out and never really applied anything. It would just go in maybe a couple of layers of the skin.

I: How many times has this happened?

PO: Maybe once or twice.

Summary and Analysis

Chips were blown away with the use of an air hose. The air hose was used to cool the body as well as to clean the worker's clothing.

Code	Unsafe Mechanical or Physical Condition
24	Hazardous arrangement, procedure, etc., in, on, or around the selected agency—unsafe processes

	Unsafe Act
31	Unsafe use of equipment

INCIDENT NO. 10

Description

PO: Every day the men handle chips with their bare hands. I do once in a while myself. Some operators take a chance and use their bare hands to handle copper or steel chips or even nickel silver. These are razor sharp.

I: How often have you seen someone get cut from handling chips?

PO: There's quite a bit of that happening, but some of the boys they just don't report that. They don't want to get that on the record. They let it slip by. I would say someone gets cut once a week or maybe less. . . . You'll see someone with a Band-Aid—ask what happened—they'll say, "Oh, I got a cut on this chip or that." If you reported all the slight injuries to the medical department, you'd be going there every day of the month. . . . We're supposed to report, but all the operators don't want to get all that on their inventory sheet and on the records. If it gets there it looks bad for us, so they don't report.

Summary and Analysis

The operators were handling metal chips with their bare hands.

> Code   Unsafe Act
> 32   Using unsafe equipment, hands instead of equipment, or equipment unsafely—using hands instead of hand tools.

INCIDENT NO. 11

Description

PO: Everybody has the tendency of not working their punch press correctly, and I myself also. A punch press is something you can't fool around with. Each part while you're putting it on you're supposed to have your foot off the pedal, which nobody does. You keep your foot on the pedal always. Because, you figure the pedal is set like this and every time you put your foot off to put the piece on you have to put your foot back on the pedal again. You're so used to leaving your foot on the pedal that you also have it half-way tripped already, so at the slightest pressure it trips. . . .

I: Have you seen any accidents from doing this?

PO: I've seen tweezers get smashed in there.

I: How often do you think this happens?

PO: Pretty often . . . put down about once a month.

Summary and Analysis

The workers were operating punch presses with their feet continuously resting on the foot pedal. This practice often results in the workers accidentally tripping the punch press.

> Code   Unsafe Act
> 50   Taking unsafe position or posture

INCIDENT NO. 12

Description

PO: An employee in the milling section will usually use stands to hold properly a small pan about so long of material. A lot of times the operator will feel he hasn't got much time, and he's trying to make up time. Maybe the last job he had he didn't do too well on it. Rather than get the proper stand to set the pan in, he'll grab a chair and pull it up. There have been cases where, as he throws material into the pan, it will tilt and fall out of the chair onto the floor. It has happened that the pan hits their feet and scratched their legs while falling.

I: About how many times have you seen someone use makeshift stands or chairs during the past two years?

PO: Sometimes it happens quite frequently. . . . It's hard to say whether it will happen eight or ten times during a year or not. I mean, it will happen quite often. A lot of times the operator can't find the stand. Rather than walk to some other area to get one or to tell the setup man about it, if the chair is available, he'll reach and grab the chair and pull it out. I would say about every week, somebody does this.

I: How many times have you seen material fall off and somebody get hurt from this?

PO: I would say about three or four times in the past two years I've seen it.

I: In the past two years?

PO: Yes.

Summary and Analysis

The operator was using a makeshift stand to support the stock pans. The stands were unstable and unable to support the loaded stock pans.

Code	Unsafe Mechanical or Physical Conditions
15	Defect of agencies—poorly contructed
20	Hazardous arrangement, procedure, etc., in, on, or around the selected agency—unsafely stored or piled tools, materials, etc.

	Unsafe Acts
31	Using unsafe equipment, hands instead of equipment, or equipment unsafely—unsafe use of equipment
43	Unsafe loading, placing, mixing, combining,—arranging or placing objects or materials unsafely

GLOSSARY OF MEASUREMENT TERMINOLOGY

Accident. An unplanned, not *necessarily* injurious or damaging event, that interrupts the completion of an activity, and is invariably preceded by an unsafe act and/or an unsafe condition or some combination of unsafe acts and/or unsafe conditions.

Alpha. The level of significance set by the evaluator for rejecting the null hypothesis (that is the point at which one can infer that the project had an effect). The alpha also specifies the probability of committing a Type I error. An alpha of .05 means there are five chances in a hundred of being wrong or rejecting the null hypothesis when it is true (see also *statistical power*).

Analysis of Variance. A statistical technique that tests for significant difference between means of sets of samples selected from more than two groups.

Arithmetic Mean. The sum of all values or scores divided by the number of values or scores. Often called simply "the mean," In symbolic form:

$$\overline{X} = \frac{\Sigma X}{N}$$

Average. A general term applied to measures of central tendency. The three most widely used averages are the arithmetic mean, median, and mode. See also *arithmetic mean, median,* and *mode.*

Bar Graph. A graph consisting of parallel bars whose lengths are proportional to quantities appearing in a set of data. See also *histogram.*

Baseline Data. Data collected prior to project operations for use in describing conditions before the treatment or project begins.

Before-After Design. An evaluation design in which measurements of variables are made before and after a project is initiated. Effectiveness is then defined in terms of the differences in measurements taken during the before and after time periods.

Before-During-After Design. An evaluation design in which measurements of variables are made before the project begins, during project activities, and after the project is completed.

Bias. An effect that deprives a statistical result of validity by systematically distorting the data.

Bottom Line. A base level at which impact is measured in terms of fatalities, injuries, and property damage.

Categorical Data. Data for which items are grouped into non-overlapping categories according to a qualitative description.

Causal-Comparative Design (ex post facto design). An evaluation design in which measurements of variables are made after a project is completed.

Causal Factor. A person, thing, or condition that contributes significantly to an accident or to a project outcome. A combination of simultaneous or sequential circumstances directly or indirectly contributing to an accident.

Chi-Square (χ^2). A test of compatibility of observed and expected frequencies based on the quantity: $\chi^2 = \dfrac{(O-E)^2}{E}$. Appropriate for ordinal scale data, where O = observed frequency and E = expected frequency. Used for testing statistical hypotheses involving independence between an observed frequency and an expected frequency or between two or more observed frequencies and the hypothetical frequencies, as in contingency tables or in problems of goodness of fit.

Class Interval. The number of units that separate one class of data from another. A grouping of possible values of a variable. For example, a variable that may be continuous from 0 to 100 may be grouped arbitrarily into class intervals 10 units wide from 0 up to 10, 10 up to 20, etc. The width of the class is sometimes called the "class interval."

Clinical Evaluation. A level of appraisal that combines objective data and information with an element of personal judgment.

Coefficient of Correlation. A measure of the degree of relationship or "going togetherness" between two or more sets of measures. The correlation coefficient most frequently used with interval data is known as the Pearson *r* (so named for Karl Pearson, originator of the method) or as the product-moment *r*, to denote the mathematical basis of its calculation. The Pearson *r* ranges from .00, denoting complete absence of relationship, to 1.00, denoting perfect correspondence, with +1.00 indicating perfect positive and −1.00 indicating perfect negative relationship.

Confidence Interval. The probability that a given interval includes a given population parameter (such as the mean).

Confidence Limits. The upper and lower limits of the confidence interval.

Confounding Factor. Characteristics or events within a plant, department, work unit or within other aspects of a countermeasure application that may influence its impact either positively or negatively, unless they are controlled by the design of the experiment or evaluative research project. If the confounding factors are not controlled in a countermeasure application, the effects of a particular countermeasure will be difficult to measure since it cannot be determined if the results obtained are truly produced by the countermeasure or by the confounding factors.

Control Group. A group with similar characteristics that is not exposed to the same countermeasure as a treatment or experimental group, used to aid in determining if the results achieved by the treatment group are a consequence of the countermeasure rather than the result of an outside influence.

Controlled Experiment. An experiment involving two or more similar groups. One, the "control" group, is held as a standard for comparison, while the other, the "test" or "experimental" group, is subjected to a procedure or countermeasure whose effect one wishes to determine. The groups are usually formed by "randomization," that is, assigning individuals to one group or the other by some means that does not involve human discrimination or bias.

Correlation. A technique that measures the degree of relationship between two or more variables. Measures the tendency of one set of data to vary concomitantly with another set of data. See also *coefficient of correlation.*

Cost-Benefit Analysis. A form of evaluation in which input is measured in terms of dollar costs and output is measured in terms of economic benefit of a project or program, that is, dollars spent compared to dollars saved.

Cost-Benefit Ratio. The ratio of a project input cost to an impact measure when the latter is assigned a dollar value, for example, dollars spent compared to dollars saved by preventing a fatal injury.

Cost-Effectiveness Analysis. A form of evaluation in which input is measured in terms of dollars and output is measured in terms of the achievement of some desired objectives.

Cost-Effectiveness Ratio. A ratio of a project or countermeasure input cost to a project or countermeasure impact measure, that is, dollars spent per accident prevented.

Countermeasure. A specific activity intended to improve one or more aspects of the safety system or contribute to the solution of a specific accident problem.

Criterion. A standard by which something can be judged or evaluated.

Data. Symbolic representation of information. Items of information collected or presented for a particular purpose. *Plural*: data; *singular*: datum.

Data Analysis. The technique or process of organizing, evaluating, and interpreting information.

Data Base. The bank of collected data that describes an area of interest and serves as the basis of an information retrieval system or an experimental evaluation.

Data Collection. The process of accumulating statistical information relating to the empirical effects of an experiment.

Data, Grouped. The result of gathering items of similar values into appropriate class intervals in order to simplify and clarify a statistical analysis by reducing the number of data categories.

Data Presentation. The final activity relating to an experiment in which the test information, analyzed and interpreted, is displayed for the purpose of describing the results of the experiment.

Data Tabulation. The process of displaying experimental results in a table (columns and/or rows) so that the information can more readily be interpreted.

Degrees of Freedom. Number of observations that are free to vary within a distribution of scores or set of data when all restrictions are imposed.

Dependent Variable. An element or condition within an experiment that is allowed to vary, subject to the influence of the independent variable(s) employed in the treatment phase of the experiment. Also referred to as the "criterion variable."

Descriptive Statistics. Standard methods used to summarize and describe the key characteristics of the total amount of raw data. The most frequently used descriptive statistics include frequency distribution, graphs or charts, arithmetic mean, median, mode, range, standard deviation, and correlation.

Deviation. The amount by which a score or other item differs from a reference value such as the mean, the norm, or the score on some other test.

Effectiveness Evaluation. A type of appraisal that examines the outcome of project activities in order to determine if project objectives have been met.

Efficiency Evaluation. An extension of effectiveness evaluation that compares the effects of a project to the cost of the project.

Error of Misplaced Precision. An incorrect assumption that results from the precise collection of data within the framework of a faulty design.

Evaluation. A comparison process that measures an item or activity against certain predetermined standards or criteria. A judgment of value or worth.

Evaluation Cycle. A measurement design that encompasses the total sequence of events relating to a program whereby problems are identified, examined, and treated. A subsequent assessment of the success or failure of projects within the program leads to another level of problem assessment, which then may initiate another evaluation sequence.

Expected Value. The mean value of a random variable.

Experimental Design. A formal procedure or method by which the independent (task) variables are manipulated to determine their effect upon dependent (criterion) variables.

Experiment Bias Effect. The partiality that an experimenter builds into a method of conducting an experiment, with the result that his method shapes the data in the direction of his foregone conclusions. Also known as the "self-fulfilling prophecy."

Exposure. The quantity of time involved and the nature (quality) of involvement with certain types of environments possessing various degrees or types of hazards. For example, the amount of time, number of drivers, or mileage driven by a certain class or type of driver exposed to certain types of hazards.

Extraneous Variable. A variable outside the control of a project design.

Extrapolation. Any process of estimating values of a function beyond the range of available data.

Factor Analysis. A technique for analyzing patterns of intercorrelations among many variables to determine which variables have similar underlying characteristics.

Frequency. The number of objects or items in a given category when objects are classified according to variations in a set of one or more specific attributes.

Frequency Distribution. A systematic arrangement of data that exhibits the division of the values of a variable into classes and that indicates the number of cases in each class.

Frequency Rate. The number of occurrences of a given type of event expressed in relation to a base unit of measure (for example, accidents per 1 million miles traveled).

Goal. An expression of a long-term target, usually in terms of ultimate success criteria, with defined units of measure that permit evaluation of progress toward accomplishment.

Halo Effect. The tendency for an irrelevant feature of a project variable to influence the relevant feature, either favorably or unfavorably.

Hawthorne Effect. A change produced in the subjects of an experiment by the very fact that they are participating in an experiment. The term was taken from the name of the Hawthorne Plant of the Western Electric Company where a series of experiments was conducted that revealed this effect.

Histogram. A graphic representation of a frequency distribution by means of rectangles or bars whose widths represent class intervals and whose heights represent the corresponding frequencies. If the class intervals are equal, the heights serve as an exact measure.

Homogeneous Population. A group having many salient characteristics in common.

Hypothesis. An assumption or proposition held to be probably true based on the knowledge of observed events, conditions, and relationships but not empirically validated.

Hypothesis Test. Evaluating randomly obtained observations or evidence to determine the probability of drawing such evidence by random sampling methods under the assumption that the hypothesis is correct.

Immediate Objective. A project target expressed in terms of desired proximate change. Action without the intervention of another object, cause, or agent.

Independent Variable. An element or condition in an experiment that is controlled by the experimental design. The item(s) that are responsible for the induced change during the treatment phase of an experiment. The factor that is manipulated in order to determine its effects on the dependent variable(s). Also referred to as the "task variable."

Inferential Statistics. A set of analytical techniques for inferring something about information obtained from a representative sample taken from a population. Methods of drawing conclusions from sample data about a larger body of population data. A body of techniques that will tell what inferences can be drawn and how safe it is to draw them. The techniques of statistical inference are designed to enable the evaluator to quantify probability statements to determine exactly the probability that a conclusion is true. Inferential statistics fall into two main categories: those that tell the margin of error when predicting or estimating a population measure such as an average and those that enable the testing of hypotheses about populations.

Information. The meaning assigned to data or a description of data. Knowledge concerning some particular fact, subject, or event in any communicable form. For purposes of documentation, it has three basic criteria: existence, availability, and semantic content.

Information System. A network of information services providing facilities by which information and data are processed and transmitted from originator to user.

Interaction Effect. The outcome generated by the combination of two or more factors in an analysis of variance. A measure of how much effect one statistical variable has upon one or more other variables.

Interaction Term. The measure of the effects of combining two or more factors in an analysis of variance.

Intermediate Objective. A project target expressed in terms of desired changes occurring between the immediate and ultimate objectives.

Interpolation. In general, any process of estimating intermediate values between two known points.

Interval. (1) Difference between two classes or categories of data. (2) In an interval scale the presence of three basic properties: (a) mutually exclusive categories, (b) categories ordered according to the amount of the attribute they represent, (c) equal differences in the attribute represented by equal differences in the numbers assigned.

Inverse Relationship. A negative relationship between variables; as one variable increases, the other decreases.

Law of the Instrument. An error that results from the indiscriminant use of certain instruments or procedures to study projects or problems.

Level of Confidence. The probability that a confidence interval will include the parameter value. See *confidence interval.*

Main Effects. An estimate of the effect of an experimental variable or treatment measured independently of other treatments that may form part of the experiment.

Mean. See *arithmetic mean.*

Measurement. The assignment of numbers to events according to rules. A field of study concerned with the development and application of methods and techniques for quantification.

Median. The point on a scale of measurement above which and below which one-half of the numbers fall. The middle value of a distribution. The point that divides a group of numbers in a distribution into two equal parts.

Mode. In a distribution of scores the most frequently occurring score (or scores). Also, the midpoint of the class interval with the greatest frequency.

Model. A mathematical and/or physical representation of real-world phenomena that serves as a plan or pattern from which these phenomena and their interrelationships can be identified, analyzed, synthesized, and altered without disturbing the real-world processes.

Monitoring. Administrative evaluation which compares actual accomplishments with established goals and assesses operational efficiency.

Multimodal. A distribution of data which has more than one concentration of frequently occurring values.

Multiple Correlation. A statistical technique that shows relationships among three or more variables.

Nominal. A scale of numbers assigned to mutually exclusive categories—that is, objects that are assigned the same number possess the same attribute. Different objects are assigned different numbers. The numbers serve only to identify the classes and do not indicate anything about the classes other than that they are different.

Normal Distribution. A frequency distribution that approximates a bell-shaped curve. In a normal distribution scores or measures are distributed symmetrically about the mean, with as many cases at various distances about the mean and decreasing in frequency the further one departs from the mean, according to a precise mathematical equation.

Null Hypothesis. The statement, tested in statistical analysis, that there is no difference between treatment group and control group in an experimental study.

Objective. An expression of a short term target in terms of the basic · planning units (subelements) of a program.

Ordinal. A data scale that includes the following properties: (1) categories are mutually exclusive, (2) categories are ordered or ranked according to the amount of the attribute they represent. The ordinal scale does not identify the exact difference in degree between any two ranks.

Output Measure. Measurement of project accomplishments or results.

Partial Correlation. A statistical technique that shows relationships among three or more variables, holding constant the effects of one or more variables.

Percentage. Individual class frequencies expressed in proportion to 100. Parts per 100.

Percentile. A point (score) in a distribution below which falls the percent of cases indicated by the given percentile number. Thus, the fifteenth percentile denotes the score or point below which 15 percent of the scores fall. "Percentile" has nothing to do with the percent of correct answers an examinee has on a test.

Performance Criteria. Operational standards for use in determining effectiveness or efficiency.

Permutation. Any one of the combinations or changes in position possible within a group. For example, the permutations of 1, 2, and 3 are 123, 132, 213, 231, 312, and 321.

PERT (Program Evaluation and Review Technique). A management tool that displays graphically the network of relationships between program objectives and the tasks, time, and resources required for their achievement.

Pie Diagram. A circular figure or chart used to show proportions of a total.

Placebo Effect. The effect of a neutral or irrelevant feature of an experiment that is intended to produce the same reaction in a subject as an important or relevant feature. An inert preparation given for its psychological effect to satisfy the subject or to act as a control in an experimental series.

Poisson Distribution. A distribution that often appears in observed events that are very improbable compared to all possible events but that do happen occasionally since so many trials occur, for example, traffic deaths, industrial accidents, and radioactive emissions. The frequency function of the Poisson distribution is of the form

$$f(x) = \frac{m^x e^{-m}}{x!}$$

where:

$x = 0, 1, 2, \ldots$

m = a parameter that is both the mean and the variance

The mean and variance of the Poisson distribution are equal.

Population. The total set of items (actual or potential) defined by some characteristic of these items.

Post Hoc Error. A faulty reasoning process that attributes the cause of a given event to another event that occurred earlier in time.

Pretest. A measure or test applied before the project activity or treatment is initiated so that scores may be statistically compared to the sample scores on the same measure or test following the project activity or treatment (post-test). A preliminary testing of an instrument or questionnaire to evaluate its validity, accuracy, and format.

Probability. The likelihood that a given event will occur. The relative frequency of occurrence of an event based on the ratio of its occurrence and the total number of cases necessary to ensure its occurrence when such cases are viewed as indefinitely extended.

Problem Area. A population of detrimental events (accidents, injuries, fatalities, etc.) identified through data analysis and possessing a common causative, contributory or geographical characteristic to be addressed through one or more countermeasure solutions.

Program. An organized directed effort which uses resources to achieve desired goals. A set of instructions in a form applicable to a computer, prepared to achieve certain results.

Project. A task or program that involves a variety of mental and physical activities related to the accomplishment of specific objectives with a specific plan or design.

Project Impact. Project effectiveness in achieving immediate inter-mediate, or ultimate objectives; also any unexpected consequences of the project.

Proportion. A part or fraction of the whole, expressed as a decimal between 0 and 1.

Proxy Measure. An indirect impact measure used in place of or in addition to the criterion impact measure.

Qualitative. The characteristic element, attribute, kind, or degree of a quality possessed by something. Refers to a characteristic (physical or non-physical, individual or typical) that constitutes the basic nature of something or is one of its distinguishing features. See *quantitative.*

Quantitative. The property of anything that can be determined by measurement. The property of being measurable in dimensions, amounts, etc., or in extensions of these that can be expressed in numbers or symbols. A quantitative statement describes "how much," while a qualitative state-ment answers the question, "What kind is it?" or "How good is it?"

Quartile. One of three points that divide the items in a distribution into four equal groups. The lower quartile, or twenty-fifth percentile, sets off the lowest fourth of the group; the middle quartile is the same as the fiftieth percentile or median; and the third quartile, or seventy-fifth percentile, marks off the highest fourth.

Quasi-experimental Study Design. A study design that approximates the conditions of the true experiment design under conditions that do not allow for the control and/or manipulation of all relevant variables.

Random Sample. Selection of cases or subjects for study in such a way that all cases in the population from which they are selected have an independent and equal chance of being included, that is, selected in a manner that precludes the operation of bias. See also *stratified random sampling.*

Range. The easiest and most quickly obtained measure of variability expressing the highest and lowest values in a distribution of scores. The difference between the highest score and the lowest score, plus one.

Rating Errors. Mistakes in evaluation that arise from rating subjects too leniently, too strictly, or too often as "average."

Ratio. (1) A fraction that expresses the relationship of one part of the whole to another. (2) As a data scale, requires that: (a) categories are mutually exclusive, (b) categories are ordered according to the amount of the attribute they measure, (c) equal differences are represented by equal dif-ferences in the numbers assigned to the categories, and (d) a fixed and true zero point exists.

Regression Toward the Mean. The tendency of extreme groups, over a period of time, to become more "average." If two measures are associated with less than perfect correlation, unusually high or low scores on one measure will tend to be associated with more average (mean) scores on the second. The lower the correlation between two measures, the more salient is the phenomenon of regression toward the mean.

Relative Frequency Distribution. A listing of the proportion of cases in each class interval. When a collection of data is separated into several categories, the number of items in a given category is the absolute frequency, and the absolute frequency divided by the total number of items is the relative frequency.

Reliability. The quality of a measure to provide consistent results upon repeated application of the measure. The probability that a system will perform satisfactorily for at least a given period of time when used under stated conditions.

Representative Sample. A sample that corresponds to or matches the population from which it is drawn, with respect to characteristics important for the purpose under investigation.

Sample. A subgroup of the population or universe selected according to some rule or plan. A finite portion of a population or universe. See also *random sample, representative sample, stratified random sampling.*

Sample Distribution. The characteristics of a given statistic or set of statistics associated with a sample drawn from a population.

Scientific Evaluation. A level of appraisal that, through careful planning and data collecting, eliminates most elements of personal judgment. An assessment of value on the basis of establishing relationships between accomplishments and criteria or proxy measures using scientific techniques designed to identify cause and effect relationships, if any, between independent (task) and dependent (criterion) variables using the "scientific method."

Skewness. The tendency of a distribution to depart from symmetry or balance around the mean.

Standard Deviation. An average of all the deviations about the mean in a sample or random variable. Arithmetically, defined as the square root of the arithmetic mean of the squared deviations of measurements from the mean. A measure or index of dispersion of a frequency distribution.

Statistical Power. The probability of accepting the null hypothesis when it is false. That is, the probability of concluding that a result had no significance when it actually did. Also referred to as *Beta* or Type II Error. (See also *Alpha*.)

Statistical Significance. One minus the probability that the difference between a treatment group and a control group is not due to chance. Deviations between hypothesis and observations that are so improbable under the hypothesis as to cause one to believe that the difference is not due to sampling errors or fluctuations. The failure of a difference to fall in the acceptable realm of sample deviations identifies it as statistically significant.

Statistics. A branch of mathematics dealing with the collection, analysis, interpretation, and presentation of numerical data.

Stratified Random Sampling. A data collection technique in which the population is divided into subclassifications, or strata, from which a certain number of cases is drawn by chance (that is, by taking random samples from each subclassification).

System. A set or arrangement of components so related or connected as to form a unity or organic whole. A set of facts, principles, rules, etc., classified or arranged in a regular, orderly form so as to show a logical plan linking the various parts. A method or plan of classification. An orderly arrangement of interdependent activities and related procedures that implements and facilitates the performance of a major activity or an organization. A set of components, man or machine or both, that has certain functions and acts and interacts, one in relation to another, to perform some task or tasks in a particular environment or environments. Any configuration of elements in which the behavior properties of the whole are functions of both the nature of the elements and the manner in which they are combined.

System Safety. An approach to accident analysis that involves the detection of deficiencies in system components and their interactions, which have an accident potential.

Theory. A principle concerned with a coordinated set of tested hypotheses that are found to be consistent with one another and with specific observed events.

Time Series Design. In statistics a set of ordered observations of a quantitative characteristic or an event taken at successive points in time.

Treatment Group (Study, Task, or Experimental Group). A group that receives or is exposed to the countermeasure established in an experiment. See *control group.*

Trend Chart. A graphical presentation showing changes in frequencies, percentages, or proportions over a period of time.

t-Test. A statistical technique for testing the null hypothesis—that the mean scores from two groups do not differ in a statistically significant way. Appropriate for interval scale data. Applicable to the test of the hypothesis that a random sample of N observations is from a normal population with mean u and with the variance unspecified.

Type I Error. Rejection of the null hypothesis when it should have been accepted (that is, when it is true). In other words, it falsely concluded that the independent or task variables had an effect. See *Alpha.*

Type II Error. Acceptance of the null hypothesis when it should have been rejected (that is, when it is false). In other words, it falsely concluded that the independent or task variable had no effect. See *statistical power.*

Validity. The quality of a test or evaluation instrument to be able to measure with accuracy what it is intended to measure.

Variable. A quantity represented by a symbol, which may take any one of a specified set of values. See also *dependent variable, extraneous variable, independent variable.*

Variance. The standard deviation squared. A measure of deviation from the average or mean within a distribution of data.

Z-*Score.* A number that expresses the distance from a particular score (X) above or below the mean ($\overline{X}$) in units of standard deviation.

$$Z = \frac{X - \overline{X}}{S}$$

BIBLIOGRAPHY

Ackoff, R. L., S. K. Gupta, and J. S. Minas. *Scientific Method: Optimizing Applied Research Decisions.* New York: Wiley, 1962.

Altman, I. and A. Ciocco. Introduction to occupational health statistics: 11—Rates. *Journal of Occupational Medicine,* October 1964, pp. 409–415. Introduction to Occupational Health Statistics: IV—Accidents and Probability, *Journal of Occupational Medicine,* May 1966, pp. 266–270.

American Institutes for Research. *Evaluation Research Strategies and Methods.* Pittsburgh: The Institute, 1970.

American National Standards Institute. *Method of Recording Basic Facts Relating to the Nature and Occurrence of Work Injuries: Z-16.2.* New York: The Institute, 1962, r. 1969.

American National Standards Institute. *Method of Recording and Measuring Work Injury Experience: Z-16.1.* New York: The Institute, 1967, r. 1973.

Attaway, D. C. *Safety Sampling Program.* DA-11-173-AMC 200 (A). Bristol, Pa.: Longhorn Division, Thiokol Chemical Corporation, February 12, 1963.

Attneane, F. *Applications of Information Theory to Psychology,* New York: Holt, Rinehart and Winston, 1959.

Baird, D. C. *Experimentation: An Introduction to Measurement Theory and Experiment Design.* Englewood Cliffs, N.J.: Prentice-Hall, 1962.

Barnes, R. M. *Work Sampling,* 2nd ed. New York: Wiley, 1957.

Barnes, R. M. *Motion and Time Study,* 4th ed. New York: Wiley, 1958.

Bernstein, I. N. (Ed.). *Validity Issues in Evaluative Research,* Beverly Hills, Calif.: Sage Publications, 1976.

Bird, Frank E., Jr. Measurement tools for safety management. *Canadian Mining Journal,* 92(9):61–62, 1971.

Bureau of Labor Statistics, U.S. Department of Labor. *Annual Releases of Injury Statistics* (Manufacturing and Non Manufacturing). Division of Injury Statistics, Office of Industrial Hazards. Washington, D.C.: U.S. Department of Labor, annually.

Bureau of Labor Statistics, U.S. Department of Labor. *Special Studies of Injury Rates and Accident Causes in Selected Industries.* Division of Accident Research, Office of Industrial Hazards. Washington, D.C.: U.S. Department of Labor, 1966.

Campbell, B. J. Highway safety program evaluation and research. *Traffic Digest and Review,* January 1970, pp. 6–11.

Campbell, D. T. Measuring the effects of social innovations by means of time series. In J. M. Tanur et al. (Eds.), *Statistics: A Guide to the Unknown*. San Francisco: Holden-Day, 1974.

Campbell, D. T., and J. Stanley. *Experimental and Quasi-Experimental Designs for Research*. Chicago: Rand McNally, 1966.

Campbell, N. R. *Symposium: Measurement and Its Importance for Philosophy*. Aristotelian Society, Suppl. Vol. 17. London: Harrison, 1938.

Caws, P. Definitions and measurement in physics. In C. W. Churchman and P. Ratoosh (Eds.), *Measurement: Definitions and Theories*. New York: Wiley, 1959.

Churchman, C. W. Why measure? In C. W. Churchman and P. Ratoosh (Eds.), *Measurement: Definitions and Theories*. New York: Wiley, 1959.

Churchman, C. W. *Prediction and Optimal Decisions*. Englewood Cliffs, N.J.: Prentice-Hall, 1961.

Churchman, C. W. *The Systems Approach*. New York: Dell, 1968 (paperback).

Churchman, C. W. and P. Ratoosh (Eds.). *Measurement: Definitions and Theories*. New York: Wiley, 1959.

Coombs, C. H. The theory and models of social measurement. In L. F. Festinger and D. Katz (Eds.), *Research Methods in the Behavioral Sciences*. New York: Dryden, 1953.

Damkot, D., and T. Pollock. *Development of Highway Safety Program and Project Evaluation Criteria: Accident Experience Characterization*. Ann Arbor, Michigan: Office of Highway Safety Research Institute, 1970.

Diamond, S. *Information and Error*. New York: Basic Books, 1959.

Duncan, A. J. *Quality Control and Industrial Statistics*. Homewood, Ill.: Irwin, 1959.

Dunlap, J. W. *Manual for the Application of Statistical Techniques for Use in Accident Control*. Stamford, Conn.: Dunlap, 1958.

Edwards, A. L. *Experimental Design in Psychological Research,* rev. ed. New York: Holt, Rinehart and Winston, 1960.

Edwards, A. L. *Statistical Analysis,* 4th ed. New York: Holt, Rinehart and Winston, 1974.

Ellis, B. *Basic Concepts of Measurement*. Cambridge, Eng.: University Press, 1966.

Elsby, O. H. Measurement of industrial safety performance: The search is on. *Transactions,* 1966 National Safety Congress. Chicago: The National Safety Council, Vol. 12, 1966, pp. 112–113.

Employers Insurance of Wausau, *Current "Costing" of Occupational Injuries*. Wausau, Wis.: Safety and Health Services, 407 Grant St., 1964.

Fine, W. T. Mathematical evaluations for controlling hazards. *Selective Readings in Safety*. Macon, Ga.: Academy Press, 1973, pp. 68–84.

Flanagan, J. C. The critical incident technique. *Psychological Bulletin, 51:*327–358, 1954.

Gilmore, C. L. *Accident Prevention and Loss Control*. New York: American Management Association, 1970.

Gilmore, C. L. and L. M. Buttery. *The use of statistics in accident prevention*. Unpublished report. Texas City, Texas: Monsanto Chemical Company, Hydrocarbons Division, 1962.

Grant, G. S. Measuring the cost of accidents. *Journal of the American Society of Safety Engineers, 11*(7):11–13, 1966.

Greenberg, L. *Quantitative Techniques in Safety and Loss Prevention with Computer Programs.* New York: Philipp Feldheim, 1972.

Grimaldi, J. V. Appraising safety effectiveness. *Journal of the American Society of Safety Engineers,* (4):57–62, 1960.

Grimaldi, J. V. Another look at stimulating safety effectiveness. *Journal of the American Society of Safety Engineers,* Vol. VII (4):20–23, 1962.

Grimaldi, J. V. Measuring safety effectiveness. In H. H. Fawcett and W. S. Wood (Eds.), *Safety and Accident Prevention in Chemical Operations.* New York: Interscience Publishers, 1965, pp. 537–553.

Grimaldi, J. V., and R. H. Simonds. *Safety Management,* 3rd ed. Homewood, Ill.: Irwin, 1975.

Hammer, W. *Handbook of System and Product Safety.* Englewood Cliffs, N.J.: Prentice-Hall, 1972.

Hansen, B. L. *Work Sampling for Modern Management.* Englewood Cliffs, N.J.: Prentice-Hall, 1960.

Heath, E. D. Tests and measurement as applied to accident prevention situations. *Journal of the American Society of Safety Engineers,* Vol. VIII, Part I, 1963, pp. 7–14; Vol. VIII, Part II, 1963, pp. 9–14.

Heiland, R. F., and W. J. Richardson. *Work Sampling.* New York: McGraw-Hill, 1957.

Helmstadter, G. C. *Principles of Psychological Measurement.* New York: Appleton-Century-Crofts, 1964.

Herzog, E. *Some Guidelines for Evaluation Research.* Washington, D.C.: U.S. Government Printing Office, 1959.

Huff, D. *How to Lie with Statistics.* New York: Norton, 1954.

Isaac, S., and W. B. Michael. *Handbook in Research and Evaluation.* San Diego, Calif.: Knapp, 1971.

Jewell, A. J. *An investigation and evaluation of the application of work sampling techniques to the safety behavior measurement problem.* Unpublished master's thesis, Ohio State University, 1958.

Johnson, W. G. *The Management Oversight and Risk Tree (MORT).* U.S. Atomic Energy Commission. Washington, D.C.: U.S. Government Printing Office, pp. 415–434, 1973.

Kerlinger, F. *Foundations of Behavioral Research,* 2nd ed. New York: Holt, Rinehart and Winston, 1973.

Kirk, R. E. *Experimental Design: Procedures for the Behavioral Sciences,* Belmont, Calif.: Brooks/Cole, 1968.

Klingel, A. R., and O. C. Haier. SOHIO serious injury index. *National Safety News* November, 1956, Vol. 1, No. 4, pp. 50–52. The Journal of ASSE (Technical Feature Section of National Safety News).

Lambrow, F. H. *Guide to Work Sampling.* New York: Rider, 1962.

Littauer, S. B., and T. S. Irby. Analytical and mathematical studies of the analysis of accident data: II—The application of statistical control methods for the reduction and control of industrial accidents. *Traffic Safety Research Review, 1*:6–15, 1957.

Malasky, S. W. *System Safety: Planning/Engineering/Management,* Rochelle Park, N.J.: Hayden Book, 1974.

Massarik, F., and P. Ratoosh (Eds.). *Mathematical Explorations in Behavioral Science.* Homewood, Ill.: Irwin, 1965.

Maynard, T. H. *Industrial Engineering Handbook,* 2nd ed. New York: McGraw-Hill, 1964.

Meyer, J. J. Statistical sampling and control for safety. *Industrial Quality Control, 19*(12):14-17, 1963.

Meyers, V. J. *Recording and Measuring Injury Experience in the Western Electric Company.* New York: Western Electric Company, 222 Broadway, 1964.

Miller, D. C. *Handbook of Research Design and Social Measurement,* New York: McKay, 1964.

National Safety Council. *Accident Facts.* Chicago: The Council, annually.

National Safety Council. Safety performance measurement in industry. Special issue. *Journal of Safety Research, 2*(3), 1970.

Newman, J. R. (Ed.). *The World of Mathematics.* New York: Simon and Schuster, 1956.

Peters, G. A. *Notes on safety performance measurement.* Unpublished paper presented at the Symposium on Measurement of Industrial Safety Performance, Chicago, May 1966.

Petersen, D. C. *Techniques of Safety Management.* New York: McGraw-Hill, 1971.

Petersen, D. C. *Safety Management—A Human Approach.* New York: McGraw-Hill, 1974.

Pollina, V. *Safety Sampling Procedure.* Milwaukee, Wis.: A. O. Smith Corporation, Office of the Corporate Director of Safety, 1966.

Recht, J. L. Systems safety analysis, reprint from *National Safety News.* Chicago: National Safety Council, 1966.

Rockwell, T. H. Safety performance measurement, *Journal of Industrial Engineering, 10*:12-16, 1959.

Rockwell, T. H. *Problems in the measurement of safety performance.* Unpublished paper presented at the Symposium on Measurement of Industrial Safety Performance, Chicago, May 1966.

Rodgers, W. P. *Introduction to System Safety Engineering.* New York: Wiley, 1971.

Rossi, P. H., and S. R. Wright. Evaluation research: An assessment of theory, practice, and politics. *Evaluation Quarterly, 1*:5-52, 1977.

Satterwhite, H. G., and R. M. LaForge. A comparison of three measures of safety performance. *Journal of the American Society of Safety Engineers, 11*(3):9-15, 1966.

Schauer, L. R., and T. S. Ryder. New approach to occupational safety and health. *Monthly Labor Review,* Vol. 95, No. 3:14-19, 1972.

Schowalter, E. J. A years's trial with a new safety measurement plan. *Journal of Occupational Medicine,* Vol. VIII, No. 6:313-316, 1966.

Schreiber, R. The development of engineering techniques for the evaluation of safety programs. *Transactions of the New York Academy of Sciences, 18*(3), 1956.

Senders, V. L. *Measurement and Statistics.* New York: Oxford, 1958.

Siegel, S. *Nonparametric Statistics for the Behavioral Sciences.* New York: McGraw-Hill, 1956.

Slonim, M. J. *Sampling.* New York: Simon and Schuster, 1966.

Solomon, H. *Mathematical Thinking in Psychology,* New York: Free Press, 1962.

Stevens, S. S. Mathematics, measurement, and psychophysics. In S. S. Stevens (Ed.), *Handbook of Experimental Psychology.* New York: Wiley, 1951.

Stevens, S. S. Measurement, psychophysics, and utility. In C. W. Churchman and P. Ratoosh (Eds.), *Measurement: Definitions and Theories.* New York: Wiley, 1959.

Stevens, S. S. Measurement, statistics, and the schemapiric view. *Science, 161*:885–888, 1968.

Struening, E. L., and M. Guttentag (Eds.), *Handbook of Evaluation Research,* Vols. I and II. Beverly Hills, Calif.: Sage Publications, 1975.

Suchman, E. A. *Evaluative Research: Principles and Practice in Public Service and Social Action Programs.* New York: Russell Sage Foundation, 1970.

Sudman, S. *Applied Sampling.* New York: Academic Press, 1976.

Surrey, J. *Industrial Accident Research: A Human Engineering Approach.* Toronto, Canada: University of Toronto, 1968.

Tarrants, W. E. *An evaluation of the critical incident technique as a method for identifying industrial accident causal factors.* Unpublished doctoral dissertation, New York University, 1963. (Also available from University Microfilm, Inc., Ann Arbor, Michigan.)

Tarrants, W. E. Applying measurement concepts to the appraisal of safety performance. *Journal of the American Society of Safety Engineers, 10*(5):15–22, 1965.

Tarrants, W. E. Research in safety performance measurement. *Transactions,* 1966 National Safety Congress. Chicago: The National Safety Council, Vol. 12, pp. 124–127.

Tarrants, W. E. *The Evaluation of Safety Program Effectiveness.* Washington, D.C.: National Highway Traffic Safety Administration, U.S. Department of Transportation, 1972.

Tarrants, W. E. (Ed.). *Research Concepts and Methods.* Proceedings of a Symposium on Safety Research sponsored by the Research Projects Committee, National Safety Council. Chicago: The Council, 1976.

Tarrants, W. E., and H. Veigel (Eds.). *The Evaluation of Highway Traffic Safety Programs: A Manual for Managers.* Washington, D.C.: U.S. Government Printing Office, 1977.

Thiokol Chemical Corporation. *Safety performance indicator.* Unpublished report. Bristol, Pa.: The Corporation, January 1966.

Thrall, R. M., C. H. Coombs, and R. L. Davis. *Decision Processes.* New York: Wiley, 1954.

Torgerson, W. S. *Theory and Methods of Scaling.* New York: Wiley, 1958.

U.S. Department of Labor. *What Every Employer Needs to Know About OSHA Recordkeeping,* rev. Report 412. Washington, D.C.: Bureau of Labor Statistics, 1973.

U.S. Department of Labor. *Recordkeeping Requirements Under the Williams–Steiger Occupational Safety and Health Act of 1970,* rev. With amendments October 1972, and January 1973. Washington, D.C.: Occupational Safety and Health Administration, 1975.

Voland, L. J. Beyond frequency and severity. *National Safety News 86*(6) December 1962, p. 28.

Wallis, A. W., and H. V. Roberts. *Statistics: A New Approach,* New York: Free Press, 1956.

Weiss, C. H. (Ed.). *Evaluating Action Programs: Readings in Social Action and Education,* Boston: Allyn and Bacon, 1972.

Weiss, C. H. *Evaluation Research,* Englewood Cliffs, N.J.: Prentice-Hall, 1972.

Wert, J. E., C. O. Neidt, and J. S. Ahman. *Statistical Methods in Educational and Psychological Research.* New York: Appleton-Century-Crofts, 1954.

Widner, J. T. (Ed.). *Selected Readings in Safety.* Macon, Ga.: Academy Press, 1973.